Dynamical Systems, Number Theory and Applications

**A Festschrift in Honor of
Armin Leutbecher's 80th Birthday**

Dynamical Systems, Number Theory and Applications

**A Festschrift in Honor of
Armin Leutbecher's 80th Birthday**

Editors

Thomas Hagen (University of Memphis, USA)

Florian Rupp (German University of Technology, Oman)

Jürgen Scheurle (Technische Universität München, Germany)

 **World Scientific**

NEW JERSEY · LONDON · SINGAPORE · BEIJING · SHANGHAI · HONG KONG · TAIPEI · CHENNAI · TOKYO

Published by

World Scientific Publishing Co. Pte. Ltd.

5 Toh Tuck Link, Singapore 596224

USA office: 27 Warren Street, Suite 401-402, Hackensack, NJ 07601

UK office: 57 Shelton Street, Covent Garden, London WC2H 9HE

Library of Congress Cataloging-in-Publication Data
Names: Leutbecher, Armin. | Hagen, Thomas, 1972–   editor. | Rupp, Florian, 1978–   editor. |
  Scheurle, Jürgen, 1951–   editor.
Title: Dynamical systems, number theory and applications : a festschrift in honor of
  Armin Leutbecher's 80th birthday / edited by Thomas Hagen (University of Memphis, USA),
  Florian Rupp (German University of Technology, Oman), Jürgen Scheurle (Technische Universität
  München, Germany).
Description: New Jersey : World Scientific, 2016. | Includes bibliographical references.
Identifiers: LCCN 2015041750 | ISBN 9789814699860 (hardcover : alk. paper)
Subjects: LCSH: Number theory. | Dynamics. | Topological algebras.
Classification: LCC QA241 .D965 2016 | DDC 512.7--dc23
LC record available at http://lccn.loc.gov/2015041750

British Library Cataloguing-in-Publication Data
A catalogue record for this book is available from the British Library.

Printed in Singapore

Preface

Deep and subtle links connect Dynamical Systems and Number Theory. The past few years have witnessed the solution of several problems on the frontiers of mathematical research, obtained by combining methods and insights from these two independent fields.

Having evolved from elementary Number Theory and first considerations about prime numbers, modern Number Theory aims to develop unifying results, like the structure of rational points on varieties. Methods from Complex Analysis, Group Theory and Algebraic Geometry, above all, shed new light on number-theoretic structures and lead to the improvement of classical methods beyond the seminal works of the 18th, 19th and 20th century. As we have seen in the proof of Fermat's Last Theorem, achievements are based on connections between different mathematical fields and on the transfer of methods and results from one field to another.

As an example in Number Theory, the classical and still interesting case is the upper complex half plane, which, on the one hand, can be viewed as a set of points or the domain of definition for interesting geometric curves and corresponding generating or zeta functions, or, on the other hand, as a group of Möbius transformations (modulo some stabilizer). Hence, numbers and groups immediately get connected, and interpretations of, say, the zeta function for groups become a subject of research. Moreover, generalizations of the thus derived group structure are natural with properties of zeta functions on these more general domains of definition arising as exciting and non-trivial research objects among many other topics of interest. Thus, groups and their representation, complex functions, connections between different algebraic structures as well as their applications in Crystallography, Quantum Physics, Coding Theory and other disciplines are areas of number-theoretic focus. In this way, for instance, recent applications link lattices and the zeta function to Physics, cf. [1].

The mathematical theory of Dynamical Systems evolved as a geometric theory of phase space where a discrete or continuous time-parameterized propagation of states is viewed as the action of a (semi-)group. Such groups are, for instance, generated by the solutions of iterated function systems or initial (-boundary) value problems associated to ordinary or partial differential equations. Above all, questions of stability, long-term behavior, singularities and blow-up, and attraction by stationary solutions are of great interest as they may reflect real-world phenomena visible in

Fig. 1. Armin Leutbecher at the colloquium on the occasion of his 80th birthday, held at TU München on 17 October 2014.

physical experiments. At the core of classical Dynamical Systems Theory lies its geometric character and its objective to study the (long-)time behavior of all (or at least sufficiently large) subsets in phase space. Here, curves and their properties play an essential role. Methods of Algebraic Geometry are widely used in this context. Moreover, it is clear that such considerations must be supported by reductions of a problem's complexity, often connected to underlying symmetries of the equation of motion (cf. Noether's Theorem) where groups and their stabilizers are central. Modern computational methods and novel techniques from the theory of Evolution Equations, Differential Geometry and Algebraic Topology add to the expanding field of methodology embraced and often appropriated by Dynamical Systems Theory. This includes the development of efficient algorithms that preserve the underlying geometric structures of the dynamics, and the consideration of physical or geometric characteristics that are present in evolution models of real-world processes. It is not surprising that one can find many important and innovative applications of Dynamical Systems in Engineering and both the Natural and Social Sciences.

Although working in seemingly completely different fields, researchers in Number Theory, Dynamical Systems, Evolution Equations and several other, closely related disciplines share common interests concerning algebraic, geometric and analytic concepts, among them groups and semi-groups, spectral and operator-theoretic methods, harmonic functions, trajectories/orbits (in space) or curves (in the complex plane) and their properties. It is worth noting that such common interests are not restricted to those just mentioned. Hence, we would like to refer, for instance, to [2], [3] and [6] for additional examples.

This volume, dedicated to Professor Dr. Armin Leutbecher on the occasion of his 80th birthday, intends to broaden the mutual understanding of researchers from the fields of Number Theory and Dynamical Systems and some closely connected disciplines in Pure and Applied Mathematics. Therefore, a carefully chosen selection of genuine research papers as well as survey articles from these fields is presented here in a way experts and non-experts in the respective disciplines will benefit from. All research contributions were diligently refereed. Of course, only selected highlights could be chosen, mostly based on Armin Leutbecher's personal preferences and scientific activities during a long and still ongoing career as a scholar and academic teacher of exquisite taste and wisdom. Let us just mention Complex Analysis as the mathematical subject where Armin Leutbecher's seminal lectures [4] and [5] are foundational for many high-impact works, covering topics in Number Theory and Dynamical Systems.

Reflecting the wide range of Armin Leutbecher's interests, the articles presented cover a broad spectrum of topics in both the fields of Number Theory and Dynamical Systems and thus serve as an excellent overview of methods relevant in those two fields of study. To be more specific, these articles cover topics dealing with groups and symmetries, discrete and continuous dynamical systems, problems in Coding Theory, questions in Differential Geometry, evolution equations from Material Science and Fluid Dynamics, computational aspects of Dynamical Systems, connections to Number Theory, and some historical developments. The spectrum of topics in Number Theory, Dynamical Systems and adjacent areas of mathematics is broad by design. The application-oriented nature of the works included may serve as a blueprint for further research on related questions and stimulate the exchange of insights and ideas between these central disciplines of modern mathematics.

Apart from a biographical note on Armin Leutbecher and a historical article on mechanical integration devices, all other contributions are ordered alphabetically by the surname of the first author. As editors we would like to thank the contributors, the reviewers and the publishing company for their excellent work, critiques and comments, and suggestions and editorial assistance, respectively.

Memphis and Munich,
August 2015

Th. Hagen, F. Rupp & J. Scheurle

References

[1] A. CONNES & M. MARCOLLI (2004): From Physics to Number Theory via Noncommutative Geometry. Part I: Quantum Statistical Mechanics of Q-lattices, arXiv:math/0404128.

[2] A. KNAUF (1998): Number Theory, Dynamical Systems and Statistical Mechanics, lecture notes, Max-Planck Institut für Mathematik in den Naturwissenschaften, Leipzig.

[3] S. KOLYADA, Y. MANIN, M. MÖLLER, P. MOREE, & T. WARD (Eds., 2010): Dynamical Numbers: Interplay Between Dynamical Systems and Number Theory, Contemporary Mathematics, AMS.

[4] A. LEUTBECHER (1990): Vorlesungen zur Funktionentheorie I. Mathematisches Institut der Technischen Universität München.

[5] A. LEUTBECHER (1991): Vorlesungen zur Funktionentheorie II. Mathematisches Institut der Technischen Universität München.

[6] J.H. SILVERMAN (2007): The Arithmetic of Dynamical Systems, Graduate Texts in Mathematics, 241, Springer.

Biographical Note on Armin Leutbecher

S. Walcher

Lehrstuhl A für Mathematik, RWTH Aachen University, D-52056 Aachen
E-mail: walcher@rwth-aachen.de

Armin Leutbecher was born on August 11, 1934 in Nancheng (China). He studied Mathematics and Physics from 1954 to 1961 in Göttingen, Tübingen and finally Münster, finishing the "Staatsexamen für das Höhere Lehramt" in 1961. In 1963 he received his doctoral degree at the University of Münster, with Hans Petersson as principal advisor. He proceeded there with postdoctoral work, culminating in his "Habilitation" in 1968. In 1970 he accepted an offer for a position at TU München, where he remained and was promoted to Professor in 1975. Even after his retirement in 1999 he continued to be active in research and teaching, and served as president of the Hurwitz-Gesellschaft, an alumni organization which he co-founded in 1996.

Notable in Armin Leutbecher's research are his broad range of mathematical knowledge and activity, and the excellent quality of the articles which satisfy his own high standards for publication.

The range of his work reaches from automorphic forms to algebra and algebraic number theory. Several of his manuscripts (including some work on Dynamical Systems) were circulated only privately or in newsletters, although those who read them agreed that they deserved publication in a journal. His research combined deep mathematical insight with elegance and rigor, and has continuing impact.

At TU München, Armin Leutbecher was renowned for giving excellent lectures which were characterized by originality, rigor and high standards, and succeeded in conveying the beauty of mathematics to many generations of students. In particular, his lectures on Linear Algebra and Analysis for first- and second-year students had a lasting influence. He led a number of students to mathematical research and advised seven doctoral students at TU München.

The following selection of his publications gives an impression of Armin Leutbecher's work:

- A. Leutbecher (1967): *Über die Heckeschen Gruppen $G(\lambda)$*. Abh. Math. Semin. Univ. Hamb. 31, 199-205.
- A. Leutbecher (1978): *Euklidischer Algorithmus und die Gruppe GL_2*.

Math. Ann. 231, 269-285.

- A. Leutbecher & J. Martinet (1982): *Lenstra's constant and Euclidean number fields.* Astérisque **94**: 87-131.
- A. Leutbecher & G. Niklasch (1989): *On cliques of exceptional units and Lenstra's construction of Euclidean fields.* In: Lecture Notes in Math. 1380, Springer, Heidelberg, 150-178.
- A. Leutbecher (1996): Zahlentheorie – Eine Einführung in die Algebra, Springer, Heidelberg.

Contents

Chapter 1

Das Jahr 1934 …

Joachim Fischer[a]

Technische Universität München, Fakultät für Mathematik,
D-85747 Garching, Germany
E-mail: `jea.fischer@t-online.de`

Contents

1. Einleitung

Im Geburtsjahr 1934 von Armin Leutbecher erblickten unter anderem auch die Welt: Giorgio Armani, Brigitte Bardot, Alfred Biolek, Sophia Loren, Shirley MacLaine und Sidney Pollack; oder, wenn wir das Gebiet der Prominenten und Stars einmal etwas hinter uns lassen und uns dafür den wirklich und langfristig bedeutenden Personen zuwenden: der Logiker Paul Cohen, der Schriftsteller Harlan Ellison, der Kosmonaut Juri Alexejewitsch Gagarin, die Verhaltensforscherin Jane Goodall, der Astronom Carl Sagan, oder die Mathematiker bzw. Informatiker Tony Hoare, Robin Milner, Gilbert Strang und Niklaus Wirth … um nur einige aufzuzählen. Man sieht also insbesondere schon: Der Leutbechersche Jahrgang 1934 war ein *guter* Jahrgang!

Das gilt nicht nur für Personen, sondern auch für manche Teilgebiete menschlicher Aktivitäten. Es wird also niemanden verwundern – und den Jubilar zu allerletzt –, wenn ich mich auf dem Terrain der "Mechanischen Integration", dem ich seit fast 30 Jahren meine (knappen) Mußestunden widme, ebenfalls im Jahr 1934 umsehe und finde, daß sich dort einige höchst bedeutsame Entwicklungen oder Teilschritte dazu festmachen lassen … Ich hatte ja die Ehre und das Vergnügen, zur Inaugurations-Feier der von Armin Leutbecher ins Leben gerufenen Hurwitz-Gesellschaft am 20. März 1998 den Festvortrag über "Instrumente zum Integrieren,

---

[a]Diplomarbeit August 1973 an der TU München: "Dirichletreihen mit Funktionalgleichung"; Aufgabensteller: Prof. Dr. Armin Leutbecher.

oder: Mathematik und Feinmechanik. Ein Streifzug durch knapp zwei Jahrhunderte (verbunden mit der Vorführung ausgewählter historischer Geräte)" halten zu dürfen:

Fig. 1.   V.l.n.r.: Reinsch, Leutbecher, Scheurle, Schleicher, Vachenauer, Fischer, Hartl, Kredler.

Und die wenigen, die sich auf diesem Gebiet sonst noch tummeln, werden mit mir die Leidenschaft teilen, daß die Kombination von (teilweise sehr guter) Mathematik und Feinmechanik – meist sogar Feinstmechanik – so reizvoll ist, daß es schwerfällt, sich davon zu lösen.– Bevor wir einen Blick in das Jahr 1934 werfen, schließen wir kurz einen Bogen zu heute mit dem Hinweis, daß sich 2014 auch die Erfindung des exakt messenden, rein mechanischen Planimeters zum 200sten Male gejährt hat ... Also: Jubiläen allenthalben!

2. Vor 1934

Was war der Stand der Mechanischen Integration im Jahr 1934? Erstaunlicherweise gibt es allein schon viele runde Jubiläen, die man 1934 hätte begehen können. So hatte 120 Jahre zuvor, 1814, Johann Martin Hermann (1785-1841) das erste theoretisch exakt messende Planimeter erfunden und damit eigentlich das Gebiet der mechanischen Integration eröffnet. Leider veröffentlichte er wohl aufgrund einer einsetzenden Krankheit nichts dazu, und die, die von seiner Erfindung, von der Möglichkeit exakten mechanischen Integrierens und sogar von dem erfolgreichen Test eines Prototyps wußten, trugen es nicht weiter; die Erfindung wurde rasch vergessen und der Prototyp 1848 verschrottet. Erhalten blieb neben fragmen-

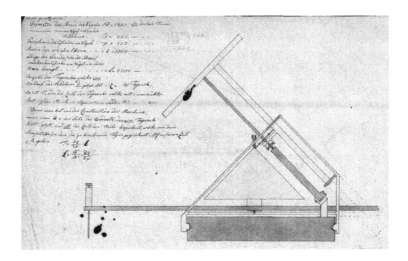

Fig. 2. Seitenriß des Hermannschen Planimeters (einzig erhaltene Zeichnung des 1848 verschrotteten Prototyp-Instruments); Nachlaß Bauernfeind, Deutsches Museum München, Archiv NL 49/14.

tarischen Dokumenten nur ein kolorierter Seitenriß mit Maßangaben, der immerhin eine Rekonstruktion erlaubt (mehr dazu in Fischer 2014):

40 Jahre nach Hermann, also 1854 und damit 80 Jahre vor dem Jahr 1934, machte Jakob Amsler (1823-1912) mit seiner Erfindung des Polarplanimeters Furore. Dieser einfachste Mechanismus, der nur aus einem zweiarmigen Gelenk und einer passend montierten Meßrolle bestand, trug wesentlich dazu bei, daß die mechanische Integration aufgrund der Einfachheit und dadurch der preiswerten Herstellung des Instruments Einzug hielt in die Büros und Kontore aller, die mit Flächenberechnung zu tun hatten: Katasterbüros, Vermesser, aber bald auch Bauingenieure, Maschinenbauer, Elektrotechniker.

Über kurz oder lang gab es fast kein Gebiet im technisch-naturwissenschaftlich-mathematischen Bereich, in dem nicht planimetriert wurde. Aber Amsler blieb nicht dort stehen; er skizzierte schon in seiner ersten Veröffentlichung zum Polarplanimeter weitere Anwendungen für andere Integrale, die nicht in erster Linie Flächeninhalte darstellten: Integratoren, später meist zutreffender Momentenplanimeter oder Potenzplanimeter genannt, zur Integration nicht nur über cartesisches $f(x)$, sondern zugleich – jedoch ohne zeichnerisch-rechnerischen Zwischenschritt – über $f(x)^2$, $f(x)^3$ und/oder $f(x)^4$; oder Harmonische Analysatoren zur Bestimmung der Fourier-Koeffizienten (Amsler 1856).

1884, 50 Jahre vor 1934, legte Amsler nochmals nach und stellte einerseits die Übertragung der Flächenmessung in der Ebene auf die Kugel mittels des wiederum äußerst einfachen Sphärischen Polarplanimeters vor, und gab andererseits eigene Ideen zu den sogenannten Präzisionsplanimetern an die Öffentlichkeit (Amsler 1884), nachdem Gottlieb Coradi in Zürich ein paar Jahre davor mit seinen – in

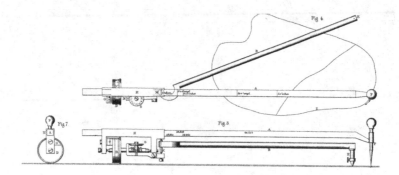

Fig. 3. Polarplanimeter von Jakob Amsler; Figuren 4, 7 und 8 aus Amsler 1856.

Zusammenarbeit mit Friedrich Hohmann entstandenen – entsprechenden Instrumenten debütiert hatte (Fischer 1995; Fischer 2002).

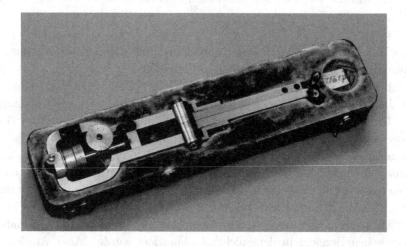

Fig. 4. Sphärisches Polarplanimeter im Etui (einziges derzeit bekanntes Exemplar; ein zur Beschwerung dienendes Polgewicht ist wohl leider abhanden gekommen), um 1883/84; Musée des arts et métiers, Paris, Inv.-Nr. 11617.

Das sphärische Polarplanimeter entsteht ganz einfach aus dem ebenen Polarplanimeter, indem die Arme des Planimetergelenks so nach unten abgeknickt werden, daß die Bewegung auf einer Kugeloberfläche möglich wird; die Theorie bleibt dabei wesentlich dieselbe.

1894, 40 Jahre vor 1934, sah dann die ersten kommerziell hergestellten Harmonischen Analysatoren; sie wurden aber nicht nach den ursprünglichen Amslerschen Ideen, sondern nach denen von Olaus Magnus Friedrich Erdmann Henrici (1840-1918) bei Coradi gebaut:

Fig. 5. Harmonischer Analysator Henrici-Coradi, Modell 50; Abbildung aus einem Coradi-Prospekt um 1935. Integrierelemente sind hier die fünf mit höchster Präzision (1/1000 mm) mattgeschliffenen Glaskugeln.

1914, im letzten halben Friedensjahr vor dem Ersten Weltkrieg (und 20 Jahre vor 1934), lag die Herstellung von Integrierinstrumenten fest in der Hand einer weniger Firmen: Jakob Amsler hatte gleich im Jahr der Erfindung des Polarplanimeters, 1854, in Schaffhausen mit der Herstellung begonnen und war 1914 bei etwa 55000 Instrumenten aller Arten – Polarplanimeter verschiedenster Modelle, Momentenplanimeter etc. – angelangt. Ihm folgte mit respektablem Abstand Gottlieb Coradi (1847-1929), der sich 1880 selbständig gemacht hatte und bald schon mit der patentierten Herstellung von Kompensationspolarplanimetern eine lange bekannte Fehlerquelle bei den Polarplanimetern, die sogenannte Rollenachsenschiefe, beseitigte; er hielt die Konkurrenz bis zum Auslaufen des Patentschutzes um 1905 in Schach. (Die Theorie des Polarplanimeters besagt, daß die Achse der Meßrolle und die Ideallinie des Fahrarms vom Planimetergelenk zum Fahrstift genau parallel sein müssen; sind sie das nicht, arbeitet das Instrument ungenau bzw. schlichtweg falsch. Durch "Kompensationsmessungen" in zwei zueinander symmetrischen Ausgangsstellungen und Mittelwertbildung läßt sich dieser Fehler beheben; dazu muß jedoch der Fahrarm "durchgeschlagen" werden können, was Coradi mit einer auf O. Lang zurückgehenden Konstruktion, die einfach als die Trennbarkeit des Planimetergelenks beschrieben werden kann, erreicht hatte.) Damit war es ihm immerhin gelungen, etwa 25000 Instrumente herzustellen, darunter Präzisionsplanimeter, die erwähnten Harmonischen Analysatoren, aber auch Momentenplanimeter und Integraphen. Die Firma der Gebrüder Haff, 1835 in Pfronten gegründet, die als erste nach Amsler schon im Jahr 1862 mit der Fabrikation eigener Planimeter begonnen hatte, war ab 1867 durch einen mündlichen oder Handschlag-Vertrag mit der Firma Keuffel & Esser (der bis 1986 Bestand hatte!) zum nahezu exklusiven

Lieferanten für die USA geworden, einem durchaus beachtlichen Absatzgebiet; sie hatte daher still und leise mit geschätzten 20000 Instrumenten Platz 3 unter den Herstellern erobert. (Eine Eigenheit dieses "Vertrags" war, daß Haff seine über Keuffel & Esser vertriebenen Instrumente nicht signierte; lediglich der später gesetzlich vorgeschriebene, gravierte Vermerk "Germany" oder "Made in Germany" ließ einen partiellen, nicht aber direkt auf Haff als Hersteller führenden Schluß zu.) Die 1873 gegründete Firma von Albert Ott (1847-1895) in Kempten hatte in den letzten Jahren vor und während des Ersten Weltkriegs einen Großauftrag des russischen Zarenreiches über allein schon 5400 Kompensationspolarplanimeter akquiriert, was sie insgesamt auf etwa 15000 Instrumente brachte. Schließlich die 1862 gegründete Firma Dennert & Pape in Altona (später sehr viel bekannter unter dem ab den 1930ern geläufigen Namen "ARISTO" und ihrer gleichnamigen Rechenstabfabrikation), bei der die Anfänge der Planimeterherstellung im Dunkeln liegen und nur eine Aussage wie "zwischen 1862 und 1872" sicher ist, hatte sich mit etwa 4500 Instrumenten als kleiner, aber qualitätvoller Hersteller etabliert. Und, meist vergessen – weil wie Haff in den U.S.A. ein überwiegend anonym fungierender Hersteller –, die mindestens seit 1879 produzierende Firma A. Blankenburg in Berlin, der wir sage und schreibe rund 7500 Instrumente (also mehr als Dennert & Pape) bis zum Ersten Weltkrieg verdanken. Die sechs Firmen Amsler, Blankenburg, Coradi, Dennert & Pape, Haff und Ott hatten damit 1914 mit rund 130000 Instrumenten schätzungsweise mehr als 95% der Gesamtproduktion an mechanischen Integrierinstrumenten hergestellt (vgl. Fischer 2002).

3. 1934

Nochmals 20 Jahre später sind wir endlich im Jahr 1934 angekommen. Hier war das Produktspektrum an Geräten zur mechanischen Integration schon klar auf dem Weg zu seinem Höhepunkt, der in der Zeit des Zweiten Weltkriegs und, nach dem zu erwartenden Einbruch, nochmals kurz danach erreicht wurde. Noch aber war es nicht soweit; doch die Schritte zur Vollendung des Programms wurden schneller. Das beginnt mit einer späten Perfektionierung der Polarplanimetertheorie durch Evert Johannes Nyström (1895-1960) im Jahr 1934:

Nullkurven beim Polarplanimeter. Das zentrale Bauteil bei fast allen mechanischen Integrierinstrumenten ist eine Meßrolle, deren Gesamtdrehung am Schluß der Befahrung einer Kurve oder meist des Randes einer Fläche zum gewünschten Integral proportional ist. Es gibt nun aber Kurven, bei deren Befahrung die Meßrolle eines Planimeters nur gleitet, aber nicht rollt; solche Kurven heißen daher in der Literatur meist Gleitkurven oder (selten, aber besser, auch wenn diese Bezeichnung sich nicht mehr durchsetzte) nach Groeneveld 1927a, 1927b Nullkurven, da die Meßrollenablesung sich während der Befahrung einer solchen Kurve nicht verändert, die Änderung dU der Meßrollenablesung U also $\equiv 0$ ist. Eine Analyse dieser Kurven ist jedoch nur dann sinnvoll (und einigermaßen in geschlossener Form machbar), wenn man gleich den Spezialfall des Polarplanimeters mit festen Armlängen l und f

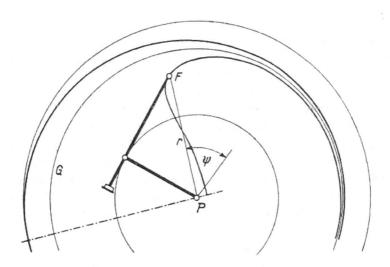

Fig. 6.   Verschiedene Nullkurven (aus Willers 1943, 114, Bild 84a, oder Willers 1951, 142, Abb. 116).

betrachtet. Man erhält für die Nullkurve(n) in Polarkoordinaten r und φ den nach φ' bzw. φ_0 aufgelösten Zusammenhang

$$\varphi' = \frac{(f^2 + l^2 - r^2)(r^2 + f^2 - l^2) + 2fg(r^2 + l^2 - f^2)}{r(f^2 + l^2 - r^2 - 2fg)\sqrt{4f^2 l^2 - (f^2 + l^2 - r^2)^2}}\, r'$$

bzw.

$$\varphi - \varphi_0 = \int_{r_0}^{r_1} \frac{(f^2 + l^2 - r^2)(r^2 + f^2 - l^2) + 2fg(r^2 + l^2 - f^2)}{r(f^2 + l^2 - r^2 - 2fg)\sqrt{4f^2 l^2 - (f^2 + l^2 - r^2)^2}}\, dr$$

(Willers 1943, 114 bzw. 1951, 143, der Nyström 1934 zitiert, ohne aber selbst die Herleitung zu geben – die auch ich hier unterdrücke, wenngleich sie nicht allzu schwierig ist ...).

r und φ (in der Abbildung mit ψ bezeichnet) sind, wie gesagt, die Polarkoordinaten; l ist die Länge des Polarms, also die Entfernung vom Pol P zum Gelenkpunkt des Planimeters; f ist die Länge des Fahrarms, also die Entfernung vom Gelenkpunkt zum Fahrstift F; und g ist die (vorzeichenrichtig gerechnete, in der Abbildung also negativ zu nehmende) Entfernung des Meßrollenauflagepunkts zum Gelenkpunkt.

Die Abbildung zeigt verschiedene Nullkurven. Gemäß des oben hergeleiteten Ausdrucks handelt es sich um zwei Kurvenscharen (sie entstehen, weil es für φ' bzw. φ zwei Vorzeichenmöglichkeiten gibt), wobei jede Schar ihrerseits erzeugt wird, indem eine ausgewählte Nullkurve der Schar um den Ursprung gedreht wird (d. h. φ_0 wird kontinuierlich variiert). Beide Nullkurvenscharen schmiegen sich asymptotisch einem Kreis an, der selbst auch eine Lösung der Differentialgleichung ist. Dieser Kreis G heißt Grundkreis; als solcher war er aber natürlich schon seit Beginn der Polarplanimetertheorie durch Amsler 1856 bekannt. Noch weitergehende Untersuchungen zu den Nullkurven verdankt man dann neben Groeneveld 1927a, 1927b vor allem eben Nyström 1934.

Momentenplanimeter I. 1934 kam auch eine Entwicklung zu ihrem Abschluß, die 1929 begonnen hatte: Damals suchte die Firma Ott nach Möglichkeiten, das Integratoren- oder Momentenplanimeterprinzip von Amsler, das (im Gegensatz zum Vorschlag in seiner Erstpublikation 1856) bei der Integration über $f(x)^2$, $f(x)^3$ oder $f(x)^4$ nun auf der Winkelvervielfachung mittels Zahnrädern basierte, durch eine einfachere und damit preiswertere Konstruktion zu ersetzen. Der Erfindungsreichtum von Heinz Adler (1908-19??), zu jener Zeit Student u. a. bei Alwin Walther (1898-1967) in Darmstadt, führte in kurzer Zeit zu Gelenkkonstruktionen, die ebenso für das Quadrieren wie auch – mit einer unwesentlichen Modifikation – für das Quadratwurzelziehen geeignet waren, siehe Figur 7.

Damit war die Integration über cartesisches $f(x)^n$ zunächst einmal für $n = 1/2$ und $n = 2$ möglich; eine ähnlich einfache Konstruktion für $n = 3$ schien jedoch zunächst außer Reichweite – und damit wohl auch eine Konkurrenz für die Amslerschen Geräte mit Winkelvervielfachung per Zahnrädern, die selbstverständlich auch

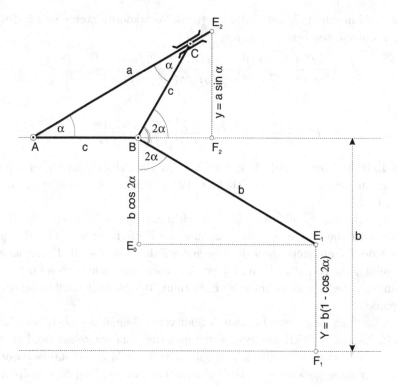

Fig. 7.   Schema eines Quadrier- und Radiziermechanismus nach Heinz Adler, 1929/30: Ist $E_2 F_2 = y$, so ist bei geeigneter Wahl der Größen a, b (und trivialerweise $a > 2c$) $E_1 F_1 = y^2$.

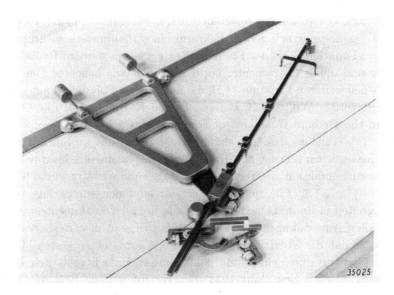

Fig. 8. 3-Rollen-Integrator Werkmeister-Ott, für Flächeninhalt, statisches und Trägheitsmoment einer Fläche; die Berechnung erfolgt simultan. Abbildung eines kurz vor 1934 entstandenen Instruments (Ott-Bild 35025).

diesen Fall $n = 3$ meisterten, weit entfernt. Doch 1934 konnte Paul Werkmeister (1878-1944) ein positives Ergebnis präsentieren (Werkmeister 1934): den 3-Rollen-Integrator Werkmeister-Ott – mitbenannt nach der Herstellerfirma, vor allem aber auch nach der Person von Ludwig Ott (1883-1946), der einer frühen Werkmeisterschen Lösung des Problems die endgültige, konstruktiv ebenso einwandfreie wie elegante Form gab, siehe Figur 8.

Diese ursprüngliche Lösung war von Werkmeister schon 1932 angegeben worden, erwies sich aber als konstruktiv zu ungelenk. Das wiederum stachelte den Ehrgeiz von Heinz Adler an, dessen allererste Lösung auf dem Papier funktionierte, aber einer konstruktiven Ausführung noch größere Hürden in den Weg gestellt hätte. Jetzt kam er auf die gleiche Idee, wie sie Amsler schon 1856 beschrieben, aber niemals gebaut hatte – und prompt war sie in Vergessenheit geraten. Adler gelang damit also die Wiedererfindung eines bereits von Amsler angegebenen Mechanismus; letztlich aber setzte sich die Ottsche Modifikation der Werkmeisterschen Idee durch (vgl. Fischer 2011, Kapitel 1 und 2).

Momentenplanimeter II. Um die gleiche Zeit, zu der die wiederentdeckten Gelenkmechanismen die bislang verwendeten Rädergetriebe zur Winkelvervielfachung ergänzten, kam aber auch noch eine dritte Lösung ins Spiel, die bis dato meist nur empirisch verwendet worden war: Steuerkurven. Nach einer überwiegend experimentell verlaufenen Phase, die schon um 1910 begonnen und zu einigen Instrumenten geführt hatte, erfolgte 1926 bei den Radialfunktionsplanimetern ein großer Schritt vorwärts mit dem Patent US 1650490 (eingereicht am 13.

September 1926, erteilt am 22. November 1927). Steuerkurven sind Vorschriften, nach denen der ursprünglich gerade Fahrarm eines Planimeters zu krümmen ist, damit andere Funktionen als der Flächeninhalt gemessen werden; Radialfunktionsplanimeter sind daher Instrumente, bei denen über eine beliebige Funktion $u(.)$ einer gegebenen weiteren Funktion $f(\varphi)$, die nun aber in Polarkoordinaten gegeben ist, integriert wird. "Willis C. Brown and Leland K. Spink, of Tulsa, Oklahoma, assignors to the Foxboro Company, of Foxboro, Massachusetts, a corporation of Massachusetts" erhielten das sehr detaillierte Patent, das ausführlich die Konstruktion eines solchen allgemeinen Radialfunktionsplanimeters beschreibt, wobei die Quadratwurzelfunktion ausdrücklich als gängige und wichtige Variante erwähnt wird, also $u(.) = \sqrt{(.)}$. (Sie spielt bei der Durchflußmengenmessung von Gasen oder Flüssigkeiten in Rohren eine herausragende Rolle.) Die Momentenberechnung in den üblichen Anwendungen findet jedoch überwiegend in cartesischen Koordinaten statt, so daß die Übertragung auf dieses Gebiet noch zu leisten blieb. Dies gelang, nun ebenfalls durch ein Patent dokumentiert, Fritz Lorenz. Sein deutsches Patent DE 610715 aus dem Jahr 1934 (ein Vorläuferpatent 1933 in der Tschechoslowakei; die allgemeine Publikation ist dann Lorenz 1935) enthält die Gleichung der Steuerkurve "für Flächenmomente n-ter Ordnung" nebst ihrer Herleitung; sie lautet

$$y = \sqrt{\sqrt[n]{(R^{n-1}x)^2} - x^2}\,.$$

R ist eine Instrumentenkonstante und beschreibt die größte x-Abszisse, die mit dem Gerät erreichbar ist. Für $n = 2$ erhält man z. B. als – in der Lorenzschen Diktion – Flächenmoment 2. Ordnung die folgende Steuerkurvengleichung:

$$y = \sqrt{Rx - x^2} \quad \Leftrightarrow \quad x^2 - Rx + y^2 = 0 \quad \Leftrightarrow \quad \left(x - \tfrac{R}{2}\right)^2 + y^2 = \left(\tfrac{R}{2}\right)^2,$$

also einen (Halb-)Kreis um $\left(\tfrac{R}{2}, 0\right)$ mit Radius $\tfrac{R}{2}$.– Ein Instrument für $n = 3$ zeigt die folgende Abbildung, vermutlich ein von Lorenz selbst gefertigter Prototyp, siehe Figur 9.

Vom Planimeter zum Integrimeter. Im Jahr 1934 befaßte sich Evert J. Nyström auch mit der Frage, wann und gegebenenfalls mit welchen Hilfsmitteln Planimeter als Integrimeter verwendet werden können, also zu jedem Zeitpunkt der Befahrung einer Kurve den richtigen Integralwert liefern, und nicht erst am Ende einer vollständigen Umfahrung[1]. Sowohl für Linear- als auch für Polarplanimeter erzielte er weitreichende, theoretisch wie praktisch interessante Ergebnisse. Am einfachsten kann seine Vorgehensweise bei der Erweiterung des Linearplanimeters zum Linearintegrimeter unter Benutzung einer einfach herzustellenden Hilfsskala gezeigt werden. (Auch wenn die nachstehende Abbildung ein sogenanntes Kugelrollplanimeter zeigt, ist die Methode natürlich nicht auf diese Form von Instrumenten beschränkt.)

---

[1] Nyström 1934.– Nyströms Artikel wurde geschrieben kurz bevor das Wort "Integrimeter" aufkam (das war 1935); er verwendet stattdessen den Allgemeinbegriff "Integrator", beruft sich dabei auf Willers 1926, und betitelt seine Arbeit daher auch "Anwendung des Planimeters als Integrator".

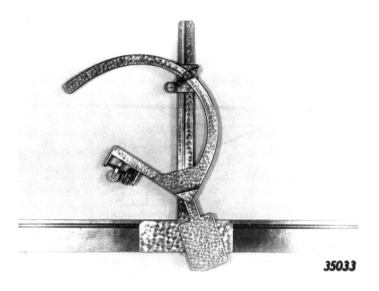

Fig. 9. Prototyp(?) des Lorenzschen Potenzplanimeters (Eigenbau Lorenz?), hier für $n = 3$; Ott-Bild 35033.

Nyström betrachtet Linearplanimeter (also solche, bei denen die Kreisführung des Polarplanimeters durch eine Geradführung ersetzt ist), bei denen die Meßrollenebene durch den Gelenkpunkt geht, oder vergleichbare Instrumente, bei denen ein Ausschlag des Fahrarms bei im übrigen festgehaltenem Instrument ebenfalls zu keiner Abwicklung der Meßrolle führt. Das nämlich ist die von Nyström im weiteren Verlauf benötigte Eigenschaft, und sie gilt in der Tat für viele Linearplanimeter, handele es sich um Linearplanimeter mit Schienen-, Spurwagen- oder Walzenführung, und unabhängig davon, ob als Integriermechanismus Ebene/Rolle, Scheibe/Rolle, Kugel/Rolle oder Kugel/Zylinder (wie in dem von ihm gewählten Beispielinstrument) verwendet werden. Diese Eigenschaft ist jedoch nicht vorhanden bei denjenigen Linearplanimetern, die durch Hinzunahme eines Geradführungsmechanismus aus einem Kompensationspolarplanimeter entstehen: denn hier schlägt die andere Montierung der Meßrolle auch auf das so entstandene Linearplanimeter durch, und Drehungen des Fahrarms um seinen Gelenkpunkt führen sehr wohl zur Meßrollendrehungen – was Nyström ausschließt.

Das Linearplanimeter sei so aufgestellt, daß die Leitgerade mit der x-Achse zusammenfällt. Beginnt man die Umfahrung – mit Anfangsablesung 0 – am Fußpunkt x_0 der Anfangsordinate, folgt dieser bis zum Punkt F_0, wo der Graph der Funktion $y = f(x)$ erreicht wird, und folgt dann diesem bis zu einem beliebigen Kurvenpunkt $F(x, y)$, an dem man den richtigen Integralwert benötigt, so müßte man jetzt zuerst zur Vervollständigung der Umfahrung (denn sie erst liefert ja diesen richtigen Wert) noch der Ordinate y des Kurvenpunktes $F(x, y)$ zurück zur x-Achse folgen und dann auf letzterer zum Anfangspunkt x_0 zurückkehren.

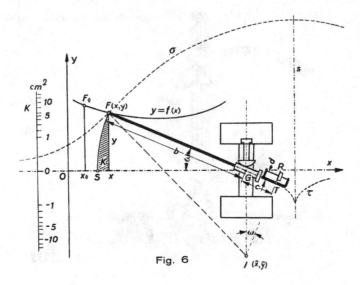

Fig. 10. Zur Verwendung des Linearplanimeters als Linearintegrimeter (Nyström 1934, 7, Fig. 6).

Die letzterwähnte Teilbefahrung verändert aber die Meßrollenablesung nicht mehr, so daß also eigentlich nur die Befahrung der Ordinate y zurück zur x-Achse zum Erreichen des richtigen Integralwerts erforderlich wäre.

Betrachtet man nun die schraffierte Fläche K, die von dieser Ordinate (in der Skizze: "y" oder "Fx"), einem Teilstück der x-Achse ("xS") und einem Kreisbogen ("SF") mit Radius = Fahrarmlänge (hier "b" genannt) zurück zum Kurvenpunkt begrenzt ist, so sieht man, daß die Befahrung des Randes von K ebenfalls nur längs der Ordinate y zu Veränderungen der Meßrollenabwicklung führt; denn sowohl auf dem Teilstück der x-Achse als auch auf dem Kreisbogen (gemäß der eingangs erwähnten Eigenschaft) findet keinerlei Meßrollenabwicklung statt. Der Flächeninhalt von K, der durch diese – gedanklich bleibende – einfache Befahrung einer Ordinate geliefert würde, stellt genau den Korrekturterm (bei Nyström: die "Korrektion") dar, der in einem beliebigen Kurvenpunkt zur dortigen Ablesung der Meßrolle hinzuzufügen ist, um das Linearplanimeter zum Linearintegrimeter zu machen; natürlich ist er am besten in der Meßeinheit des jeweiligen Instruments anzugeben.

Die Bestimmung des Korrekturterms kann empirisch, aber auch analytisch-geometrisch erfolgen (Nyström 1934, 8 9). Am einfachsten und unmittelbarsten ist seine geometrische Deutung: denn K ist offensichtlich die Hälfte eines Kreissegments mit Radius (der Fahrarmlänge) b und zur Sehne $2y$ (bzw. zum Zentriwinkel 2ω). Mit $y = b \sin(\omega)$ erhält man aus einer der bekannten Formeln für die (halbe) Fläche eines solchen Kreissegments nach kurzer Rechnung z. B.

$$K(\omega) = \tfrac{1}{2}b^2\left(\omega - \sin(\omega)\cos(\omega)\right) .$$

In dieser Form wird K auch von Nyström angegeben. Als Funktion von y ergibt sich (in Nyström 1934, 9 nur angedeutet, aber nicht mehr ausgeführt)

$$K(y) \;=\; \tfrac{1}{2}b^2 \arcsin(y/b) - \tfrac{1}{2}y\sqrt{b^2 - y^2}\,.$$

K kann nun also tabelliert werden; und wo auch immer man dann die Befahrung von $f(x)$ anhält, ergibt sich durch Addition des tabellierten Korrekturterms rasch der "richtige" Integralwert. Das Verfahren läßt sich übrigens, ohne daß dies hier noch ausgeführt werden soll, auf Polarplanimeter übertragen, wofür ebenfalls eine Korrektur-Skala eingesetzt wird.

Der Harmonische Analysator nach Harvey. John Harvey (dessen Lebensdaten erstaunlicherweise unbekannt sind, obwohl er immerhin Professor am City and Guilds College in London war) befaßte sich spätestens seit Ende der 1920er Jahre mit mechanischer Harmonischer Analyse. Zu seinen ersten Versuchen auf diesem Gebiet zählt ein Harmonischer Analysator, der jedoch prinzipiell nur eine Näherung liefert; sie beruht auf der Auswertung einer 2π-periodischen Funktion an 12 gleichabständigen Stützstellen. Er ließ sich das ein bißchen wie eine Waage aussehende und auch funktionierende Instrument mit dem englischen Patent GB 335969 (eingereicht 2. Juli 1929, akzeptiert 2. Oktober 1930) als "A Harmonic Analyser" patentieren. Das Gerät wurde 1929 von E. A. Nehan im *Mechanics and Mathematics Laboratory, City and Guilds (Engineering) College*, South Kensington, gebaut, und gelangte 1933 als Geschenk Harveys an das *Science Museum*, London (Inv. Nr. 1933 439).

Das Gebiet der exakten Mechanischen Harmonischen Analyse betrat Harvey nur wenige Jahre später mit einem Instrument, das – zumindest größtenteils – als eine geschickte, dafür aber aufwendigere Modifikation des Yuleschen Harmonischen Analysators von 1894 angesehen werden muß (George Udny Yule, 1871-1951). Eine Patentierung von Harvey zu diesem neuen Instrument ist, einer Patentrecherche zufolge, jedoch nicht mehr erfolgt.

In zwei Veröffentlichungen des Jahres 1934 (Harvey 1934a, Harvey 1934b) stellte Harvey dann seinen Harmonischen Analysator der Öffentlichkeit vor. Meyer zur Capellen 1941 bzw. 1949 und Willers 1943 bzw. 1951 weisen ausdrücklich darauf hin, daß es sich "im Wesentlichen ⟨um⟩ eine andere Form des bereits seit langem bekannten Analysators von Yule" handele (Meyer zur Capellen 1941, 224) bzw. daß Harveys Analysator "ein Konstruktionsprinzip ⟨verwende⟩, das schon in dem Apparat von Yule benutzt wurde" (Willers 1943, 176). In der Tat liegt der wesentliche Unterschied in der technischen Ausführung: während der Yulesche Harmonische Analysator als ein Zusatzgerät für ein Planimeter konzipiert war (oder andersherum: ein Planimeter als Zusatzgerät benötigte), ist der Harveysche Harmonische Analysator als in sich autonomes Instrument gestaltet, das zudem auch noch die Möglichkeit zur Verwendung als Planimeter, ja: sogar als Momentenplanimeter erlaubt.– Mehr als viele Worte sagt eine Prinzipskizze bzw. eine schematische Darstellung des Instruments, hier nur für die Hauptanwendung als Harmonischer Analysator:

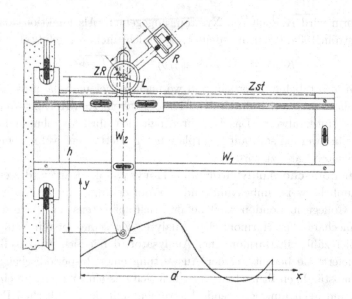

Fig. 11. Schematische Darstellung des Harmonischen Analysators nach Harvey (hier aus Willers 1951, 201, Abb. 181; identisch so schon in Willers 1943, 176, dort Bild 149).

Bei einer Bewegung des Fahrstifts F nur in y-Richtung erfährt der Wagen W_1 – und mit ihm also das gesamte Instrument – eine Parallelverschiebung in y-Richtung. Bei einer Bewegung des Fahrstifts F nur in x-Richtung bleibt W_1 in Ruhe, jedoch erfährt der Wagen W_2 eine entsprechende Verschiebung in x-Richtung; dabei dreht sich der Arm LR um den Punkt L. Technisch geschieht dies durch das Eingreifen des Zahnrades ZR in die Zahnstange Zst. Folgt der Fahrstift F einer Kurve, so überlagern sich diese beiden Bewegungen. Die Bewegung von W_2 ist durch zwei (nicht eingezeichnete) Anschläge auf eine Strecke der Länge d, der Periodenlänge, begrenzt. Macht man den Umfang des Zahnrades gleich d/n, wird sich das Zahnrad – und mit ihm der Arm LR – bei einer Bewegung des Wagens W_2 vom linken zum rechten Anschlag genau n Mal drehen. Die Anfangslage von LR (der Wagen W_2 liegt dabei am linken Anschlag an) wird folgendermaßen geregelt: für die Bestimmung der Koeffizienten a_n muß LR in Richtung der positiven y-Achse weisen; entsprechend hat LR in Richtung der negativen x-Achse zu weisen, wenn man b_n erhalten will.– Es handelt es sich nach dieser Beschreibung ersichtlich um einen Harmonischen Analysator – allerdings mit vorgegebener Periodenlänge d, so daß zu analysierende Kurven gegebenenfalls umgezeichnet werden müssen.

Ein Prototyp wurde 1933/1934 von E. A. Nehan im Mechanics and Mathematics Laboratory, City and Guilds (Engineering) College, South Kensington, gebaut, also von demselben Mechaniker, der bereits den waageähnlichen, approximativen Harmonischen Analysator Harveys hergestellt hatte:

Auch dieses Instrument ist heute noch erhalten; es gelangte 1953 als Geschenk Harveys ebenfalls an das Science Museum.– Der Analysator nach Harvey ist,

Fig. 12. Harmonischer Analysator nach Harvey. Prototyp, gebaut von E. A. Nehan, um 1933/34 (*Science Museum*, London, Inv.-Nr. 1953-318, Negativ-Nr. 706/53).

wie schon erwähnt, in jeweils leicht veränderter Konfiguration nicht nur als Harmonischer Analysator, sondern auch als Planimeter und überdies auch noch als Momentenplanimeter (bis $n = 3$) einsetzbar. Die Basis- oder Periodenlänge d beträgt übrigens environ 24 cm (Harvey 1934b, 554). Der anschließend von der Firma Amsler übernommenen Herstellung war jedoch kein allzu großer Erfolg mehr beschieden; hemmend wirkte sich vermutlich vor allem die feste Periodenlänge des Instruments aus.

4. Epilog

So weit ein kurzer Blick in das Jahr 1934 ... Man hat anhand dieser Beispiele gesehen, daß es ein Jahr war, in dem immer noch eine größere Anzahl von Überlegungen, Theorien oder Erfindungen selbst innerhalb eines so überschaubaren Gebiets wie der Mechanischen Integration stattfand. Man sieht daraus auch, daß zu diesem Zeitpunkt, 120 Jahre nach den ersten Gehversuchen, die Mechanische Integration ein durchaus lebendiger Bereich war, in dem es eben auch lange Zeit nach seiner ersten Grundlegung immer noch reichlich Aufgaben zu lösen gab und der damals eine nicht zu unterschätzende Bedeutung hatte. Das ist auch daran abzulesen, daß die weniger als 10 Jahre später erscheinenden und grundlegenden Werke über "Mathematische Instrumente" von Walther Meyer zur Capellen 1941 und Friedrich Adolf Willers 1943 dem Thema der Mechanischen Integration jeweils mehr als zwei Drittel des Umfangs ihrer Bücher einräumen – denn die Digitaltechnik steckte, mit Ausnahme der (elektro-)mechanischen Rechenmaschinen und einiger weniger

Lochkartenmaschinen, die immerhin schon etwas komplexere Abläufe verwirklichen können, noch in den Kinderschuhen. Und Konrad Zuse (1910-1995) dachte gerade erst ab 1935 über seine dann 1937 fertiggestellte und in Maßen schon universelle Rechenanlage Z1 nach ...

Heute, 2014 und damit nochmals 80 Jahre später, ist das alles Geschichte; nur noch ein einziger Hersteller – die Gebrüder Haff in Pfronten – fertigt rein mechanisch arbeitende Planimeter in geringer Stückzahl; aber alle weiterführenden Integrierinstrumente, also z. B. Präzisionsplanimeter, Potenzplanimeter, Integrimeter, Harmonische Analysatoren, Stieltjes-Planimeter oder Integraphen, haben nur noch museale Bedeutung ...

... aber das soll uns, zumindest aber mich, nicht davon abhalten, diesem Gebiet weiterhin treu zu bleiben: denn es beinhaltet – auch wenn es in der täglichen Praxis so gut wie keine Rolle mehr spielt – doch faszinierende und teilweise überraschende Einblicke. Vielleicht ist dem einen oder der anderen auch noch mein oben erwähnter Festvortrag in Erinnerung; von Armin Leutbecher weiß ich das. Als er im Sommersemester 1970, aus Münster kommend, die Vorlesung "Funktionentheorie" (Analysis IV in heutiger Zählung) hielt, begann für mich das Studium einen neuen Reiz und wirkliche Inhalte zu bekommen. Und wenn die Zeit mich von der Analytischen Zahlentheorie (Diplom) über Wahrscheinlichkeitstheorie und Statistik (Promotion) hin zur Geschichte der Mathematik (Habilitation) geführt hat, dann ist das nur scheinbar ein weiter Weg oder gar umwegig – ebensowenig, wie ich meine Tätigkeiten als Programmierer, Akademischer Rat, Konservator am Deutschen Museum oder dann 26 Jahre in der Kunst- und Kulturförderung als um- oder gar abwegig bezeichnen würde. Denn bei allem war Mathematik in den verschiedensten Erscheinungsformen mir nützlich; und auch in meinen Berliner Jahren (bei der Kulturstiftung der Länder) blieb der Kontakt mit Armin Leutbecher bestehen. Dafür danke ich ihm, und wünsche ihm jetzt zum 80. Geburtstag *Alles Gute!*

References

[Amsler 1856] Amsler, Jakob: Ueber die mechanische Bestimmung des Flächeninhaltes, der statischen Momente und der Trägheitsmomente ebener Figuren, insbesondere über einen neuen Planimeter. Schaffhausen: A. Beck und Sohn 1856 [Zuerst erschienen in der Vierteljahresschrift der naturforschenden Gesellschaft in Zürich; anschließend in dieser separaten Ausgabe]

[Amsler 1884] Amsler-Laffon, Jakob: Neuere Planimeter-Constructionen. ZfI [Zeitschrift für Instrumentenkunde] 4 (1884) 11-24

[Fischer 1995] Fischer, Joachim: Instrumente zur Mechanischen Integration. Ein Zwischenbericht. In: Brückenschläge (H.-W. Schütt und B. Weiss Hrsg.). 25 Jahre Lehrstuhl für Geschichte der exakten Wissenschaften und der Technik an der Technischen Universität Berlin, 1969-1994. Berlin: Verlag für Wissenschafts- und Regionalgeschichte Dr. Michael Engel 1995, 111-156

[Fischer 2002] Fischer, Joachim: Instrumente zur mechanischen Integration II. Ein (weiterer) Zwischenbericht. In: Chemie – Kultur – Geschichte (A. Schürmann und B. Weiss, Hrsg.). Festschrift für Hans-Werner Schütt anlässlich seines 65. Geburtstags.

Berlin/Diepholz: Verlag für Geschichte der Naturwissenschaft und der Technik 2002, 143-155

[Fischer 2011] Fischer, Joachim: Zur Rolle von Heinz Adler zwischen Ludwig Albert Ott und Alwin Oswald Walther. Hartmut Petzold zum 65. Geburtstag. In: Rechnende Maschinen im Wandel: Mathematik, Technik, Gesellschaft (U. Hashagen und II. D. Hellige Hrsg.). Festschrift für Hartmut Petzold zum 65. Geburtstag. München: Deutsches Museum 2011 (= Preprint 3; ISSN 2190-8966 [Printausgabe], ISSN 2191-0871 [Onlineausgabe]), 33 110

[Fischer 2014] Fischer, Joachim: 200 Jahre Planimeter. Ein bayerischer Vermesser und seine geniale Idee. 1814-2014. Katalog zur gleichnamigen Ausstellung, 5. Mai bis 15. September 2014, innerhalb der Vermessungshistorischen Ausstellung des Landesamts für Digitalisierung, Breitband und Vermessung. München: Landesamt für Digitalisierung, Breitband und Vermessung 2014, 171 S.

[Groeneveld 1927a] Groeneveld, Jan: Eine neue Planimetertheorie. ZfI 47 (1927) 1-16

[Groeneveld 1927b] Groeneveld, Jan: Planimetrische Integration mit Nullkurven. ZfI 47 (1927) 185-189

[Harvey 1934a] Harvey, John: A Harmonic Analyser. Engineering 138 (1934) 667-668

[Harvey 1934b] Harvey, John: Un nouvel analyseur harmonique basé sur le principe du planimètre polaire. Génie civil 105 (1934) 552-555 (bei Willers 1943 fälschlich "analysateur")

[Lorenz 1935] Lorenz, Fritz: Vorrichtung zur mechanischen Bestimmung von Flächenmomenten beliebiger Ordnung. ZfI 55 (1935) 213-217

[Meyer zur Capellen 1941] Meyer zur Capellen, Walther: Mathematische Instrumente. Leipzig: Akademische Verlagsgesellschaft 1941 [= Mathematik und ihre Anwendungen in Physik und Technik, Reihe B, Band 1]

[Meyer zur Capellen 1949] Meyer zur Capellen, Walther: Mathematische Instrumente. Leipzig: Akademische Verlagsgesellschaft 1949 [3. erweiterte Auflage; eine geplante 2. Auflage 1944 ist nicht erschienen]

[Nyström 1934] Nyström, Evert Johannes: Anwendung des Planimeters als Integrator. Societas Scientiarum Fennica, Commentationes Physico-Mathematicae VII.10 (1934)

[Werkmeister 1934] Werkmeister, Paul: Ein Dreirollen-Momentenplanimeter. ZfI 54 (1934) 410-412 [auch als Sonderdruck Dd 362 der Fa. A. Ott]

[Willers 1926] Willers, Friedrich Adolf: Mathematische Instrumente. Berlin und Leipzig: de Gruyter 1926 [= Sammlung Göschen 922]

[Willers 1943] Willers, Friedrich Adolf: Mathematische Instrumente. München und Berlin: Oldenbourg 1943

[Willers 1951] Willers, Friedrich Adolf: Mathematische Maschinen und Instrumente. Berlin: Akademie-Verlag 1951 [erweiterte Ausgabe von Willers 1943]

Chapter 2

Explicit Expressions for the Iwasawa Factors, the Metric and the Monodromy Matrices for Minimal Lagrangian Surfaces in $\mathbb{C}P^2$

Josef F. Dorfmeister & Hui Ma[a]

Technische Universität München, Fakultät für Mathematik,
D-85747 Garching, Germany
E-mail: dorfm@ma.tum.de

Department of Mathematical Sciences, Tsinghua University,
Beijing 100084, P.R. China
E-mail: hma@math.tsinghua.edu.cn

In this chapter we continue our study of equivariant minimal Lagrangian surfaces in $\mathbb{C}P^2$, characterizing the rotationally equivariant cases and providing explicit formulae for relevant geometric quantities of translationally equivariant minimal Lagrangian surfaces in terms of Weierstrass elliptic functions.

Contents

---

[a]The second author is partially supported by NSFC grant No. 11271213.

1. Introduction

Classical differential geometry was about the construction of specific examples of certain surfaces. For certain classes of surfaces, methods from the theory of integrable systems have been particularly useful. This way surfaces of finite type, and in particular many tori have been constructed. For surfaces in three-dimensional space forms one could also implement these methods on a computer and thus draw a large amount of pictures of the surfaces of interest.

A fairly general "integrable system like" method is the "loop group method" (see [8]), which allows to construct all CMC surfaces in $\mathbb{R}^3$, all minimal Lagrangian surfaces in $\mathbb{C}P^2$ etc. The method starts from some holomorphic or meromorphic data ("potentials") involving a parameter λ. Then one solves an ODE and performs an Iwasawa decomposition of this ODE solution, considered as a function of λ. The unitary factor ("extended frame") then is a "moving frame" for the surface one wants to construct. The actual surface is easily derived from the extended frame by a differentiation for λ or just by projection, called Sym-Bobenko type formula.

The main problem left is to find potentials such that the resulting surface has the properties one wants to have specifically.

Clearly, the Iwasawa decomposition is the crucial step in the construction via the loop group method. It is therefore natural to look for examples where one can carry out this decomposition explicitly. Where this is possible one does not only construct one surface, but its whole associated family.

In the case of translationally equivariant CMC surfaces in $\mathbb{R}^3$ one considers a simple generalization of the classically investigated Delaunay surfaces. Burstall and Kilian [2] have given a nice description of the loop group method of these surfaces and have given an explicit Iwasawa decomposition. Using this explicit Iwasawa decomposition they have shown that in each associate family of a Delaunay surface there are infinitely many non-congruent CMC cylinders with screw-motion symmetry. Also, even without using the explicit Iwasawa decomposition, one can prove that CMC surfaces with screw-motion symmetry all are in the associate family of some Delaunay surface, i.e. a result by Dacjer and Do Carmo (see [13]). Finally, the potentials of translationally equivariant CMC surfaces are the building blocks of CMC trinoids, i.e. CMC spheres with three properly embedded annular ends (see [10]).

A goal for the future is to construct minimal Lagrangian "trinoids" and to investigate symmetry properties of minimal Lagrangian surfaces. For this an explicit Iwasawa decomposition should be very helpful. This paper is a continuation of [6] on the discussion of this subject via the loop group method.

The paper is organized as follows: in Section 2, we recall the basic setup for minimal Lagrangian surfaces in $\mathbb{C}P^2$. In Section 3, we obtain that any vacuum can be deformed to the potential of the Clifford torus by an isometric transformation and a coordinate change. In Section 4, we characterize the rotationally equivariant minimal Lagrangian surfaces in $\mathbb{C}P^2$. In Section 5, we present the details of the computation for the Iwasawa decomposition of translationally equivariant minimal Lagrangian surfaces and we also give explicit solutions for the metrics and the associated family of immersions of such surfaces in terms of the Weierstrass $\wp$−functions. In Section 6, we provide explicit formulae for relevant geometric quantities of translationally equivariant minimal Lagrangian surfaces in terms of Weierstrass elliptic functions. In Section 7, we present a quite direct classification of homogeneous minimal Lagrangian surfaces into $\mathbb{C}P^2$ by using the loop group method.

2. Basic setup of minimal Lagrangian surfaces in $\mathbb{C}P^2$

We recall briefly the basic set-up for minimal Lagrangian surfaces in $\mathbb{C}P^2$. For details we refer to [6, 15] and references therein.

Let $\mathbb{C}P^2$ be the complex projective plane endowed with the Fubini-Study metric and $f : M \to \mathbb{C}P^2$ be a minimal Lagrangian immersion of an oriented surface. The induced metric on M generates a conformal structure with respect to which the metric is $g = 2e^u dz d\bar{z}$, where $z = x + iy$ is a local conformal coordinate on M and u is a real-valued function defined on M locally. For any Lagrangian immersion f, there exists a local horizontal lift $F : U \to S^5(1)$. We therefore have

$$F_z \cdot \overline{F} = F_{\bar{z}} \cdot \overline{F} = 0,$$
$$F_z \cdot \overline{F_z} = F_{\bar{z}} \cdot \overline{F_{\bar{z}}} = e^u, \quad F_z \cdot \overline{F_{\bar{z}}} = 0.$$

Thus $\mathcal{F} = (e^{-\frac{u}{2}} F_z, e^{-\frac{u}{2}} F_{\bar{z}}, F)$ is a Hermitian orthonormal moving frame globally defined on the universal cover of M. Set

$$\psi = F_{zz} \cdot \overline{F_{\bar{z}}}.$$

Then the cubic differential $\Psi = \psi dz^3$ is globally defined on M and independent of the choice of the local horizontal lift, which is called the Hopf differential of f. One can obtain the Gauss-Codazzi equations of a minimal Lagrangian surface given by

$$u_{z\bar{z}} + e^u - e^{-2u}|\psi|^2 = 0, \tag{1}$$
$$\psi_{\bar{z}} = 0. \tag{2}$$

Now let's introduce some notions on loop groups. Let σ denote the automorphism of $SL(3, \mathbb{C})$ of order 6 defined by

$$\sigma : g \mapsto P(g^t)^{-1} P^{-1}, \quad P = \begin{pmatrix} 0 & \alpha & 0 \\ \alpha^2 & 0 & 0 \\ 0 & 0 & 1 \end{pmatrix}, \quad \alpha = e^{2\pi i/3},$$

Let τ denote the anti-holomorphic involution of $SL(3, \mathbb{C})$ which defines the real form $SU(3)$,

$$\tau(g) := (\bar{g}^t)^{-1}.$$

Then the corresponding automorphism σ of order 6 and the anti-holomorphic automorphism τ of $sl(3, \mathbb{C})$ are

$$\sigma : \xi \mapsto -P\xi^t P^{-1}, \quad \tau : \xi \mapsto -\bar{\xi}^t.$$

Set

$$\Lambda SL(3, \mathbb{C})_\sigma = \{g : S^1 \to SL(3, \mathbb{C}) | g \text{ has finite Wiener norm, } g(\epsilon\lambda) = \sigma g(\lambda)\},$$
$$\Lambda^+ SL(3, \mathbb{C})_\sigma = \{g \in \Lambda SL(3, \mathbb{C})_\sigma | g \text{ extends holomorphically to } D, g(0) \in K^\mathbb{C}\},$$
$$\Lambda SU(3, \mathbb{C})_\sigma = \{g \in \Lambda SL(3, \mathbb{C})_\sigma | \tau(g(\frac{1}{\bar{\lambda}})) = g(\lambda)\},$$

where **D** denotes the interior of the unit disk. We will need the following loop group decomposition.

Theorem 1 (Iwasawa Decomposition theorem of $\Lambda SL(3, \mathbb{C})_\sigma$).
The multiplication map $\Lambda SU(3)_\sigma \times \Lambda^+ SL(3, \mathbb{C})_\sigma \to \Lambda SL(3, \mathbb{C})_\sigma$ is surjective. Explicitly, every element $g \in \Lambda SL(3, \mathbb{C})_\sigma$ can be represented in the form $g = hV_+$ with $h \in \Lambda SU(3)_\sigma$ and $V_+ \in \Lambda^+ SL(3, \mathbb{C})_\sigma$. One can assume without loss of generality that $V_+(\lambda = 0)$ has only positive diagonal entries. In this case the decomposition is unique.

Example 1. For the Clifford torus $f : \mathbb{C} \to \mathbb{C}P^2$, we have a horizontal lift $F : \mathbb{C} \to S^5(1)$ as follows

$$F(z, \bar{z}) = \frac{1}{\sqrt{3}}(e^{z-\bar{z}}, e^{\alpha z - \alpha^2 \bar{z}}, e^{\alpha^2 z - \alpha\bar{z}}),$$

where $\alpha = e^{\frac{2}{3}\pi i}$. It is easy to see that $\psi = F_{zz} \cdot \overline{F_{\bar{z}}} = -1$ and $e^u = 1$. Then it follows from Wu's formula in [7] that the normalized potential of the Clifford torus is given by

$$\eta = \lambda^{-1} \begin{pmatrix} 0 & 0 & i \\ i & 0 & 0 \\ 0 & i & 0 \end{pmatrix} dz.$$

We write $\eta = \lambda^{-1} A dz$ and verify $[A, \tau(A)] = 0$. Therefore the solution to $dC = C\eta, C(0, \lambda = 1) = I$ is given by $C(z, \lambda) = \exp(z\lambda^{-1}A)$. Moreover, we can perform the Iwasawa decomposition directly and obtain for the extended frame the expression $\mathbb{F}(z, \lambda) = \exp(z\lambda^{-1}A + \bar{z}\lambda\tau(A))$. Consider the translation

$$z \mapsto z + \delta, \qquad \text{with } \delta \in \mathbb{C}.$$

As a consequence, the monodromy matrix of the frame $F(z, \lambda)$ for this translation is given by

$$\mathbb{F}(z + \delta, \lambda) = M(\delta, \lambda)\mathbb{F}(z, \lambda),$$

where

$$M(\delta, \lambda) = \exp(\delta\lambda^{-1}A + \bar{\delta}\lambda\tau(A)).$$

As a consequence we obtain $F(z + \delta, \lambda) = M(\delta, \lambda)F(z, \lambda)$ and $f(z + \delta, \lambda) = M(\delta, \lambda)f(z, \lambda)$.

Clearly, the map $f_{\lambda_0} : \mathbb{C} \to \mathbb{C}P^2$ can be defined on $\mathbb{C}/\delta\mathbb{Z}$ if and only if $f_{\lambda_0}(z + \delta) = f_{\lambda_0}(z)$. By the above this is equivalent to $M(\delta, \lambda_0)f(z, \lambda_0) = f_{\lambda_0}(z)$ for all z. If we assume that f is "full" and that it descends to a torus, then the last relation implies that $M(\delta, \lambda_0)$ is a multiple of identity, $M(\delta, \lambda_0) - cI$, where c is a scalar. Clearly then, c needs to satisfy $c^3 = 1$.

Since the eigenvalues of A are i, $i\alpha$ and $i\alpha^2$, it follows that the closing conditions for $\lambda_0 \in S^1$ are

$$e^{i\lambda_0^{-1}\delta + i\lambda_0\bar{\delta}} = e^{i\lambda_0^{-1}\alpha\delta + i\lambda_0\alpha^2\bar{\delta}} = e^{i\lambda_0^{-1}\alpha^2\delta + i\lambda_0\alpha\bar{\delta}} = c,$$

which is

$$\mathrm{Re}(\lambda_0^{-1}\delta) = \frac{\pi + k\pi}{3} + l_1\pi, \tag{3}$$

$$\mathrm{Re}(\lambda_0^{-1}\alpha\delta) = \frac{\pi + k\pi}{3} + l_2\pi, \tag{4}$$

$$\mathrm{Re}(\lambda_0^{-1}\alpha^2\delta) = \frac{\pi + k\pi}{3} + l_3\pi, \tag{5}$$

for $k = 0, 1$ or 2 and $l_1, l_2, l_3 \in \mathbb{Z}$. Then it is easy to see that for any $\lambda_0 \in S^1$, the solutions to (3)-(5) are given by

$$\delta = \frac{2l_1 - l_2 - l_3}{3}\lambda_0\pi + i\frac{l_3 - l_2}{\sqrt{3}}\lambda_0\pi.$$

where $l_1 + l_2 + l_3 + 1 + k = 0$ for $k = 0, 1$ or 2 and $l_1, l_2, l_3 \in \mathbb{Z}$. Therefore, for arbitrary λ_0, we obtain $\delta(\lambda_0)\mathbb{Z} = \lambda_0\delta(1)\mathbb{Z}$, i.e. the lattice $\delta(\lambda_0)\mathbb{Z}$ is obtained from the lattice $\delta(1)\mathbb{Z}$ by rotation by λ_0. This implies the following

Proposition 1. *Every member in the associated family of the Clifford torus is a torus.*

3. Vacuum solutions

A "vacuum" is an extended framing whose normalized potential is given by $\eta = \lambda^{-1}Adz$ with $A \in \mathcal{G}_{-1}$ a constant matrix satisfying $[A, \tau(A)] = 0$. To clarify what this means we consider the constant matrix

$$A = \begin{pmatrix} 0 & 0 & a \\ b & 0 & 0 \\ 0 & a & 0 \end{pmatrix} \in \mathcal{G}_{-1}. \quad \text{Then } \tau(A) = \begin{pmatrix} 0 & -\bar{b} & 0 \\ 0 & 0 & -\bar{a} \\ -\bar{a} & 0 & 0 \end{pmatrix},$$

and the condition $[A, \tau(A)] = 0$ says $|a|^2 = |b|^2$.

Let's next write $a = ire^{i\theta}$ and $b = ire^{i\beta}$. Now take the following isometric transformation

$$\begin{pmatrix} e^{i\delta} & & \\ & e^{-i\delta} & \\ & & 1 \end{pmatrix} A \begin{pmatrix} e^{-i\delta} & & \\ & e^{i\delta} & \\ & & 1 \end{pmatrix} = \begin{pmatrix} 0 & 0 & ire^{i(\theta+\delta)} \\ ire^{i(\beta-2\delta)} & 0 & 0 \\ 0 & ire^{i(\theta+\delta)} & 0 \end{pmatrix}.$$

Then choose δ such that $\theta + \delta = \beta - 2\delta$, i.e., $\delta = \frac{\beta - \theta}{3}$. Thus,

$$\eta = \lambda^{-1} i r e^{i\frac{2\theta + \beta}{3}} \begin{pmatrix} 0 & 0 & i \\ i & 0 & 0 \\ 0 & i & 0 \end{pmatrix} dz.$$

Finally, choose a new coordinate: $z \mapsto w = re^{i\frac{2\theta + \beta}{3}} z$, and we obtain

$$\eta = \lambda^{-1} \begin{pmatrix} 0 & 0 & i \\ i & 0 & 0 \\ 0 & i & 0 \end{pmatrix} dw.$$

Summing up we have

Proposition 2. *Any vacuum can be deformed by an isometric transformation and a coordinate change (if necessary) to the potential of the Clifford torus.*

4. Equivariant minimal Lagrangian immersions into $\mathbb{C}P^2$

4.1. *General background*

For all classes of surfaces, the surfaces admitting some symmetries are of particular interest and beauty.

While a basic definition of a symmetry R for a surface $f(M)$ may only mean $Rf(M) = f(M)$, it is very useful to know that if the induced metric is complete, then on the universal cover $\tilde{M}$ of M one finds some automorphism γ such that

$$f(\gamma \cdot z) = Rf(z) \text{ for all } z \in \tilde{M}. \tag{6}$$

Therefore, in this paper, a "symmetry" will always be a pair $(\gamma, R) \in (Aut(M), Iso(\mathbb{C}P^2))$, such that (6) holds.

The usual transition to the associated family f_λ then produces some family $R(\lambda)$ of isometries of $\mathbb{C}P^2$ such that we have

$$f_\lambda(\gamma \cdot z) = R(\lambda)f_\lambda(z) \text{ for all } z \in \tilde{M}.$$

More details can be found in [4], [5], [9].

In this paper we will investigate minimal Lagrangian immersions for which there exists a one-parameter family $(\gamma_t, R_t) \in (Aut(M), Iso(\mathbb{C}P^2))$ of symmetries.

Definition 1. Let M be any connected Riemann surface and $f : M \to \mathbb{C}P^2$ an immersion. Then f is called equivariant, relative to the one-parameter group $(\gamma_t, R(t)) \in (Aut(M), Iso(\mathbb{C}P^2))$, if

$$f(\gamma_t \cdot p) = R(t)f(p)$$

for all $p \in M$ and all $t \in \mathbb{R}$.

By the definition above, any Riemann surface M admitting an equivariant minimal Lagrangian immersion admits a one-parameter group of (biholomorphic) automorphisms. Fortunately, the classification of such surfaces is very simple:

Theorem 2. *(Classification of Riemann surfaces admitting one-parameter groups of automorphisms, e.g. [12])*

(1) S^2,
(2) $\mathbb{C}$, $\mathbf{D}$,
(3) $\mathbb{C}^*$,
(4) $\mathbf{D}^*, \mathbf{D}_r$,
(5) $T = \mathbb{C}/\Lambda_\tau$,

where the superscript "$*$" denotes deletion of the point 0, the subscript "r" denotes the open annulus between $0 < r < 1/r$ and Λ_τ is the free group generated by the two translations $z \mapsto z + 1$, $z \mapsto z + \tau$, $\operatorname{Im}\tau > 0$.

Looking at this classification a bit more closely, one sees that after some biholomorphic transformations one obtains the following picture, including representative one-parameter groups:

Theorem 3. *(Classification of Riemann surfaces admitting one-parameter groups of automorphisms and representatives for the one-parameter groups, e.g. [12])*
   (1) S^2, group of all rotations about the z-axis,
   (2a) $\mathbb{C}$, group of all real translations,
   (2b) $\mathbb{C}$, group of all rotations about the origin 0,
   (2c) $\mathbf{D}$, group of all rotations about the origin 0,
   (2d) $\mathbf{D} \cong \mathbb{H}$, group of all real translations,
   (2e) $\mathbf{D} \cong \mathbb{H} \cong \log \mathbb{H} = \mathbb{S}$, the strip between $y = 0$ and $y = \pi$, group of all real translations,
   (3) $\mathbb{C}^*$, group of all rotations about 0,
   (4) $\mathbf{D}^*, \mathbf{D}_r$, group of all rotations about 0,
   (5) T, group of all real translations.

For later purposes we state the following

Definition 2. Let $f : M \to \mathbb{C}P^2$ be an equivariant minimal Lagrangian immersion, then f will be called "translationally equivariant", if the group of automorphisms acts by (all real) translations. It will be called "rotationally equivariant", if the group acts by (all) rotations (about the origin).

Remark 1. The case of S^2 is usually special and has been treated in the literature. For minimal Lagrangian immersions this case has been treated in (see [17]) and it has been shown that any minimal Lagrangian immersion f from a sphere to $\mathbb{C}P^2$ is totally geodesic and it is the standard immersion of S^2 into $\mathbb{C}P^2$ (see [17]). Therefore, up to a few exceptions, we will exclude the case S^2 from the discussions in this chapter. We would like to point out, however, that this case could be discussed like the general case below. In this case we would need to deal with algebraic solutions to elliptic equations listed below.

4.2. *Rotationally equivariant minimal Lagrangian immersions*

In view of Theorem 3 there are two types of equivariant surfaces, translationally equivariant surfaces and rotationally equivariant surfaces. The translationally equivariant case will be discussed in Section 5. Thus here it remains to consider rotationally equivariant minimal Lagrangian immersions.

There are essentially three types of such surfaces: those without fixed point in M, those with exactly one fixed point in M and those with two fixed points in M, i.e. $M = S^2$. Let's first consider the cases $\mathbb{C}^*, \mathbf{D}^*, \mathbf{D}_r$, which do not contain the fixed point of the group of rotations.

Using the covering map $w \to exp(iw)$, we see that the rotationally symmetric minimal Lagrangian immersion f is obtained from some translationally equivariant minimal Lagrangian immersion $\tilde{f}$ defined on some strip $\mathbb{S}$. Obviously, the condition of descending to the given rotationally equivariant minimal Lagrangian immersion is equivalent to $\tilde{f}$ being $2\pi-$periodic in the variable corresponding to the group of translations. Since $\tilde{f}$ is actually defined on $\mathbb{C}$ (see Proposition 4.4, [6]) and real analytic, it is clear that $\tilde{f}$ is $2\pi-$periodic on $\mathbb{C}$ and thus descends to a rotationally equivariant minimal Lagrangian immersion on $\mathbb{C}^*$.

Theorem 4. *Consider a rotationally equivariant minimal Lagrangian immersion $f : M \to \mathbb{C}P^2$ defined on $M = \mathbb{C}^*, \mathbf{D}^*, \mathbf{D}_r$. Then f can be extended without loss of generality to $\mathbb{C}^*$ and can be obtained from some $2\pi-$periodic translationally equivariant minimal Lagrangian immersion defined on $\mathbb{C}$ by projection.*

Next we consider the rotationally equivariant minimal Lagrangian immersions defined on $\mathbb{C}$ or $\mathbf{D}$. In these cases we remove the fixed point 0 and obtain rotationally equivariant minimal Lagrangian immersions without fixed point. The last theorem shows that we only need to consider the case $M = \mathbb{C}$. Clearly this is a special case of a rotationally equivariant minimal Lagrangian immersion defined on $\mathbb{C}^*$. Finally, considering S^2, we can assume without loss of generality that the group acts by rotations about the $z-$axis. Then any rotationally equivariant minimal Lagrangian immersion on S^2 is a special case of a rotationally equivariant minimal Lagrangian immersion defined on $\mathbb{C}$.

4.3. *Rotationally equivariant minimal Lagrangian immersions defined on $\mathbb{C}$*

Let $f : \mathbb{C} \to \mathbb{C}P^2$ be a rotationally equivariant minimal Lagrangian immersion and $\mathbb{F}(z, \lambda)$ an extended frame. We normalize $\mathbb{F}$ by $\mathbb{F}(z = 0, \lambda) = I$. Then the equivariance is reflected by the equation

$$\mathbb{F}(e^{it}z, \lambda) = \chi(e^{it}, \lambda)\mathbb{F}(z, \lambda)\mathcal{K}(e^{it}, z),$$

where $\chi(e^{it}, \lambda) \in SU(3)$ and $\mathcal{K}(e^{it}, z) \in K = U(1)$. For more details see Section 4 of [6].

Setting $z = 0$ shows $\chi(e^{it}, \lambda)\mathcal{K}(e^{it}, 0) = I$. As a consequence, χ is independent of λ and a one-parameter group in K. Hence

$$\chi(e^{it}, \lambda) = \exp(itr\delta),$$

where $r \in \mathbb{R}$, $t \in \mathbb{R}$ and $\delta = \mathrm{diag}(1, -1, 0)$.

Performing a Birkhoff splitting of $\mathbb{F}$, $\mathbb{F} = \mathbb{F}_- V_+$, we derive

$$\mathbb{F}_-(e^{it}z, \lambda) = \exp(itr\delta)\mathbb{F}_-(z, \lambda)\exp(-itr\delta).$$

For the Maurer-Cartan form $\eta_- = \mathbb{F}_-^{-1}d\mathbb{F}_-$ of $\mathbb{F}_-$ we then obtain

$$(e^{it})^*\eta_- = \exp(itr\delta)\eta_-\exp(-itr\delta).$$

Hence the normalized potential η_- has the form $\eta_- = \lambda^{-1}Adz$, where

$$A = \begin{pmatrix} 0 & 0 & az^{m-1} \\ bz^{-2m-1} & 0 & 0 \\ 0 & az^{m-1} & 0 \end{pmatrix}$$

with $m \in \mathbb{Z}$ and certain complex numbers a and b.

Since we had normalized everything at $z = 0$, the normalized potential is holomorphic at $z = 0$. Hence $m \geq 1$ and $b = 0$, which implies that the cubic Hopf differential Ψ vanishes. It follows from (1) that the Gauss curvature satisfies $K = -u_{z\bar{z}}e^{-u} = 1$. Then from the Gauss equation we obtain that $\frac{S}{2} = 1 - K$ vanishes, where S is the norm square of the second fundamental form of the surface. Therefore f is totally geodesic, hence the image of this minimal Lagrangian immersion f lies in $\mathbb{R}P^2$ up to isometries of $\mathbb{C}P^2$. We thus have reproved part of Corollary 3.9 of [11]. As a consequence we obtain

Theorem 5. *Any minimal Lagrangian immersion f from $\mathbb{C}$ or S^2 into $\mathbb{C}P^2$ which is rotationally equivariant has a vanishing cubic Hopf differential, and therefore is totally geodesic in $\mathbb{C}P^2$ and its image is, up to isometries of $\mathbb{C}P^2$, contained in $\mathbb{R}P^2$.*

Remark 2. This result can also be obtained by using the explicit Iwasawa decomposition discussed below. Also see Remark 4.

5. Explicit discussion of translationally equivariant minimal Lagrangian immersions

For the convenience of the reader, we recall some basic notions from [6].

5.1. *Application of a result by Burstall and Kilian for translationally equivariant minimal Lagrangian immersions*

We now consider translationally equivariant minimal Lagrangian immersions defined on some strip $\mathbb{S}$ with values in $\mathbb{C}P^2$, $f : \mathbb{S} \to \mathbb{C}P^2$, i.e. minimal Lagrangian immersions for which there exists a one-parameter subgroup $R(t)$ of $SU(3)$ such that

$$f(t + z, t + \bar{z}) = R(t)f(z, \bar{z})$$

for all $z \in \mathbb{S}$ and $t \in \mathbb{R}$. Following the approach of [1], we have shown

Theorem 6 (see [6]). *For the extended frame $\mathbb{F}$ of any translationally equivariant minimal Lagrangian immersion we can assume without loss of generality $\mathbb{F}(0, \lambda) = I$ and*

$$\mathbb{F}(t + z, \lambda) = \chi(t, \lambda)\mathbb{F}(z, \lambda),$$

with $\chi(t, \lambda) = e^{tD(\lambda)}$ for some $D(\lambda) \in \Lambda su(3)_\sigma$.

Note that $\mathbb{F}$ satisfies

$$\mathbb{F}^{-1}\mathbb{F}_z = \frac{1}{\lambda}\begin{pmatrix} 0 & 0 & ie^{\frac{u}{2}} \\ -i\psi e^{-u} & 0 & 0 \\ 0 & ie^{\frac{u}{2}} & 0 \end{pmatrix} + \begin{pmatrix} \frac{u_z}{2} & & \\ & -\frac{u_z}{2} & \\ & & 0 \end{pmatrix}$$

$$:= \lambda^{-1}U_{-1} + U_0,$$

$$\mathbb{F}^{-1}\mathbb{F}_{\bar{z}} = \lambda\begin{pmatrix} 0 & -i\bar{\psi}e^{-u} & 0 \\ 0 & 0 & ie^{\frac{u}{2}} \\ ie^{\frac{u}{2}} & 0 & 0 \end{pmatrix} + \begin{pmatrix} -\frac{u_{\bar{z}}}{2} & & \\ & \frac{u_{\bar{z}}}{2} & \\ & & 0 \end{pmatrix}$$

$$:= \lambda V_1 + V_0.$$

and we can assume

$$\mathbb{F}(x + iy, \lambda) = e^{(x+iy)D(\lambda)}U_+(y, \lambda)^{-1} = e^{xD(\lambda)}\mathbb{F}(iy, \lambda), \qquad (7)$$

with $U_+(y, \lambda) \in \Lambda^+ SL(3, \mathbb{C})_\sigma$.

5.2. *The basic set-up for an explicit Iwasawa decomposition*

We have seen above in subsection 5.1 that every translationally equivariant minimal Lagrangian immersion can be obtained from some potential of the form

$$\eta = D(\lambda)dz,$$

where

$$D(\lambda) = \lambda^{-1}D_{-1} + D_0 + \lambda D_1 \in \Lambda su(3)_\sigma.$$

The general loop group approach requires to consider the solution to $dC = C\eta$, $C(0, \lambda) = I$. This is easily achieved by $C(z, \lambda) = \exp(zD)$.

Next one needs to perform an Iwasawa splitting. In general this is very complicated and difficult to carry out explicitly. But, for translationally equivariant minimal Lagrangian surfaces in $\mathbb{C}P^2$, one is able to carry out an explicit Iwasawa decomposition of $\exp(zD)$.

In view of equation (7) we obtain

$$\mathbb{F}(iy, \lambda) = e^{iyD}U_+(y, \lambda)^{-1}. \tag{8}$$

Using (8) we obtain for the Maurer-Cartan form $\alpha = \mathbb{F}^{-1}d\mathbb{F} = Adx + Bdy$ of $\mathbb{F}$ the equations

$$A_\lambda(y) = U_+(y, \lambda)DU_+(y, \lambda)^{-1}, \tag{9}$$

$$B_\lambda(y) = U_+(y, \lambda)iDU_+(y, \lambda)^{-1} - \frac{d}{dy}U_+(y, \lambda)U_+(y, \lambda)^{-1}.$$

Writing, on the other hand, $\alpha = U+V$ with U a $(1,0)-$form and V a $(0,1)-$form, we obtain

$$U_+(y, \lambda)DU_+^{-1}(y, \lambda) = \lambda^{-1}U_{-1} + U_0 + \lambda V_1 + V_0 =: \Omega, \tag{10}$$

$$\frac{d}{dy}U_+(y, \lambda)U_+(y, \lambda)^{-1} = 2i(\lambda V_1 + V_0). \tag{11}$$

The above two equations are the basis for an explicit computation of the Iwasawa decomposition of $\exp(zD(\lambda))$.

It is important to note that because U_+ only depends on y and Ω is of the form

$$\Omega = \begin{pmatrix} \frac{u_z - u_{\bar{z}}}{2} & -i\lambda\bar{\psi}e^{-u} & i\lambda^{-1}e^{\frac{u}{2}} \\ -i\lambda^{-1}\psi e^{-u} & -\frac{u_z - u_{\bar{z}}}{2} & i\lambda e^{\frac{u}{2}} \\ i\lambda e^{\frac{u}{2}} & i\lambda^{-1}e^{\frac{u}{2}} & 0 \end{pmatrix},$$

both u and ψ also only depend on y.

Lemma 1. *If f is a translationally equivariant minimal Lagrangian immersion into $\mathbb{C}P^2$ with respect to translations in x-direction, then the metric only depends on y and the cubic Hopf differential has a constant coefficient.*

There will be two steps for the computation of the Iwasawa decomposition of $\exp(zD(\lambda))$.

Step 1: Solve equation (10) in any way one pleases by some matrix Q. Then U_+ and Q satisfy

$$U_+ = QE, \quad \text{where } E \text{ commutes with } D.$$

Step 2: Solve equation (11). This will generally only mean to carry out two integrations in one variable.

5.3. *Evaluation of the characteristic polynomial equations*

Step 1 mentioned above actually consists of two sub-steps. First of all one determines Ω from D and then one computes a solution W to **Step 1:** Solve the equation (10).

In this section we will discuss the first sub-step. In our case we observe that D and Ω are conjugate and therefore have the same characteristic polynomials. Using the explicit form of Ω stated just above and writing D in the form

$$D = \begin{pmatrix} \alpha & -\lambda \bar{b} & \lambda^{-1}a \\ \lambda^{-1}b & -\alpha & -\lambda \bar{a} \\ -\lambda \bar{a} & \lambda^{-1}a & 0 \end{pmatrix} \in \Lambda_1 \subset \Lambda su(3)_\sigma,$$

where α, a and b are constants, (10) is equivalent to

$$2e^u + |\psi|^2 e^{-2u} + \frac{1}{4}(u')^2 = -\alpha^2 + 2|a|^2 + |b|^2 =: \beta, \qquad (12)$$

$$\psi = -ia^2 b, \qquad (13)$$

where α, a, b and ψ are constants.

Remark 3. It has been noticed long time ago that the cases $\psi = 0$ and $u =$ constant are related with very special surfaces. As pointed out above, the case $\psi = 0$ implies that the surface is an open portion of $\mathbb{R}P^2$. The case $u =$ constant yields a flat surface thus is, up to isometries, an open portion of the Clifford torus (see [14]). Both cases can be treated like the general case below. In the first case, as already mentioned, one needs to use hyperbolic solutions and in the second case one needs to use constant solutions to the elliptic equations occurring in this context. At any rate, from here on (unless stated explicitly otherwise) we will assume that ψ is not 0 and u is not constant. In particular, we will assume $u' \not\equiv 0$.

5.4. *Explicit solutions for metric and cubic form in terms of Weierstrass $\wp$—functions*

We start by noticing that (12) is a first integral of the Gauss equation

$$\frac{1}{4}u'' + e^u - |\psi|^2 e^{-2u} = 0. \qquad (14)$$

Making the change of variables $w = e^u$ in (12), we obtain equivalently

$$(w')^2 + 8w^3 - 4\beta w^2 + 4|\psi|^2 = 0. \qquad (15)$$

Set $w(y) = \frac{\beta}{6} - \frac{v(y)}{2}$. We obtain the fundamental differential equation

$$(v')^2 = 4v^3 - g_2 v - g_3$$

of the Weierstrass function $\wp(z) = \wp(z; g_2, g_3)$ with

$$g_2 = \frac{4}{3}\beta^2, \qquad g_3 = 16|\phi|^2 - \frac{8}{27}\beta^3.$$

Thus the general non-constant solution to (15) can be given by

$$w(y) = \frac{\beta}{6} - \frac{\wp(y - y_0; g_2, g_3)}{2}$$

for a constant y_0.

Since the beginning of section 5.2 we discuss the construction of translationally equivariant minimal Lagrangian immersions, starting from some Delaunay type potential. It is clear that the metrics of such surfaces only depends on y and is bounded for $y \in \mathbb{R}$ (cf. [6]). Hence the periodic function $w(y)$ above is bounded along the real line. This implies at one hand that y_0 is not real and on the other hand that there exists a point, where the derivative of w vanishes. We change coordinates such that

$$w'(0) = 0. \tag{16}$$

This convention in combination with (12) implies

$$2a_1 + \frac{|\psi|^2}{a_1^2} = \beta,$$

where $a_1 := e^{u(0)} > 0$. Considering β as a function of a_1, one can easily see that $\beta^3 \geq 27|\psi|^2$. Thus the discriminant of the cubic equation

$$4v^3 - g_2 v - g_3 = 0 \tag{17}$$

satisfies

$$\Delta = g_2^3 - 27g_3^2 = 256|\psi|^2(\beta^3 - 27|\psi|^2) > 0, \tag{18}$$

if and only if $\psi \neq 0$ and $\beta^3 \neq 27|\psi|^2$.

Remark 4. When $\psi = 0$, $g_2 = \frac{4\beta^2}{3} > 0$, $g_3 = -\frac{8}{27}\beta^3 < 0$ and the roots of the Weierstrass equation (17) are

$$e_1 = e_2 = \frac{\beta}{3}, \quad e_3 = -\frac{2\beta}{3},$$

and the Weierstrass elliptic function $\wp(z; g_2, g_3)$ is reduced to (cf. 18.12.3, [16])

$$\wp(z; g_2, g_3) = \frac{\beta}{3} + \beta[\sinh(\sqrt{\beta}z)]^{-2}.$$

Thus the initial condition (16) gives the following solution to (15)

$$w(y) = \frac{\beta}{2\cosh^2(\sqrt{\beta}y)}.$$

This is nothing but the metric of the real projective plane in $\mathbb{C}P^2$.

When $\beta^3 = 27|\psi|^2$, $g_2 = \frac{4\beta^2}{3} > 0$, $g_3 = \frac{8}{27}\beta^3 > 0$ and there also are two equal real roots of (17) given by (cf. 18.12.25, [16])

$$e_1 = \frac{2\beta}{3}, \quad e_2 = e_3 = -\frac{\beta}{3}$$

and the Weierstrass elliptic function $\wp(z; g_2, g_3)$ is reduced to (cf. 18.12.27, [16])

$$\wp(z; g_2, g_3) = -\frac{\beta}{3} + \beta[\sin(\sqrt{\beta}z)]^{-2}.$$

Thus the initial condition (16) gives

$$w(y) = \frac{\beta}{3} - \frac{\beta}{2}\frac{1}{\cos^2(\sqrt{\beta}y)}.$$

But since $w(y)$ needs to be positive, we exclude this solution. On the other hand, we obtain the three roots of

$$8w^3 - 4\beta w^2 + 4|\psi|^2 = 0 \tag{19}$$

given by

$$a_1 = a_2 = \frac{\beta}{3}, \quad a_3 = -\frac{\beta}{6}.$$

Thus the solution of (15) is a constant function $w(y) \equiv \frac{\beta}{3}$, which corresponds to a flat minimal Lagrangian surface and has been ruled out in the beginning of our discussion.

Now for the general case, g_2 and g_3 are real and $\Delta > 0$, thus there are three distinct non-zero real roots of (17), denoted by

$$e_1 > e_2 > e_3.$$

Because of the initial condition (16), we know from (15) that

$$e^{u(0)} = w(0) =: a_1$$

is a root of (19). And now we assume that this is the largest root of (19). Recall that the half-periods ω, ω' and $\omega + \omega'$ of the Weierstrass elliptic function are related to the roots e_1, e_2 and e_3 by

$$\wp(\omega) = e_1, \qquad \wp(\omega + \omega') = e_2, \qquad \wp(\omega') = e_3.$$

Consequently, $\wp'(\omega) = \wp'(\omega') = \wp'(\omega + \omega') = 0$. The initial condition (16) thus yields the particular solution of (15) given by

$$e^{u(y)} = w(y) = \frac{\beta}{6} - \frac{\wp(y - \omega'; g_2, g_3)}{2}. \tag{20}$$

Remark that now the half-period

$$\omega = \int_{e_1}^{\infty} \frac{dt}{\sqrt{4t^3 - g_2t - g_3}}$$

is real, whereas the other half-period

$$\omega' = i \int_{-\infty}^{e_3} \frac{dt}{\sqrt{|4t^3 - g_2t - g_3|}}$$

is purely imaginary. It is easy to see that the solution $u(y)$ inherits from the Weierstrass elliptic function the following properties:

(1) $u(y + 2\omega) = u(y)$,

(2) $u(-y) = u(y)$,

(3) $u(\omega) = \log a_2$ and $u'(\omega) = 0$, where $a_2 = \frac{\beta}{6} - \frac{e_2}{2} > 0$.

In particular, $\hat{u}(y) = u(y + \omega)$ is also a solution to (14) with $\hat{u}'(0) = 0$.

Thus for any (in x−direction) translationally equivariant minimal Lagrangian surface in $\mathbb{C}P^2$, its metric conformal factor e^u is given by (20) in terms of a Weierstrass elliptic function and its cubic Hopf differential is constant and given by (13).

For our loop group setting the assumption $u'(0) = 0$ has an important consequence:

Theorem 7. *By choosing the coordinates such that the metric for a given translationally equivariant minimal Lagrangian immersion has a vanishing derivative at $z = 0$, we obtain that the generating matrix D satisfies $D_0 = 0$.*

We will therefore always assume this condition from here on.

5.5. *Solving equation* (10)

The main goal of this subsection is to find some "sufficiently nice" matrix function Q satisfying (10), i.e.

$$QDQ^{-1} = \Omega. \tag{21}$$

Recall that for translationally equivariant minimal Lagrangian surfaces, the potential matrix D coincides with $A_\lambda(0) = \Omega|_{y=0}$ of (9) so we have (including the convention above about the origin)

$$D = \begin{pmatrix} 0 & -i\lambda\bar{\psi}e^{-u(0)} & i\lambda^{-1}e^{\frac{u(0)}{2}} \\ -i\lambda^{-1}\psi e^{-u(0)} & 0 & i\lambda e^{\frac{u(0)}{2}} \\ i\lambda e^{\frac{u(0)}{2}} & i\lambda^{-1}e^{\frac{u(0)}{2}} & 0 \end{pmatrix},$$

where $\alpha = -\frac{iu'(0)}{2} = 0$, $a = ie^{\frac{u(0)}{2}}$ and $b = -i\psi e^{-u(0)}$. We may summarize the following proposition:

Proposition 3. *Up to isometries in $\mathbb{C}P^2$, any translationally equivariant minimal Lagrangian surface can be generated by a potential of the form*

$$\begin{pmatrix} 0 & -\lambda\bar{b} & \lambda^{-1}a \\ \lambda^{-1}b & 0 & -\lambda\bar{a} \\ -\lambda\bar{a} & \lambda^{-1}a & 0 \end{pmatrix} dz, \tag{22}$$

where a is purely imaginary and both a and $b = \frac{i\psi}{a^2}$ are constants.

Thus the characteristic polynomial of D in (22) is given by

$$\det(\mu I - D(\lambda)) = \mu^3 + \beta\mu - 2i\text{Re}(\lambda^{-3}\psi).$$

Remark 5. It is easy to derive from (18) that the discriminant of the above polynomial satisfies

$$\Delta = (\frac{\beta}{3})^3 - [\text{Re}(\lambda^{-3}\psi)]^2 \geq (\frac{\beta}{3})^3 - |\psi|^2 \geq 0.$$

The second equal sign holds when $\lambda^{-3}\psi$ is real and the third one holds only for special cases which we excluded. Hence for the general case when $\Delta > 0$, $D(\lambda)$ has three distinct purely imaginary roots for any choice of $\lambda^{-3}\psi$. Moreover, the root 0 occurs if and only if $\lambda^{-3}\psi$ is purely imaginary. This case can only happen for six different values of λ (See Lemma 5.3 in [6]).

Denote the eigenvalues of $D(\lambda)$ by $\mu_1 = id_1$, $\mu_2 = id_2$, $\mu_3 = id_3$.
The following relations will be frequently used later.

$$d_1 + d_2 + d_3 = 0, \quad d_1 d_2 + d_2 d_3 + d_3 d_1 = -\beta, \quad d_1 d_2 d_3 = -2\text{Re}(\lambda^{-3}\psi).$$

Now take

$$Q_0 = \text{diag}(ia^{-1}e^{\frac{u}{2}}, -iae^{-\frac{u}{2}}, 1), \tag{23}$$

such that

$$\hat{\Omega} = Q_0^{-1}\Omega Q_0 = \begin{pmatrix} -\frac{iu'}{2} & i\lambda\bar{\psi}a^2 e^{-2u} & \lambda^{-1}a \\ \lambda^{-1}b & \frac{iu'}{2} & -\lambda a^{-1}e^u \\ -\lambda a^{-1}e^u & \lambda^{-1}a & 0 \end{pmatrix}$$

has the same coefficients at λ^{-1} as $D(\lambda)$.

Consider now a 3×3 matrix $\hat{Q}$ given by

$$\hat{Q} = \begin{pmatrix} A & \gamma \\ 0 & c \end{pmatrix}, \quad A = \begin{pmatrix} p & q \\ s & t \end{pmatrix}, \quad \gamma = \begin{pmatrix} v_1 \\ v_2 \end{pmatrix}, \tag{24}$$

where c is a scalar. Put

$$D = \begin{pmatrix} E & \xi \\ -\bar{\xi}^t & 0 \end{pmatrix}, \quad \hat{\Omega} = \begin{pmatrix} \Omega' & \eta \\ \zeta & 0 \end{pmatrix},$$

where E and Ω' are 2×2 matrices, ξ and η are 2×1 matrices, and ζ is a 1×2 matrix.

Then $\hat{Q}D\hat{Q}^{-1} = \hat{\Omega}$ is equivalent to the following equations

$$AEA^{-1} - \gamma\bar{\xi}^t A^{-1} = \Omega', \tag{25}$$

$$-(AEA^{-1}\gamma - \gamma\bar{\xi}^t A^{-1}\gamma) + A\xi = c\eta, \tag{26}$$

$$-c\bar{\xi}^t A^{-1} = \zeta, \tag{27}$$

$$\bar{\xi}^t A^{-1}\gamma = 0. \tag{28}$$

Inserting (27) into (28) gives $\zeta\gamma = 0$, which implies that

$$\gamma = \begin{pmatrix} \lambda^{-1}a^2 \\ \lambda e^u \end{pmatrix} h,$$

where h is an arbitrary factor. Noticing that (25) is equivalent to $AE - \gamma \bar{\xi}^t = \Omega' A$ and inserting the assumption (24), we obtain the following equivalent equations

$$\frac{iu'}{2}p + \lambda^{-1}bq - i\lambda a^2 \bar{\psi}e^{-2u}s - |a|^2 ah = 0, \tag{29}$$

$$-\lambda \bar{b}p + \frac{iu'}{2}s - i\lambda a^2 \bar{\psi}e^{-2u}t + \lambda^{-2}a^3 h = 0, \tag{30}$$

$$-\lambda^{-1}bp - \frac{iu'}{2}s + \lambda^{-1}bt - \lambda^2 \bar{a}e^u h = 0, \tag{31}$$

$$-\lambda^{-1}bq - \lambda \bar{b}s - \frac{iu'}{2}t + ae^u h = 0. \tag{32}$$

Multiplying both sides of (29) by $-\frac{iu'}{2}$, (30) by $\lambda^{-1}b$ and adding them together, we infer

$$(\frac{(u')^2}{4} - |b|^2)p + i\lambda a^2 \bar{\psi}e^{-2u}\frac{iu'}{2}s - ia^2 b\bar{\psi}e^{-2u}t$$
$$+ (\lambda^{-3}a^3 b + |a|^2 a\frac{iu'}{2})h = 0.$$

Multiplying (31) by $i\lambda a^2 \bar{\psi}e^{-2u}$ and adding the above equation to eliminate s and t, we obtain

$$p = \frac{|a|^2 a(-\frac{iu'}{2} + i\lambda^3 \bar{\psi}e^{-u}) - \lambda^{-3}a^3 b}{\frac{(u')^2}{4} - |b|^2 + |\psi|^2 e^{-2u}}h.$$

Similarly, multiplying both sides of (29) by $\lambda \bar{b}$, (30) by $\frac{iu'}{2}$ and adding them together, we get

$$(|b|^2 - \frac{(u')^2}{4})q - i\lambda^2 a^2 \bar{b}\bar{\psi}e^{-2u}s - i\lambda a^2 \bar{\psi}e^{-2u}\frac{iu'}{2}t$$
$$+ (-\lambda|a|^2 a\bar{b} + \lambda^{-2}a^3 \frac{iu'}{2})h = 0.$$

Multiplying (32) by $i\lambda a^2 \bar{\psi}e^{-2u}$ and subtracting the above equation to eliminate s and t, we conclude

$$q = \frac{\lambda^{-2}\frac{iu'}{2}a^3 - \lambda|a|^2 a\bar{b} - i\lambda a^3 \bar{\psi}e^{-u}}{\frac{(u')^2}{4} - |b|^2 + |\psi|^2 e^{-2u}}h.$$

Multiplying (31) by $\frac{iu'}{2}$, (32) by $\lambda^{-1}b$, and adding them together yields

$$-\lambda^{-1}b\frac{iu'}{2}p - \lambda^{-2}b^2 q + (\frac{(u')^2}{4} - |b|^2)s$$
$$+ (-\lambda^2 \bar{a}e^u \frac{iu'}{2} + ae^u \lambda^{-1}b)h = 0.$$

Multiplying (29) by $\lambda^{-1}b$ and adding the above equation to eliminate p and q results in

$$s = \frac{\lambda^2 \bar{a}e^u \frac{iu'}{2} - \lambda^{-1}abe^u + \lambda^{-1}|a|^2 ab}{\frac{(u')^2}{4} - |b|^2 + |\psi|^2 e^{-2u}}h.$$

Finally, multiplying (31) by $\lambda \bar{b}$, (32) by $-\frac{iu'}{2}$ and adding them together, we arrive at

$$- |b|^2 p + \lambda^{-1} b \frac{iu'}{2} q + (|b|^2 - \frac{(u')^2}{4}) t + (-\lambda^3 \bar{a} \bar{b} e^u - a e^u \frac{iu'}{2}) h = 0.$$

Multiplying (30) by $\lambda^{-1} b$ and subtracting the above equation to eliminate p and q, we obtain

$$t = \frac{-a e^u \frac{iu'}{2} - \lambda^3 \bar{a} \bar{b} e^u - \lambda^{-3} a^3 b}{\frac{(u')^2}{4} - |b|^2 + |\psi|^2 e^{-2u}} h.$$

Due to (25), the equation (26) is equivalent to $-\Omega' \gamma + A\xi = c\eta$. Moreover, (27) is equivalent to $-c\bar{\xi}^t = \zeta A$. Substituting (24) into these two equations, we arrive at the following system of equations

$$\lambda^{-1} ap - \lambda \bar{a} q + (\lambda^{-1} \frac{iu'}{2} - i\lambda^2 \bar{\psi} e^{-u}) a^2 h = \lambda^{-1} ac, \qquad (33)$$

$$\lambda^{-1} as - \lambda \bar{a} t - (\lambda^{-2} a^2 b + \lambda \frac{iu'}{2} e^u) h = -\lambda a^{-1} e^u c, \qquad (34)$$

$$-\lambda a^{-1} e^u p + \lambda^{-1} as = -\lambda \bar{a} c, \qquad (35)$$

$$-\lambda a^{-1} e^u q + \lambda^{-1} at = \lambda^{-1} ac. \qquad (36)$$

Note, since $a \neq 0$, equation (36) yields $c = t - \lambda^2 a^{-2} e^u q$. Substituting the expression for q derived above, we obtain

$$c = \frac{ia(\lambda^3 \bar{\psi} - \lambda^{-3} \psi - e^u u')}{\frac{(u')^2}{4} - |b|^2 + |\psi|^2 e^{-2u}} h.$$

By a direct computation, we see that (35) is an identity and (33) and (34) are both equivalent to (12).

In view of equations (12) and (13), we obtain

$$\hat{Q} = \frac{iah}{2(|a|^2 - e^u)} \check{Q}, \quad \check{Q} = \begin{pmatrix} \check{p} & \check{q} & \check{v}_1 \\ \check{s} & \check{t} & \check{v}_2 \\ 0 & 0 & \check{c} \end{pmatrix}, \qquad (37)$$

where

$$\check{p} = -|a|^2 \frac{u'}{2} + \lambda^3 \bar{\psi} |a|^2 e^{-u} - \lambda^{-3} \psi,$$

$$\check{q} = \frac{\lambda^{-2} a}{\bar{a}} [\frac{u'}{2} |a|^2 - \lambda^3 \bar{\psi} e^{-u} (|a|^2 - e^u)],$$

$$\check{s} = \frac{\lambda^2}{a^2} [|a|^2 \frac{u'}{2} e^u + \lambda^{-3} \psi (|a|^2 - e^u)],$$

$$\check{t} = \frac{1}{|a|^2} (-|a|^2 \frac{u'}{2} e^u + \lambda^3 \bar{\psi} e^u - \lambda^{-3} \psi |a|^2),$$

$$\check{v}_1 = -2i\lambda^{-1} a (|a|^2 - e^u),$$

$$\check{v}_2 = -2i\lambda a^{-1} e^u (|a|^2 - e^u),$$

$$\check{c} = \lambda^3 \bar{\psi} - \lambda^{-3} \psi - e^u u'.$$

$$(38)$$

So far we did not impose any restrictions on $\hat{Q}$. In particular, we ignored possible poles in λ and in z. It is easy to verify that all matrix entries of $\hat{Q}$ are defined for sufficiently small $\lambda \in \mathbb{C}^*$. In addition, we would like to impose now the condition for $\hat{Q}$ to have determinant 1. Computing this determinant we obtain

$$(\breve{p}\breve{t} - \breve{q}\breve{s})\breve{c} = (\lambda^3\bar{\psi} - \lambda^{-3}\psi - e^u u')^2(\lambda^3\bar{\psi} - \lambda^{-3}\psi) \in \mathbb{C}.$$

For small $\lambda \in \mathbb{C}^*$, we define

$$\tilde{Q} = \frac{\lambda^3}{\kappa}\hat{Q}, \tag{39}$$

where $\kappa = (\lambda^6\bar{\psi} - \psi - \lambda^3 e^u u')^{2/3}(\lambda^6\bar{\psi} - \psi)^{1/3}$. Then $\det \tilde{Q} = 1$ and $\tilde{Q}(0, \lambda) = Q_0(0, \lambda) = I$ due to $a = ie^{\frac{u(0)}{2}}$. Moreover, $\tilde{Q}$ is holomorphic in λ in a small disk about $\lambda = 0$. If λ is small, the denominator of the coefficient of $\tilde{Q}$ single-valued. Altogether we have found a solution to equation (21) by $Q = Q_0\tilde{Q}$.

5.6. *Solving equation* (11)

Since also U_+ has the same properties as Q, we obtain that $E = Q^{-1}U_+$ has determinant 1, attains the value I for $z = 0$, is holomorphic for all small λ and satisfies $[Q^{-1}U_+, D] = 0$.

By Remark 5 we can assume without loss of generality that $D = D(\lambda)$ is regular semi-simple for all but finitely many values of λ. Therefore, for all z and small λ we can write $E = \exp(\mathcal{E})$, where $[\mathcal{E}, D] = 0$.

Since, in the computation of Q, we did not worry about the twisting condition, the matrix $\mathcal{E}$ is possibly an untwisted loop matrix in $SL(3, \mathbb{C})$. But since $SL(3, \mathbb{C})$ has rank 2, for any regular semi-simple matrix $D = D(\lambda)$, the commutant of $D(\lambda)$ is spanned by $D(\lambda)$ and one other matrix.

Lemma 2. *Every element in the commutant* $\{X \in \Lambda sl(3, \mathbb{C})_\sigma : [X, D] = 0\}$ *of* $D(\lambda)$ *has the form* $X(\lambda) = \kappa_1(\lambda)D(\lambda) + \kappa_2(\lambda)L_0(\lambda)$ *with* $\kappa_1(\epsilon\lambda) = \kappa_1(\lambda)$, $\kappa_2(\epsilon\lambda) = -\kappa_2(\lambda)$, *where* $L_0 = D^2(\lambda) - \frac{1}{3}\mathrm{tr}(D^2)I$.

Hence, the matrix $Q^{-1}U_+$ has the form

$$Q^{-1}U_+ = \exp(\beta_1 D + \beta_2 L_0),$$

where β_1 and β_2 are functions of y and λ near 0. Thus equation (11) leads to

$$\beta_1' D + \beta_2' L_0 = -Q^{-1}\frac{d}{dy}Q + 2iQ^{-1}(V_0 + \lambda V_1)Q.$$

Recalling $Q = Q_0\tilde{Q}$, we obtain

$$\beta_1'\tilde{Q}D + \beta_2'\tilde{Q}L_0 = (-Q_0^{-1}\frac{dQ_0}{dy} + 2iQ_0^{-1}VQ_0)\tilde{Q} - \frac{d\tilde{Q}}{dy}. \tag{40}$$

A direct computation shows

$$\tilde{Q}D = \frac{\lambda^3}{\kappa}\begin{pmatrix} \lambda^{-1}b\check{q} - \lambda\bar{a}\check{v}_1 & -\lambda\bar{b}\check{p} + \lambda^{-1}a\check{v}_1 & \lambda^{-1}a\check{p} - \lambda\bar{a}\check{q} \\ \lambda^{-1}b\check{t} - \lambda\bar{a}\check{v}_2 & -\lambda\bar{b}\check{s} + \lambda^{-1}a\check{v}_2 & \lambda^{-1}a\check{s} - \lambda\bar{a}\check{t} \\ -\lambda\bar{a}\check{c} & \lambda^{-1}a\check{c} & 0 \end{pmatrix},$$

$$L_0 = \begin{pmatrix} \frac{|a|^2-|b|^2}{3} & \lambda^{-2}a^2 & \lambda^2\bar{a}b \\ \lambda^2\bar{a}^2 & \frac{|a|^2-|b|^2}{3} & \lambda^{-2}ab \\ \lambda^{-2}ab & \lambda^2\bar{a}\bar{b} & -\frac{2}{3}(|a|^2-|b|^2) \end{pmatrix},$$

where

$$\lambda^3 L_0 \in \Lambda SL(3,\mathbb{C})_\sigma, \quad \text{and} \quad \hat{\sigma}(L_0(\lambda)) := \sigma(L_0(\varepsilon^{-1}\lambda)) = -L_0(\lambda)$$

and

$$\tilde{Q}L_0 = \frac{\lambda^3}{\kappa}\cdot\begin{pmatrix} \frac{|a|^2-|b|^2}{3}\check{p} + \lambda^2\bar{a}^2\check{q} + \lambda^{-2}ab\check{v}_1 & * & * \\ \frac{|a|^2-|b|^2}{3}\check{s} + \lambda^2\bar{a}^2\check{t} + \lambda^{-2}ab\check{v}_2 & * & * \\ \lambda^{-2}ab\check{c} & \lambda^2\bar{a}\bar{b}\check{c} - \frac{2}{3}(|a|^2-|b|^2)\check{c} \end{pmatrix}.$$

On the other hand,

$$-Q_0^{-1}\frac{dQ_0}{dy} + 2iQ_0VQ_0 = \begin{pmatrix} 0 & -2\lambda\bar{\psi}a^2e^{-2u} & 0 \\ 0 & 0 & -\frac{2i\lambda e^u}{a} \\ -\frac{2i\lambda e^u}{a} & 0 & 0 \end{pmatrix},$$

and

$$\left(-Q_0^{-1}\frac{dQ_0}{dy} + 2iQ_0VQ_0\right)\tilde{Q} = \frac{\lambda^3}{\kappa}\begin{pmatrix} -2\lambda\bar{\psi}a^2e^{-2u}\check{s} & -2\lambda\bar{\psi}a^2e^{-2u}\check{t} & -2\lambda\bar{\psi}a^2e^{-2u}\check{v}_2 \\ 0 & 0 & -\frac{2i\lambda e^u}{a}\check{c} \\ -\frac{2i\lambda e^u}{a}\check{p} & -\frac{2i\lambda e^u}{a}\check{q} & -\frac{2i\lambda e^u}{a}\check{v}_1 \end{pmatrix},$$

and

$$\frac{d\tilde{Q}}{dy} = \frac{\lambda^3}{\kappa}\left(\check{Q}' + \frac{\frac{2}{3}\lambda^3e^u[(u')^2+u'']}{\lambda^6\bar{\psi}-\psi-\lambda^3e^uu'}\check{Q}\right).$$

Substituting this into (40) we obtain 9 equations for β_1' and β_2'. In particular, we obtain

$$-\beta_1'\lambda\bar{a} + \beta_2'\lambda^{-2}ab = -\frac{2i\lambda e^u}{a}\frac{\check{p}}{\check{c}} \tag{41}$$

$$\beta_1'\lambda^{-1}a + \beta_2'\lambda^2\bar{a}\bar{b} = -\frac{2i\lambda e^u}{a}\frac{\check{q}}{\check{c}}. \tag{42}$$

Solving (41) and (42) and integrating yields

$$\begin{aligned}\beta_1(y) &= \int_0^y \frac{2i\lambda^3\bar{\psi} - iu'e^u}{\lambda^3\bar{\psi} - \lambda^{-3}\psi - e^uu'}ds, \\ \beta_2(y) &= \int_0^y \frac{2e^u}{\lambda^3\bar{\psi} - \lambda^{-3}\psi - e^uu'}ds.\end{aligned} \tag{43}$$

Since we already assume $u' \neq 0$, we find that the solutions β_1' and β_2' also satisfy the other 7 equations.

Putting everything together we obtain

Theorem 8 (Explicit Iwasawa decomposition). *The extended frame* $\mathbb{F}$*, satisfying* $\mathbb{F}(0, \lambda) = I$*, for the translationally equivariant minimal Lagrangian surface in* $\mathbb{C}P^2$ *generated by the potential* $D(\lambda)dz$ *with vanishing diagonal, and satisfying* $ab \neq 0$*, is given by*

$$\mathbb{F}(z, \lambda) = \exp(zD - \beta_1(y, \lambda)D - \beta_2(y, \lambda)L_0)Q^{-1}(y, \lambda),$$

with β_1, β_2 *as in (43) and* $Q = Q_0\tilde{Q}$ *as in (23), (37), (38), (39) and* u *as in (20).*

Remark 6. In the proof of the last theorem we have derived the equation $U_+ = Q \exp(\beta_1 D + \beta_2 L_0)$. In this equation each separate term is only defined for small λ and a possibly restricted set of y's. However, due to the globality and the uniqueness of the Iwasawa splitting, the matrix U_+ is defined for all λ in $\mathbb{C}$ and all $z \in \mathbb{C}$.

5.7. *Explicit expressions for minimal Lagrangian immersions*

We know from Remark 5 that except for special cases (which we have excluded) the matrix D has three different eigenvalues. Since D is skew-Hermitian, the corresponding eigenvectors are automatically perpendicular. Therefore there exists a unitary matrix L such that $D = L\text{diag}(id_1, id_2, id_3)L^{-1}$. As a consequence, for the extended lift F we thus obtain

$$F = \mathbb{F}e_3 = L\exp(z\Lambda - \beta_1\Lambda - \beta_2(\Lambda^2 - \frac{\text{tr}\Lambda^2}{3}I))L^{-1}Q^{-1}e_3,$$

where $\Lambda = \text{diag}(id_1, id_2, id_3)$. Set $L = (l_1, l_2, l_3)$. Altogether we have shown

Theorem 9 (see [6]). *Every translationally equivariant minimal Lagrangian immersion generated by the potential* $D(\lambda)dz$ *has a canonical lift* $F = F(z, \lambda)$ *of the form*

$$F(z, \lambda) = \sum_{j=1}^{3} \exp\{izd_j(\lambda) - i\beta_1(y, \lambda)d_j(\lambda) + \beta_2(y, \lambda))(d_j(\lambda)^2 - \frac{2\beta}{3})\} \tag{44}$$
$$\langle Q^{-1}e_3, l_j\rangle l_j.$$

Along the ideas of the paper [3] by Castro-Urbano we have obtained in [6]

Theorem 10.

(1) *When the cubic differential* $\lambda^{-3}\Psi$ *of an translationally equivariant minimal Lagrangian immersion* f *is not real, the canonical lift* F *of* f *has the form*

$$F(x, y, \lambda) = \sum_{j=1}^{3} h_j(y)e^{id_jx+iG_j(y)}\hat{l}_j, \tag{45}$$

where

$$h_j(y) = \left(\frac{d_je^u - \text{Re}(\lambda^{-3}\psi)}{d_j^3 - \text{Re}(\lambda^{-3}\psi)}\right)^{\frac{1}{2}}, \quad G_j(y) = \int_0^y \frac{d_j\text{Im}(\lambda^{-3}\psi)}{d_je^u - \text{Re}(\lambda^{-3}\psi)}ds. \tag{46}$$

(2) When $\lambda^{-3}\Psi$ is real, the canonical lift F of f has the form

$$F(x, y, \lambda) = \sum_{j=1}^{3} \epsilon_j (\frac{\beta}{3} - e_j) \sqrt{\frac{e_j - \wp(y - \omega')}{8|\psi|^2 - (\frac{\beta}{3} - e_j)^3}} e^{id_j x} \hat{l}_j$$

with $\epsilon_1 = -1, \epsilon_2 = \epsilon_3 = 1$.

Since both l_j and $\hat{l}_j$ are orthonormal eigenvectors of $D(\lambda)$ with respect to the eigenvalue id_j and are independent of z, there exists a phase factor α_j which is also independent of z such that $l_j = \hat{l}_j \alpha_j$. Then we can check straightforwardly at the point $y = 2\omega$ that the two horizontal lifts (44), obtained by using the loop group method and (45) obtained by using the idea of Castro-Urbano, are the same up to a constant factor of length 1.

We can also prove the following theorem directly, which appears as Theorem 7.1 in [6].

Theorem 11. *For every translation $z \mapsto z + p + im2\omega$ for $p \in \mathbb{R}$, $m \in \mathbb{Z}$, and $j = 1, 2, 3$, the equation*

$$G_j(2\omega, \lambda) + \mathrm{Re}\beta_1(2\omega, \lambda)d_j(\lambda) + \mathrm{Im}\beta_2(2\omega, \lambda)(-d_j(\lambda)^2 + \frac{2\beta}{3}) = 0.$$

holds.

Proof. Since

$$\begin{aligned}
\beta_1(y) &= i \int_0^y \frac{e^u u' - 2\lambda^3 \bar{\psi}}{e^u u' + 2i\mathrm{Im}(\lambda^{-3}\psi)} ds \\
&= i \int_0^y \frac{(e^u u')^2 - 2\mathrm{Re}(\lambda^{-3}\psi)e^u u' + 4i\lambda^3 \bar{\psi}\mathrm{Im}(\lambda^{-3}\psi)}{(e^u u')^2 + 4[\mathrm{Im}(\lambda^{-3}\psi)]^2} ds \\
&= iy - 2i\mathrm{Re}(\lambda^{-3}\psi) \int_0^y \frac{e^u u'}{(e^u u')^2 + 4[\mathrm{Im}(\lambda^{-3}\psi)]^2} ds \\
&\quad - 4\mathrm{Im}(\lambda^{-3}\psi)\mathrm{Re}(\lambda^{-3}\psi) \int_0^y \frac{ds}{(e^u u')^2 + 4[\mathrm{Im}(\lambda^{-3}\psi)]^2},
\end{aligned} \tag{47}$$

and

$$\begin{aligned}
\beta_2(y) &= -\int_0^y \frac{2e^u}{e^u u' + 2i\mathrm{Im}(\lambda^{-3}\psi)} ds \\
&= -2\int_0^y \frac{e^{2u} u'}{(e^u u')^2 + 4[\mathrm{Im}(\lambda^{-3}\psi)]^2} ds + 4i\mathrm{Im}(\lambda^{-3}\psi) \int_0^y \frac{e^u}{(e^u u')^2 + 4[\mathrm{Im}(\lambda^{-3}\psi)]^2} ds,
\end{aligned} \tag{48}$$

thus we have

$$\mathrm{Re}\beta_1(y) = -4\mathrm{Im}(\lambda^{-3}\psi)\mathrm{Re}(\lambda^{-3}\psi) \int_0^y \frac{ds}{(e^u u')^2 + 4[\mathrm{Im}(\lambda^{-3}\psi)]^2} ds,$$

$$\mathrm{Im}\beta_2(y) = 4\mathrm{Im}(\lambda^{-3}\psi) \int_0^y \frac{e^u}{(e^u u')^2 + 4[\mathrm{Im}(\lambda^{-3}\psi)]^2} ds.$$

Hence,

$$G_1(y) + \text{Re}\beta_1(y)d_1 + \text{Im}\beta_2(y)(-d_1^2 + \frac{2\beta}{3})$$

$$= \text{Im}(\lambda^{-3}\psi) \int_0^y \{\frac{d_1}{d_1 e^u - \text{Re}(\lambda^{-3}\psi)} + \frac{-4d_1\text{Re}(\lambda^{-3}\psi) + 4(-d_1^2 + \frac{2\beta}{3})e^u}{(e^u u')^2 + 4[\text{Im}(\lambda^{-3}\psi)]^2}\}ds.$$

Denote $w(y) = e^{u(y)}$ as before. Recall that

$$d_j^3 - \beta d_j + 2\text{Re}(\lambda^{-3}\psi) = 0,$$
$$w'' + 12w^2 - 4\beta w = 0, \quad \text{if } w' \neq 0.$$

As a consequence, we have

$$d_1\{(w')^2 + 4[\text{Im}(\lambda^{-3}\psi)]^2\} + [-4d_1\text{Re}(\lambda^{-3}\psi) + 4(-d_1^2 + \frac{2\beta}{3})w][d_1 w - \text{Re}(\lambda^{-3}\psi)]$$

$$= -8d_1 w^3 + 4(-d_1^3 + \frac{5\beta}{3}d_1)w^2 - \frac{8}{3}\beta\text{Re}(\lambda^{-3}\psi)w$$

$$= \frac{2}{3}[d_1 w - \text{Re}(\lambda^{-3}\psi)]w''.$$

Therefore,

$$G_1(y) + \text{Re}\beta_1(y)d_1 + \text{Im}\beta_2(y)(-d_1^2 + \frac{2\beta}{3})$$

$$= \frac{2}{3}\text{Im}(\lambda^{-3}\psi) \int_0^y \frac{w''}{(w')^2 + 4[\text{Im}(\lambda^{-3}\psi)]^2}ds$$

$$= \frac{1}{3}\arctan\frac{w'(y)}{2\text{Im}(\lambda^{-3}\psi)},$$

and

$$G_1(2\omega) + \text{Re}\beta_1(2\omega)d_1 + \text{Im}\beta_2(2\omega)(-d_1^2 + \frac{2\beta}{3}) = 0.$$

The proof for the cases $j = 2$ and 3 is analogous. □

6. Explicit expressions for β_j and G_j in terms of Weierstrass elliptic functions

In this section, we present completely explicit expressions for the quantities $\beta_1(2\omega)$, $\beta_2(2\omega)$, and $G_j(2\omega)$.

From (47) we obtain

$$\beta_1(2\omega) = 2\omega i - 16\text{Im}(\lambda^{-3}\psi)\text{Re}(\lambda^{-3}\psi) \int_0^{2\omega} \frac{dy}{[\wp'(y - \omega')]^2 + 16[\text{Im}(\lambda^{-3}\psi)]^2},$$

$$= 2\omega i - 16\text{Im}(\lambda^{-3}\psi)\text{Re}(\lambda^{-3}\psi) \int_0^{2\omega} \frac{dy}{4[\wp(y - \omega')]^3 - g_2\wp(y - \omega') - \tilde{g}_3}, \quad (49)$$

where $\tilde{g}_3 = g_3 - 16[\text{Im}(\lambda^{-3}\psi)]^2$.

It is easy to derive from (18) that the cubic equation $4\wp^3 - g_2\wp - \tilde{g}_3 = 0$ has three distinct roots which we will order to satisfy $\tilde{e}_1 > \tilde{e}_2 > \tilde{e}_3$. Thus

$$\int_0^{2\omega} \frac{dy}{\wp(y-\omega')^3 - \frac{g_2}{4}\wp(y-\omega') - \frac{\tilde{g}_3}{4}}$$
$$= \int_0^{2\omega} \left[\frac{1}{\tilde{H}_1^2}\frac{1}{\wp(y-\omega')-\tilde{e}_1} + \frac{1}{\tilde{H}_2^2}\frac{1}{\wp(y-\omega')-\tilde{e}_2} + \frac{1}{\tilde{H}_3^2}\frac{1}{\wp(y-\omega')-\tilde{e}_3}\right]dy, \tag{50}$$

where

$$\tilde{H}_i^2 := (\tilde{e}_i - \tilde{e}_j)(\tilde{e}_i - \tilde{e}_k) \tag{51}$$

for distinct $i, j, k \in \{1, 2, 3\}$.

If $\lambda^{-3}\psi$ is not real, then $\tilde{g}_3 \neq g_3$. Let α_j denote the uniquely determined element in the fundamental parallelogram satisfying $\wp(\alpha_j) = \tilde{e}_j$ for $j = 1, 2$ or 3. By using formula (18.7.3) of [16]

$$\int \frac{du}{\wp(u) - \wp(\alpha)} = \frac{1}{\wp'(\alpha)}\left[\ln\frac{\sigma(u-\alpha_1)}{\sigma(u+\alpha)} + 2u\zeta(\alpha)\right]$$

for $\wp(\alpha) \neq e_1, e_2$ or e_3, and (18.2.20)

$$\sigma(z + 2m\omega + 2n\omega') = (-1)^{m+n+mn}\sigma(z)\exp[(z + m\omega + n\omega')(2m\eta + 2n\eta')],$$

where $\eta = \zeta(\omega)$, $\eta' = \zeta(\omega')$, we obtain

$$\int_0^{2\omega} \frac{dy}{\wp(y-\omega') - \tilde{e}_j} = \frac{1}{\wp'(\alpha_j)}[4\omega\zeta(\alpha_j) - 4\alpha_j\eta]. \tag{52}$$

Here $\sigma(z) = \sigma(z; g_2, g_3)$ and $\zeta(z) = \zeta(z; g_2, g_3)$ are the Weierstrass σ-function and the Weierstrass ζ-function, respectively. Therefore, substituting (50), (51) and (52) into (49) we derive

$$\beta_1(2\omega) = 2\omega i - 4\mathrm{Im}(\lambda^{-3}\psi)\mathrm{Re}(\lambda^{-3}\psi)\sum_{j=1}^3 \frac{1}{\tilde{H}_j^2}\frac{1}{\wp'(\alpha_j)}[4\omega\zeta(\alpha_j) - 4\alpha_j\eta].$$

Similarly, from (48) and (52) we derive

$$\beta_2(2\omega) = 4i\mathrm{Im}(\lambda^{-3}\psi)\int_0^{2\omega} \frac{\frac{\beta}{6} - \frac{\wp(y-\omega')}{2}}{4[\wp(y-\omega')]^3 - g_2\wp(y-\omega') - \tilde{g}_3}dy,$$
$$= i\mathrm{Im}(\lambda^{-3}\psi)\sum_{j=1}^3 \int_0^{2\omega} \frac{\frac{\beta}{3} - \tilde{e}_j}{2\tilde{H}_j^2}\frac{1}{\wp(y-\omega') - e_j}dy$$
$$= i\mathrm{Im}(\lambda^{-3}\psi)\sum_{j=1}^3 \frac{\frac{\beta}{3} - \tilde{e}_j}{2\tilde{H}_j^2}\frac{1}{\wp'(\alpha_j)}[4\omega\zeta(\alpha_j) - 4\alpha_j\eta].$$

Notice that actually $\frac{\beta}{3} + d_2 d_3 = \tilde{e}_1$ holds. From (46) and (52) we compute

$$
\begin{aligned}
G_1(2\omega) &= 2\mathrm{Im}(\lambda^{-3}\psi) \int_0^\omega \frac{1}{2e^{u(y)} + d_2 d_3} dy \\
&= -2\mathrm{Im}(\lambda^{-3}\psi) \int_0^\omega \frac{dy}{\wp(y - \omega') - (\frac{\beta}{3} \mid d_2 d_3)} \\
&= -2\mathrm{Im}(\lambda^{-3}\psi) \frac{4\omega\zeta(\alpha_1) - 4\alpha_1\eta}{\wp'(\alpha_1)}.
\end{aligned}
$$

Similarly, we obtain

$$
G_j(2\omega) = -2\mathrm{Im}(\lambda^{-3}\psi) \frac{4\omega\zeta(\alpha_j) - 4\alpha_j\eta}{\wp'(\alpha_j)}
$$

for $j = 2, 3$.

Remembering that $\wp(\alpha_j) = \tilde{e}_j$ is a root of $4\wp^3 - g_2\wp - \tilde{g}_3 = 0$, we obtain

$$
[\wp'(\alpha_j)]^2 = 4\wp(\alpha_j)^3 - g_2\wp(\alpha_j) - g_3 = \tilde{g}_3 - g_3 = -16[\mathrm{Im}(\lambda^{-3}\psi)]^2.
$$

We now consider the case where $\mathrm{Im}(\lambda^{-3}\psi)$ is close to 0. Since $\alpha_j = \alpha_j(\lambda)$ was chosen in the fundamental parallelogram, it follows that α_j approaches ω_j if λ approaches any of the six solutions of the equation $\mathrm{Im}(\lambda^{-3}\psi) = 0$. Here we use the usual notation: $\omega_1 = \omega$, $\omega_2 = \omega + \omega'$ and $\omega_3 = \omega'$. From 18.8 of [16], we know

$$
\wp'(\alpha_j) = 4i|\mathrm{Im}(\lambda^{-3}\psi)|
$$

Therefore,

$$
G_j(2\omega) = 2i \, \mathrm{sgn}(\mathrm{Im}(\lambda^{-3}\psi))[\omega\zeta(\alpha_j) - \alpha_j\eta].
$$

Recalling Legendre's relation (18.3.37 in [16])

$$
\zeta(\omega)\omega' - \zeta(\omega')\omega = \frac{\pi}{2}i,
$$

we see that as $\mathrm{Im}(\lambda^{-3}\psi)$ approaches 0,

$$
G_1(2\omega) \to 0, \quad G_2(2\omega) \to \mathrm{sgn}(\mathrm{Im}(\lambda^{-3}\psi))\pi, \quad G_3(2\omega) \to \mathrm{sgn}(\mathrm{Im}(\lambda^{-3}\psi))\pi.
$$

Combining with Theorem 11 and (42) in [6], we obtain tat for $\lambda^{-3}\psi \in \mathbb{R}$ the monodromy $M(\lambda)$ of the translation $2\omega i$ becomes $\mathrm{diag}(1, -1, -1)$. This matches with the immersion formula we got in (2) of Theorem 10 in the sprit of Castro-Urbano in [6].

7. Homogeneous minimal Lagrangian immersions into $\mathbb{C}P^2$

In this section we will consider homogeneous minimal Lagrangian immersions into $\mathbb{C}P^2$ from the point of view of the loop group method.

7.1. *Basic results*

In our context we consider minimal Lagrangian immersions for which the group of symmetries acts transitively.

More precisely, we consider minimal Lagrangian immersions $f : M \to \mathbb{C}P^2$, where M is a Riemann surface, and consider their group of symmetries

$$\Gamma_S^f = \{(\gamma, R) \in Aut(M) \times SU(3), f(\gamma \cdot z) = R \cdot f(z) \text{ for all } z \in M\}.$$

We will also consider its group Γ_M^f of projections onto the first component

$$\Gamma_M^f = \{\gamma \in Aut(M); \text{there exists some } R \in SU(3) \text{ such that}$$
$$(\gamma, R) \in \Gamma_S^f\}.$$

It is easy to see that the following statements hold

Lemma 3. *The group Γ_S^f is closed in the Lie group $Aut(M) \times SU(3)$ and the group Γ_M^f is closed in $Aut(M)$. In particular, both groups are Lie groups.*

Definition 3. A minimal Lagrangian immersion $f : M \to \mathbb{C}P^2$ is homogeneous, if the Lie group Γ_M^f acts transitively in M.

Remark 7. (1) If a minimal Lagrangian immersion $f : M \to \mathbb{C}P^2$ is homogeneous, then it is clear that also its lift to the universal cover $\tilde{M}$ of M is homogeneous.

(2) We can replace, without loss of generality, all groups considered so far by their connected components containing the identity element I.

(3) Since homogeneity implies that the groups under consideration are Lie groups, a homogeneous minimal Lagrangian immersion is in particular equivariant.

7.2. *Homogeneous minimal Lagrangian surfaces defined on simply-connected Riemann surfaces*

As pointed out in (3) of the remark above, the notion of "homogeneous" implies "equivariant", but, obviously, is much stronger. This will show up clearly in the discussion below.

Since there are exactly three simply connected Riemann surfaces, namely the Riemann sphere S^2, the unit disk $\mathbb{D}$, and the complex plane $\mathbb{C}$, we will separate the discussion accordingly.

7.2.1. *The case $M = S^2$*

The case of S^2 is very special, since then the cubic differential ψ on S^2 vanishes identically. As pointed out in [7], this implies that the normalized potential of such an immersion is nilpotent and only depends on one function. As a consequence, the normalized potential can be assumed (almost everywhere, at least locally after some change of coordinates) to be constant, and, since $\psi = 0$, to be nilpotent. Therefore, as stated explicitly in [7], one can carry out the loop group method explicitly and obtains that the image of f is contained in some isometric image of $\mathbb{R}P^2$. The same result has been obtained before by classical differential geometric methods in [11, 18].

We thus obtain:

Theorem 12. *Every minimal Lagrangian immersion $f : S^2 \to \mathbb{C}P^2$ is homogeneous and $f(S^2)$ is, up to isometries of $\mathbb{C}P^2$ contained in $\mathbb{R}P^2$.*

7.2.2. The case $M = \mathbb{D}$

In this case the group Γ_M^f is a closed subgroup of $Aut(\mathbb{D}) = SL(2, \mathbb{R})$ which acts transitively on $\mathbb{D}$. Now it is easy to see, by performing a Levi decomposition of Γ_M^f and then an Iwasawa decomposition of its semi-simple part that Γ_M^f actually contains (up to conjugation) the group of upper triangular matrices Δ in $SL(2, \mathbb{R})$ for which the diagonal elements are positive. But the image of the connected solvable group Δ in $SU(3)$ under the monodromy representation needs to be unitary. Therefore, the kernel of the monodromy representation has at least dimension 1 and the image of f would be one-dimensional.

Theorem 13. *There does not exist any homogeneous, minimal Lagrangian immersion $f : \mathbb{D} \to \mathbb{C}P^2$.*

7.2.3. The case $M = \mathbb{C}$

This case is slightly more complicated. The basic result is

Theorem 14. *If $f : \mathbb{C} \to \mathbb{C}P^2$ is a homogeneous minimal Lagrangian immersion, the the group Γ_M^f contains, up to conjugation, the subgroup of all translations of $\mathbb{C}$.*

Proof. We note that any element X of the Lie algebra of $Aut(\mathbb{C})$ can be represented in the form

$$X = \begin{pmatrix} u & v \\ 0 & 0 \end{pmatrix}$$

Choosing the base point $z = 0$, then the transitivity of the action means that all $v \in \mathbb{C}$ will occur (with a certain $u = u(v)$).

If there is a transitive subgroup such that all elements of its Lie algebra have $u = 0$, then this subalgebra consists of translations only and the claim follows.

Assume now there exists some X with $u \neq 0$. Then it is straightforward to see that there exists some $z_0 \in \mathbb{C}$ such that $\exp(tX) \cdot z_0 = z_0$ for all $t \in \mathbb{R}$, where z_0 does not depend on t.

Hence, after a change of the base point we can assume that X is of the form $v = 0$. But then, it is clear, that the commutators of such an X with an arbitrary Y of the

Lie algebra of some transitive group under consideration form a two-dimensional Lie algbera and only consist of translations. Hence we obtain an abelian, transitive group of translations.   □

Now we obtain (also see [11, 18])

Theorem 15. *Every homogeneous minimal Lagrangian immersion* $f : \mathbb{C} \to \mathbb{C}P^2$ *is isometrically isomorphic with the Clifford torus.*

Proof. Every homogeneous minimal Lagrangian immersion is a doubly equivariant immersion. Therefore the metric and the cubic differential both are constant. As a consequence, the Maurer-Cartan form of the frame $\mathbb{F}$ is constant. Writing $\mathbb{F}^{-1}d\mathbb{F} = Xdz + \tau(X)d\bar{z}$, we observe that the integrability condition implies $[X, \tau(X)] = 0$. Since X is of the form $X = \lambda^{-1}X_{-1} + X_0'$ and $\mathbb{F}(z, \bar{z}, \lambda) = \exp(zX)\exp(\bar{z}\tau(X))$ we see that the immersion is generated by the potential X. A straightforward computation shows that $[X, \tau(X)] = 0$, which implies $X_0 = 0$. Therefore the potential X is a vacuum, whence the immersion is isometrically isomorphic to the Clifford torus (see Section 3).   □

Remark 8. The classification of all homogeneous minimal surfaces and the classification of all homogeneous surfaces in $\mathbb{C}P^2$ have been discussed in [11, 18], respectively. Here we just consider the classification of homogeneous minimal Lagrangian surfaces using the loop group method. For the surfaces under consideration the proof is quite simple and direct.

Acknowledgement

The second author would like to express her sincere gratitude to Professor Robert Conte for his guidance on elliptic functions.

References

[1] F.E. Burstall and M. Kilian, *Equivariant harmonic cylinders,* Q. J. Math. **57** (2006), 449–468.

[2] F.E. Burstall and F. Pedit, *Dressing orbits of harmonic maps,* Duke Math. J. **80** (1995), 353–382.

[3] I. Castro and F. Urbano, *New examples of minimal Lagrangian tori in the complex projective plane,* Manuscripta Math. **85**(1994), no.3-4, 265–281.

[4] J. Dorfmeister and G. Haak, *On symmetries of constant mean curvature surfaces, part I: general theory,* Tohoku Math. J. **50** (1998), 437–154.

[5] J. Dorfmeister and G. Haak, *On symmetries of constant mean curvature surfaces. II. Symmetries in a Weierstraß-type representation,* Int. J. Math. Game Theory Algebra **10** (2000), no. 2, 121–146

[6] J. Dorfmeister and H. Ma, *A new look at equivariant minimal Lagrangian surfaces in* $\mathbb{C}P^2$, arXiv: 1502.04877v1, 2015.

[7] J. Dorfmeister and H. Ma, *Minimal Lagrangian surfaces in* $\mathbb{C}P^2$ *via the loop group method.* in preparation.

[8] J. Dorfmeister, F. Pedit and H. Wu, *Weierstrass type representation of harmonic maps into symmetric spaces,* Comm. Anal. Geom. **6** (1998), 633–668.

[9] J. Dorfmeister and P. Wang, *On symmetric Willmore surfaces in spheres I: the orientation preserving case,* preprint, arXiv:1404.4278v1, 2014.

[10] J. Dorfmeister and H. Wu, *Construction of constant mean curvature n-noids from holomorphic potentials,* Math. Z. **258** (2008), 773–803.

[11] J.-H. Eschenburg, I.V. Guadalupe and R. de A. Tribuzy, *The fundamental equations of minimal surfaces in CP^2,* Math. Ann. **270** (1985), no. 4, 571–598

[12] H.M. Farkas and I. Kra, Riemann Surfaces, Springer, Berlin, Heidelberg, New York, 1991.

[13] G. Haak, *On a theorem by do Carmo and Dajczer,* Proc. Amer. Math. Soc. **126** (1998), no. 5, 1547–1548.

[14] G.D. Ludden, M. Okumura and K. Yano, *A totally real surface in $\mathbb{C}P^2$ that is not totally geodesic,* Proc. Amer. Math. Soc. **53** (1975), 186–190.

[15] H. Ma and Y. Ma, *Totally real minimal tori in $\mathbb{C}P^2$,* Math. Z. **249** (2005), 241–267.

[16] T. Sotjthar, Weierstrass Elliptic and Related Functions, Handbook of Mathematical Functions with Formulas, Graphs, and Mathematical Tables, M. Abramowitz and I.A. Stegun, eds., New York: Dover Publications. 1972.

[17] S.T. Yau, *Submanifolds with constant mean curvature. I.* Amer. J. Math. **96** (1974), 346–366.

[18] C.P. Wang, *The classification of homogeneous surfaces in $\mathbb{C}P^2$.* Geometry and topology of submanifolds, X (Beijing/Berlin, 1999), World Sci. Publishing, River Edge, NJ, 2000, pp. 303314

Chapter 3

Rational Parameter Rays of the Multibrot Sets

Dominik Eberlein, Sabyasachi Mukherjee & Dierk Schleicher

Mayor-Huber-Weg 7, 82140 Olching
E-mail: dominik.eberlein@kabelmail.de

Jacobs University Bremen, Campus Ring 1, Bremen 28759, Germany
E-mail: s.mukherjee@jacobs-university.de
New Address: Stony Brook, Department of Mathematics
E-mail: Sabyasachi.Mukherjee@stonybrook.edu

Jacobs University Bremen, Campus Ring 1, Bremen 28759, Germany
E-mail: d.schleicher@jacobs-university.de

We prove a structure theorem for the multibrot sets, which are the higher degree analogues of the Mandelbrot set, and give a complete picture of the landing behavior of the rational parameter rays and the bifurcation phenomenon. Our proof is inspired by previous works of Schleicher and Milnor on the combinatorics of the Mandelbrot set; in particular, we make essential use of combinatorial tools such as orbit portraits and kneading sequences. However, we avoid the standard global counting arguments in our proof and replace them by local analytic arguments to show that the parabolic and the Misiurewicz parameters are landing points of rational parameter rays.

Contents

1. Introduction

The dynamics of quadratic polynomials and their parameter space have been an area of extensive study in the past few decades. The seminal papers of Douady and Hubbard [5, 6] laid the foundation of subsequent works on the topological and combinatorial structures of the Mandelbrot set, which is indeed one of the most complicated objects in the study of dynamical systems.

In this article, we study the multibrot sets $\mathcal{M}_d := \{c \in \mathbb{C} : \text{The Julia set of } z^d + c \text{ is connected}\}$, which are the immediate generalizations of the Mandelbrot set.

The principal goal of this paper is twofold.

(1) We give a new proof of the structure theorem for the Mandelbrot set consisting of a complete description of the landing properties of the rational parameter rays and the bifurcation of hyperbolic components.

The classical proofs of the structure theorem of the Mandelbrot set can be found in the work of Douady and Hubbard [5, 6], Schleicher [23], Milnor [15]. While Douady-Hubbard's proof involves a careful analysis of the Fatou coordinates and perturbation of parabolic points, more elementary and combinatorial proofs were given by Schleicher and Milnor. The non-trivial part of the structure theorem consists of showing that every parabolic and Misiurewicz parameter is the landing point of the required number of rays, which can be detected by looking at the corresponding dynamical plane. Both the combinatorial proofs are carried out by establishing bounds on the total number of parabolic parameters with given combinatorics and showing that at least (Milnor's proof) or at most (Schleicher's proof) two parameter rays at periodic angles can land at a parabolic parameter. The present proof follows a suggestion from Milnor and avoids the global counting argument. This is replaced by a combination of the combinatorial techniques of [23] and Milnor [15] (kneading sequences and orbit portraits, respectively) together with a monodromy argument.

(2) We write the proof of the structure theorem in more generality, namely for the multibrot sets. Due to existence of a single critical orbit, the passage from degree two to higher degrees does not add further technical complicacies. However, one needs to look at the combinatorics more carefully and modify some of the proofs in the higher degree unicritical case.

There has been a growth of interest in the dynamics and parameter spaces of degree d unicritical polynomials (Avila, Kahn and Lyubich [1, 12], Milnor [17], Chéritat [3]) in the recent years. A different proof of landing of rational parameter rays of the multibrot sets can be found in [21]. Combinatorial classifications of post-critically finite polynomials in terms of external dynamical rays were given in [2, 22]. To our knowledge, there is no written account of the structure theorem for the multibrot sets in the literature and this paper aims at bridging that gap.

It is also worth mentioning that the parameter spaces of unicritical antiholomorphic polynomials were studied [10, 11, 18, 19] by some of the authors and several combinatorial and topological differences between the connectedness loci of unicritical polynomials and unicritical anti-polynomials (e.g. discontinuity of landing points of dynamical rays, bifurcation along arcs, non-local connectivity of the connectedness loci, non-trivial accumulation of parameter rays etc.) have been discovered. Hence, the present paper also serves as a precise reference for the corresponding properties in the holomorphic setting which facilitates the comparison between these two families.

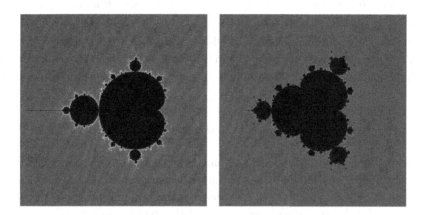

Fig. 1.　Left: The Mandelbrot set, Right: The multibrot set $\mathcal{M}_4$.

Let us now review some background material and fix our notations before stating the main theorem of this paper. For a general introduction into the field, see [16]. In the following, let $d \geq 2$ be an integer, fixed for the whole paper. Let $f_c(z) = z^d + c$ be a unicritical polynomial of degree d (any unicritical polynomial of degree d can be affinely conjugated to a polynomial of the above form). The set of all points which remain bounded under all iterations of f_c is called the Filled-in Julia set $K(f_c)$. The boundary of the Filled-in Julia set is defined to be the Julia set $J(f_c)$ and the complement of the Julia set is defined to be its Fatou set $F(f_c)$.

We measure angles in the fraction of a whole turn, i.e., our angles are elements of $\mathbb{S}^1 \cong \mathbb{R}/\mathbb{Z}$. We define for two different angles $\theta_1, \theta_2 \in \mathbb{R}/\mathbb{Z}$, the interval $(\theta_1, \theta_2) \subset \mathbb{R}/\mathbb{Z}$ as the open connected component of $\mathbb{R}/\mathbb{Z} \setminus \{\theta_1, \theta_2\}$ that consists of the angles we traverse if we move on $\mathbb{R}/\mathbb{Z}$ in anti-clockwise direction from θ_1 to θ_2. Finally, we denote the length of an interval $I_1 \subset \mathbb{R}/\mathbb{Z}$ by $\ell(I_1)$ such that $\ell(\mathbb{S}_1) = 1$.

It is well-known that there is a conformal map ϕ_c near ∞ such that $\lim_{z \to \infty} \phi_c(z)/z = 1$ and $\phi_c \circ f_c(z) = \phi_c(z)^d$. ϕ_c extends as a conformal isomorphism to an equipotential containing 0, when $c \notin \mathcal{M}_d$, and extends as a biholomorphism from $\hat{\mathbb{C}} \setminus K(f_c)$ onto $\hat{\mathbb{C}} \setminus \overline{\mathbb{D}}$ when $c \in \mathcal{M}_d$. The dynamical ray $\mathcal{R}_t^c$ of f_c at an angle t is defined as the pre-image of the radial line at angle t under ϕ_c.

The dynamical ray $\mathcal{R}_t^c$ at angle $t \in \mathbb{R}/\mathbb{Z}$ maps to the dynamical ray $\mathcal{R}_{dt}^c$ at angle dt under f_c. We say that the dynamical ray $\mathcal{R}_t^c$ of f_c lands if $\overline{\mathcal{R}_t^c} \cap K(f_c)$ is a singleton, and this unique point, if exists, is called the landing point of $\mathcal{R}_t^c$. It is worth mentioning that for a complex polynomial (of degree d) with connected Julia set, every dynamical ray at a periodic angle (under multiplication by d) lands at a repelling or parabolic periodic point, and conversely, every repelling or parabolic periodic point is the landing point of at least one periodic dynamical ray [16, §18].

Every $\mathcal{M}_d$ is simply connected and there is a biholomorphic map Φ from $\mathbb{C} \setminus \mathcal{M}_d$ onto the complement of the closed unit disk $\mathbb{C} \setminus \overline{\mathbb{D}}$ (see [6] for a proof in the Mandelbrot set case). The *parameter ray with angle* θ is defined as the set $\mathcal{R}_\theta := \{\Phi^{-1}\left(r^{2\pi i\theta}\right), r > 1\}$ If $\lim_{r \to 1^+} \Phi^{-1}\left(r^{2\pi i\theta}\right)$ exists, we say that $\mathcal{R}_\theta$ lands. The parameter rays of the multibrot sets have been profitably used to reveal the combinatorial and topological structure of the multibrot sets. We refer the readers to [13] for a combinatorial model of the Mandelbrot set in terms of the external parameter rays.

For a periodic orbit $\mathcal{O} = \{z_1, z_2, \cdots, z_p\}$ denote by $\lambda(f, \mathcal{O}) := \lambda(f, z) := \frac{d}{dz} f^{\circ p}(z)$ the *multiplier of* $\mathcal{O}$. In the case $f = f_c$ we write $\lambda(c, \mathcal{O})$ and $\lambda(c, z)$ instead of $\lambda(f_c, \mathcal{O})$ and $\lambda(f_c, z)$.

A parameter c in $\mathcal{M}_d$ is called parabolic if the corresponding polynomial has a (necessarily unique) parabolic cycle. A parabolic parameter c is called essential if at least two dynamical rays land at each point of the (unique) parabolic cycle of f_c. On the other hand, c is called a non-essential parabolic parameter if exactly one dynamical ray lands at each point of the (unique) parabolic cycle of f_c.

The main theorem of this paper is the following:

Theorem 1 (Structure Theorem for Multibrot Sets). *For the Multibrot set $\mathcal{M}_d$ and the associated parameter rays the following statements hold:*

(1) Every parameter ray at a periodic angle lands at a parabolic parameter of $\mathcal{M}_d$.

(2) Every essential (resp. non-essential) parabolic parameter of $\mathcal{M}_d$ is the landing point of exactly two (resp. one) parameter ray(s) at periodic angle(s).

(3) A parameter ray at a periodic angle θ lands at a parabolic parameter c if and only if, in the dynamics of f_c, the dynamical ray at angle θ lands at the parabolic orbit and is one of its characteristic rays.

(4) Every parameter ray at a pre-periodic angle lands at a post-critically pre-periodic parameter of $\mathcal{M}_d$.

(5) Every post-critically pre-periodic parameter is the landing point of at least one parameter ray at a pre-periodic angle.

(6) A parameter ray at a pre-periodic angle θ lands at a Misiurewicz parameter c if and only if, in the dynamics of c, the dynamical ray at angle θ lands at the critical value.

(7) Every hyperbolic component of period greater than one of $\mathcal{M}_d$ has exactly one root and $d - 2$ co-roots and the period one hyperbolic component has exactly $d - 1$ co-roots and no root.

Let us now spend a few words on the organization of the paper. In Section 2, we discuss the basic properties of orbit portraits, which is a combinatorial tool to describe the pattern of all periodic dynamical rays landing at different points of a periodic cycle. This includes a complete description of the orbit portraits and a realization theorem for 'formal orbit portraits'. This allows us to define 'wakes', which play an important role in the combinatorial structure of the multibrot sets. In Section 3, we explore the duality between parameter rays and dynamical rays. Subsection 3.1 contains some basic dynamical and topological properties of Hubbard trees, which contain crucial information about post-critically finite polynomials. This is followed by a preliminary description of centers, roots and co-roots of hyperbolic components and the structure of landing patterns of parameter rays at periodic angles. In Section 4 and Section 5, we complete the proof of Theorem 1 by giving an exact count of the number of parameter rays at periodic angles landing at parabolic parameters. In Section 6, we collect the landing properties of parameter rays at pre-periodic angles; the proofs of these results are direct generalizations of the corresponding results for the Mandelbrot set, so we omit the proofs here.

We thank John Milnor for fruitful discussions and useful advice. The second author gratefully acknowledges the support of Deutsche Forschungsgemeinschaft DFG during this work. The paper is based on the diploma thesis of the first author in spring 1999 at TUM [7].

2. Orbit Portraits

Orbit portraits of quadratic complex polynomials were first introduced by Lisa Goldberg and John Milnor in [8, 9, 15] as a combinatorial tool to describe the pattern of all periodic external rays landing at different points of a periodic cycle. Milnor proved that any collection of finite subsets of $\mathbb{Q}/\mathbb{Z}$ satisfying some conditions indeed occur as the orbit portrait of some quadratic complex polynomial. The usefulness of orbit portraits stems from the fact that these combinatorial objects contain substantial information about the connection between the dynamical and the parameter planes.

2.1. *Definitions and Properties*

In this section, we define orbit portraits for unicritical polynomials of arbitrary degree and prove their basic properties. Finally, we prove an analogous realization theorem for these generalized orbit portraits.

Definition 1 (Orbit portraits). Let $\mathcal{O} = \{z_1, z_2, \cdots, z_p\}$ be a periodic cycle of some unicritical polynomial f. If a dynamical ray $\mathcal{R}_t^f$ at a rational angle $t \in \mathbb{Q}/\mathbb{Z}$ lands at some z_i; then for all j, the set $\mathcal{A}_j$ of the angles of all the dynamical rays landing at z_j is a non-empty finite subset of $\mathbb{Q}/\mathbb{Z}$. The collection $\{\mathcal{A}_1, \mathcal{A}_2, \cdots, \mathcal{A}_p\}$ will be called the *Orbit Portrait* $\mathcal{P}(\mathcal{O})$ of the orbit $\mathcal{O}$ corresponding to the polynomial f.

An orbit portrait $\mathcal{P}(\mathcal{O})$ will be called trivial if only one ray lands at each point of $\mathcal{O}$; i.e. $|\mathcal{A}_j| = 1$, for all j. Otherwise, the orbit portrait will be called non-trivial. The portrait $\mathcal{P} = \{\{0\}\}$ is also called non-trivial.

Lemma 1 (Orientation Preservation). *For any polynomial f, if the external ray $\mathcal{R}_t$ at angle t lands at a point $z \in J(f)$, then the image ray $f(\mathcal{R}_t) = \mathcal{R}_{dt}$ lands at the point $f(z)$. Furthermore, multiplication by d maps every $\mathcal{A}_j$ bijectively onto $\mathcal{A}_{j+1}$ preserving the cyclic order of the angles around $\mathbb{R}/\mathbb{Z}$.*

Proof. Since the ray $\mathcal{R}_t$ lands at z, it must not pass through any pre-critical point of f, hence the same is true for the image ray $\mathcal{R}_{dt}$. Therefore, the image ray is well-defined all the way to the Julia set and continuity implies that it lands at $f(z)$. For the second part, observe that f is a local orientation-preserving diffeomorphism from a neighborhood of z_i to a neighborhood of z_{i+1}. Hence, it sends the set of rays landing at z bijectively onto the set of rays landing at $f(z)$ preserving their cyclic order. $\square$

Lemma 2 (Finitely Many Rays). *If an external ray at a rational angle lands at some point of a periodic orbit $\mathcal{O}$ corresponding to a polynomial, then only finitely many rays land at each point of $\mathcal{O}$. Moreover, all the rays landing at the periodic orbit have equal period and the ray period can be any multiple of p.*

Remark. An angle $t \in \mathbb{R}/\mathbb{Z}$ (resp. a ray $\mathcal{R}_t$) is periodic under multiplication by d (resp. under f) if and only if $t = a/b$ (in the reduced form), for some $a, b \in \mathbb{N}$ with $(b, d) = 1$. On the other hand, t (resp. $\mathcal{R}_t$) is strictly pre-periodic if and only if $t = a/b$ (in the reduced form), for some $a, b \in \mathbb{N}$ with $(b, d) \neq 1$.

Proof. This is well-known. See [16] for example. $\square$

So far all our discussions hold for general polynomials of degree d. Now we want to investigate the consequences of unicriticality on orbit portraits. The next two lemmas are essentially due to Milnor, who proves them for quadratic polynomials in [15].

Lemma 3 (The Critical Arc). *Let f be a unicritical polynomial of degree d and $\mathcal{O} = \{z_1, z_2, \cdots, z_p\}$ be an orbit of period p. Let $\mathcal{P}(\mathcal{O}) = \{\mathcal{A}_1, \mathcal{A}_2, \cdots, \mathcal{A}_p\}$ be the corresponding orbit portrait. For each $j \in \{1, 2, \cdots, p\}$, $\mathcal{A}_j$ is contained in some arc of length $1/d$ in $\mathbb{R}/\mathbb{Z}$. Thus, all but one connected component of $(\mathbb{R}/\mathbb{Z}) \setminus \mathcal{A}_j$ map bijectively to some connected component of $(\mathbb{R}/\mathbb{Z}) \setminus \mathcal{A}_{j+1}$ and the remaining complementary arc of $(\mathbb{R}/\mathbb{Z}) \setminus \mathcal{A}_j$ covers one particular complementary arc of $\mathcal{A}_{j+1}$ d-times and all others $(d-1)$-times.*

Proof. Let $\theta \in \mathcal{A}_j$. Let β be the element of $\mathcal{A}_j$ that lies in $[\theta, \theta + 1/d)$ and is closest to $(\theta + 1/d)$. Similarly, let α be the member of $\mathcal{A}_j$ that lies in $(\theta - 1/d, \theta]$

and is closest to $(\theta - 1/d)$. Note that there is no element of $\mathcal{A}_j$ in $(\beta, \beta + 1/d]$; otherwise the orientation preserving property of multiplication by d would be violated. Similarly, $[\alpha - 1/d, \alpha)$ contains no element of $\mathcal{A}_j$. Also, the arc (α, β) must have length less than $1/d$. We will show that the entire set $\mathcal{A}_j$ is contained in the arc (α, β) of length less than $1/d$.

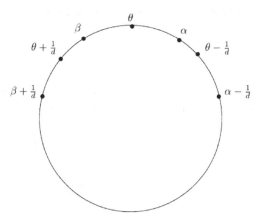

Fig. 2.   No element of $\mathcal{A}_j$ outside (α, β).

If there exists some $\gamma \in \mathcal{A}_j$ lying outside (α, β), then $\gamma \in (\beta + 1/d, \alpha - 1/d)$. Therefore, there exist at least two complementary arcs of $\mathbb{R}/\mathbb{Z} \setminus \mathcal{A}_j$ of length greater than $1/d$. Both these arcs cover the whole circle and some other arc(s) of $\mathbb{R}/\mathbb{Z} \setminus \mathcal{A}_{j+1}$ under multiplication by d. In the dynamical plane of f, the two corresponding sectors (of angular width greater than $1/d$) map to the whole plane and some other sector(s) under the dynamics. Therefore, both these sectors contain at least one critical point of f. This contradicts the unicriticality of f.

This proves that the entire set $\mathcal{A}_j$ is contained in the arc (α, β) of length less than $1/d$. □

Remark.   Following Milnor, the largest component of $(\mathbb{R}/\mathbb{Z}) \setminus \mathcal{A}_j$ (of length greater than $(1 - 1/d)$) will be called the *critical arc* of $\mathcal{A}_j$ and the complementary component of $\mathcal{A}_{j+1}$ that is covered d-times by the critical arc of $\mathcal{A}_j$, will be called the *critical value arc* of $\mathcal{A}_{j+1}$. In the dynamical plane of f, the two rays corresponding to the two endpoints of the critical arc of $\mathcal{A}_j$ along with their common landing point bound a sector containing the unique critical point of f. This sector is called a *critical sector*. Analogously, the sector bounded by the two rays corresponding to the two endpoints of the critical value arc of $\mathcal{A}_{j+1}$ and their common landing point contains the unique critical value of f. This sector is called a *critical value sector*.

Lemma 4 (The Characteristic Arc). *Among all the complementary arcs of the various $\mathcal{A}_j$'s, there is a unique one of minimum length. It is a critical value arc for some $\mathcal{A}_j$ and is strictly contained in all other critical value arcs.*

Proof. Among all the complementary arcs of the various $\mathcal{A}_j$'s, there is clearly at least one, say (t^-, t^+), of minimal length. This arc must be a critical value arc for some $\mathcal{A}_j$: else it would be the diffeomorphic image of some arc of $1/d$ times its length. Let (a, b) be a critical value arc for some $\mathcal{A}_k$ with $k \neq j$. Both the critical value sectors in the dynamical plane of f contain the unique critical value of f. Therefore, $(t^-, t^+) \bigcap (a, b) \neq \emptyset$. From the unlinking property of orbit portraits, it follows that (t^-, t^+) and (a, b) are strictly nested; i.e. the critical value sector (a, b) strictly contains (t^-, t^+). □

This shortest arc $\mathcal{I}_\mathcal{P}$ will be called the *Characteristic Arc* of the orbit portrait and the two angles at the ends of this arc will be called the *Characteristic Angles*. The characteristic angles, in some sense, are crucial to the understanding of orbit portraits.

We are now in a position to give a complete description of orbit portraits of unicritical polynomials.

Theorem 2. *Let f be a unicritical polynomial of degree d and $\mathcal{O} = \{z_1, z_2, \cdots, z_p\}$ be a periodic orbit such that at least one rational dynamical ray lands at some z_j. Then the associated orbit portrait (which we assume to be non-trivial) $\mathcal{P}(\mathcal{O}) = \{\mathcal{A}_1, \mathcal{A}_2, \cdots, \mathcal{A}_p\}$ satisfies the following properties:*

(1) Each $\mathcal{A}_j$ is a non-empty finite subset of $\mathbb{Q}/\mathbb{Z}$.

(2) The map $\theta \to d\theta$ sends $\mathcal{A}_j$ bijectively onto $\mathcal{A}_{j+1}$ and preserves the cyclic order of the angles.

(3) For each $j, \mathcal{A}_j$ is contained in some arc of length less than $1/d$ in $\mathbb{R}/\mathbb{Z}$.

(4) Each $\theta \in \bigcup_{j=1}^{p} \mathcal{A}_j$ is periodic under multiplication by d and they have a common period rp for some $r \geq 1$.

(5) For every $\mathcal{A}_i$, the translated sets $\mathcal{A}_{i,j} := \mathcal{A}_i + j/d$ $(j = 0, 1, 2, \cdots, d-1)$ are unlinked from each other and from all other $\mathcal{A}_m$.

Proof. The first four properties follow from the previous lemmas. Property (5) simply states the fact that if two rays $\mathcal{R}_\theta^c$ and $\mathcal{R}_{\theta'}^c$ land together at some periodic point z, then the rays $\mathcal{R}_{\theta+j/d}^c$ and $\mathcal{R}_{\theta'+j/d}^c$ land together at a pre-periodic point z' with $f(z') = f(z)$: the Julia set of f_c has a d-fold rotation symmetry. □

Definition 2 (Formal Orbit Portraits). A finite collection $\mathcal{P}(\mathcal{O}) = \{\mathcal{A}_1, \mathcal{A}_2, \cdots, \mathcal{A}_p\}$ of subsets of $\mathbb{R}/\mathbb{Z}$ satisfying the six properties of Theorem 2 is called a *formal orbit portrait*.

The condition (3) of Theorem 2 implies that each $\mathcal{A}_j$ has a complementary arc of length greater than $(1 - 1/d)$ (which we call the critical arc of $\mathcal{A}_j$) that, under multiplication by d covers exactly one complementary arc of $\mathcal{A}_{j+1}$ d-times (which we call the critical value arc of $\mathcal{A}_{j+1}$) and the others $(d-1)$-times. We label all the critical value arcs as $\mathcal{I}_1, \mathcal{I}_2, \cdots, \mathcal{I}_p$.

The next lemma is a combinatorial version of Lemma 4 and this is where condition (5) of the definition of formal orbit portraits comes in.

Lemma 5. *Let $\mathcal{P}(\mathcal{O}) = \{\mathcal{A}_1, \mathcal{A}_2, \cdots, \mathcal{A}_p\}$ be a formal orbit portrait. Among all the complementary arcs of the various $\mathcal{A}_j$'s, there is a unique one of minimum length. It is a critical value arc for some $\mathcal{A}_j$ and is strictly contained in all other critical value arcs.*

Proof. Among all the complementary arcs of the various $\mathcal{A}_j$'s, there is clearly at least one, say $\mathcal{I} = (t^-, t^+)$, of minimal length l. This arc must be a critical value arc of some $\mathcal{A}_j$: else it would be the diffeomorphic image of some arc of $1/d$ times its length. Let $\mathcal{I}' = (a, b)$ be the critical arc of $\mathcal{A}_{j-1}$ having length $(d-1)/d + l/d$ so that its image under multiplication by d covers (t^-, t^+) d-times and the rest of the circle exactly $(d-1)$-times. (t^-, t^+) has d pre-images $\mathcal{I}/d$, $(\mathcal{I}/d + 1/d)$, $(\mathcal{I}/d + 2/d), \cdots, (\mathcal{I}/d + (d-1)/d)$; each of them is contained in (a, b) and has length l/d. By our minimality assumption, (t^-, t^+) contains no element of $\mathcal{P}$ and hence neither do its d pre-images. Label the d connected components of

$$\mathbb{R}/\mathbb{Z} \setminus \bigcup_{r=0}^{d-1} (\mathcal{I}/d + r/d) \text{ as } C_1, C_2, \cdots, C_d \text{ with } C_1 = [b, a].$$

Clearly, $\mathcal{A}_{j-1}$ is contained in C_1 and the two end-points of a and b of C_1 belong to $\mathcal{A}_{j-1}$. Also, $C_{i+1} = C_1 + i/d$ for $0 \leq i \leq d-1$. Therefore, $\mathcal{A}_{j-1} + i/d$ is contained in C_{i+1} with the end-points of C_{i+1} belonging to $\mathcal{A}_{j-1} + i/d$. By condition (5) of the definition of formal orbit portraits, each $\mathcal{A}_{j-1} + i/d$ (for fixed j and varying i) is unlinked from $\mathcal{A}_k$, for $k \neq j-1$. This implies that any $\mathcal{A}_k$ ($k \neq j-1$) is contained in $\text{int}(C_r)$, for a unique $r \in \{1, 2, \cdots, d\}$, where r depends on k.

Since any $\mathcal{A}_k$ ($k \neq j-1$) is contained in $\text{int}(C_r)$, for a unique $r \in \{1, 2, \cdots, d\}$, all the non-critical arcs of $\mathcal{A}_k$ are contained in the interior of the same C_r. Thus all the non-critical value arcs of $\mathcal{A}_{k+1}$ ($k+1 \neq j$) are contained $\mathbb{R}/\mathbb{Z} \setminus [t^-, t^+]$. Hence the critical value arc of any $\mathcal{A}_m$ ($m \neq j$) strictly contains $\mathcal{I} = (t^-, t^+)$. The uniqueness follows. $\square$

Lemma 6. *For a formal orbit portrait $\mathcal{P} = \{\mathcal{A}_1, \mathcal{A}_2, \cdots, \mathcal{A}_p\}$, multiplication by d either permutes all the angles of $\mathcal{A}_1 \cup \mathcal{A}_2 \cup \cdots \cup \mathcal{A}_p$ or $|\mathcal{A}_j| \leq 2$ for all j and the first return map of $\mathcal{A}_j$ fixes each angle.*

Proof. We assume that the cardinality of each $\mathcal{A}_j$ is at least three and we'll show that multiplication by d permutes all the rays of $\mathcal{P}$. We can also assume that the characteristic arc $\mathcal{I}_{\mathcal{P}}$ is a critical value arc of $\mathcal{A}_1$. Since $|\mathcal{A}_1| \geq 3$, $\mathcal{A}_1$ has at least three complementary components. Let $\mathcal{I}^+$ be the arc just to the right of $\mathcal{I}_{\mathcal{P}}$ and $\mathcal{I}^-$ be the one just to the left of $\mathcal{I}_{\mathcal{P}}$. Let $\mathcal{I}^-$ be longer than $\mathcal{I}^+$; i.e. $l(\mathcal{I}^-) \geq l(\mathcal{I}^+)$. Since $\mathcal{I}^+$ is not the critical value arc of $\mathcal{A}_1$, there must exist a critical value arc $\mathcal{I}_c$ which maps diffeomorphically onto $\mathcal{I}^+$ under multiplication by d; i.e. $\mathcal{I}^+ = (d)^m \mathcal{I}_c$, for some $m \geq 1$.

We claim that $\mathcal{I}_c = \mathcal{I}_\mathcal{P}$. Otherwise, $\mathcal{I}_c$ will strictly contain the characteristic arc $\mathcal{I}_\mathcal{P}$. Since $\mathcal{I}_c$ is strictly smaller than $\mathcal{I}^+$, $\mathcal{I}_c$ cannot contain $\mathcal{I}^+$. So one end of $\mathcal{I}_c$ must lie in $\mathcal{I}^+$; but then it follows from the unlinking property that both ends of $\mathcal{I}_c$ are in $\mathcal{I}^+$. Therefore, $\mathcal{I}_c$ contains $\mathcal{I}^-$. But this is impossible because $l(\mathcal{I}^-) \geq l(\mathcal{I}^+) \gneq l(\mathcal{I}_c)$.

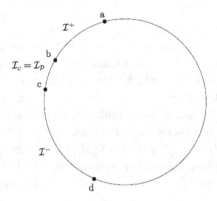

Fig. 3.   The characteristic arc $\mathcal{I}_\mathcal{P}$ maps to the shorter adjacent arc $\mathcal{I}^+$.

Therefore, $\mathcal{I}^+ = (d)^m \mathcal{I}_\mathcal{P}$. Also let, $\mathcal{I}^+ = (a, b)$ and $\mathcal{I}_\mathcal{P} = (b, c)$. Since $\mathcal{I}_\mathcal{P}$ maps to $\mathcal{I}^+$ by an orientation preserving diffeomorphism, we have: $b = d^m c$ and $a = d^m b$. Multiplication by d^m is an orientation preserving map and it sends $\mathcal{A}_1$ bijectively onto itself such that the point b is mapped to an adjacent point a. It follows that multiplication by d^m acts transitively on $\mathcal{A}_1$. Hence multiplication by d permutes all the rays of $\mathcal{P}$. $\square$

The above dichotomy leads to the following definition:

Definition 3 (Primitive and Satellite). If p is the common period of all the angles in $\mathcal{A}_1 \cup \ldots \cup \mathcal{A}_p$, i.e., if each angle is fixed by the first return map, then the portrait is called *primitive*. Otherwise, the orbit portrait is called *non-primitive* or *satellite*. In the latter case, all angles are permuted transitively under multiplication by d.

We record a few easy corollaries that follow from the proof of the previous lemma.

Corollary 1. *If $\mathcal{P}$ is a non-trivial formal orbit portrait, the two characteristic angles are on the same cycle if and only if all angles are on the same cycle.*

Corollary 2. *All angles of a formal orbit portrait are on the orbit of at least one of the characteristic angles.*

Corollary 3 (Maximality). *If $\mathcal{P} = \{\mathcal{A}_1, \mathcal{A}_2, \cdots, \mathcal{A}_p\}$ is a non-trivial (formal) orbit portrait, then it is maximal in the sense that there doesn't exist any other (formal) orbit portrait $\mathcal{P}' = \{\mathcal{A}'_1, \mathcal{A}'_2, \cdots, \mathcal{A}'_q\}$ with $\mathcal{A}_1 \subsetneq \mathcal{A}'_1$.*

Proof. Suppose there exists non-trivial orbit portraits $\mathcal{P}$ and $\mathcal{P}'$ satisfying the above properties. We will consider two cases and obtain contradictions in each of them.

Let us first assume that $\mathcal{P}$ is a primitive orbit portrait. Then $|\mathcal{A}_1| = 2$ and the two elements of $\mathcal{A}_1$ belong to different cycles under multiplication by d. By Lemma 6, $\mathcal{P}'$ must necessarily be an orbit portrait of satellite type and all angles of $\bigcup_{i=1}^{q} \mathcal{A}'_i$, in particular, the two elements of $\mathcal{A}_1$ must lie in the same cycle under multiplication by d: which is a contradiction.

Now suppose $\mathcal{P}$ is an orbit portrait of satellite type; i.e. multiplication by d permutes all the angles of $\mathcal{P}$ transitively. Let, $|\mathcal{A}_1| = v$ (so that the common period of all the angles of $\mathcal{P}$ is pv) and $|\mathcal{A}'_1| = v'$ (so that the common period of all the angles of $\mathcal{P}'$ is qv'). Since the two portraits have angles in common, it follows that $pv = qv'$. The hypothesis $\mathcal{A}_1 \subsetneq \mathcal{A}'_1$ implies that $v' > v$. Therefore, $q < p$. Clearly, both the sets $\mathcal{A}_1$ and $\mathcal{A}_{1+q}$ are contained in $\mathcal{A}'_1$; hence multiplication by d^p would map these two sets onto themselves preserving their cyclic order (multiplication by d^p maps $\mathcal{A}'_1$ onto itself). This forces multiplication by d^p to be the identity map on $\mathcal{A}_1$: a contradiction to the transitivity assumption. $\qquad\square$

The next lemma gives a necessary and sufficient condition for the characteristic rays of a formal holomorphic orbit portrait to co-land and is a mild generalization of the corresponding result proved in [15].

Lemma 7 (Outside The Multibrot Sets). *Let $\mathcal{P}$ be a formal holomorphic orbit portrait for $d \geq 2$ and (t^-, t^+) be its characteristic arc. For some c not in $\mathcal{M}_d$, the two dynamical rays $\mathcal{R}^c_{t^-}$ and $\mathcal{R}^c_{t^+}$ (where, $f_c = z^d + c$) land at the same point of $J(f_c)$ if the external angle $t(c) \in (t^-, t^+)$.*

Proof. We retain the terminology of Lemma 5. If $c \notin \mathcal{M}_d$, all the periodic points of $f_c = z^d + c$ are repelling and the Julia set is a cantor set.

Let the external angle of c in the parameter plane be $t(c)$. Label the connected components of $\mathbb{R}/\mathbb{Z} \setminus \{t(c)/d, t(c)/d + 1/d, \cdots, t(c)/d + (d-1)/d\}$ counter-clockwise as $L_0, L_1, \cdots, L_{d-1}$ such that the component containing the angle '0' gets label L_0. The $t(c)$-itinerary of an angle $\theta \in \mathbb{R}/\mathbb{Z}$ is defined as a sequence $(a_n)_{n \geq 0}$ in $\{0, 1, \cdots, d-1\}^{\mathbb{N}}$ such that $a_n = i$ if $d^n \theta \in L_i$. All but a countably many θ's (the ones which are not the iterated pre-images of $t(c)$ under multiplication by d) have a well-defined $t(c)$-itinerary.

Similarly, in the dynamical plane of f_c, the d external rays $\mathcal{R}^{f_c}_{t(c)/d}$, $\mathcal{R}^{f_c}_{t(c)/d+1/d}$, $\cdots$, $\mathcal{R}^{f_c}_{t(c)/d+(d-1)/d}$ land at the critical point 0 and cut the dynamical plane into

d sectors. Label these sectors counter-clockwise as $L'_0, L'_1, \cdots, L'_{d-1}$ such that the component containing the external ray $\mathcal{R}_0^{f_c}$ at angle '0' gets label L'_0. Any point $z \in J(f_c)$ has an associated symbol sequence $(a_n)_{n \geq 0}$ in $\{0, 1, \cdots, d-1\}^{\mathbb{N}}$ such that $a_n = i$ if $f_c^{\circ n}(z) \in L'_i$. Clearly, a dynamical ray $\mathcal{R}_\theta^{f_c}$ at angle θ lands at z if and only if the $t(c)$-itinerary of θ coincides with the symbol sequence of z defined above.

If $t(c) \in \mathcal{I} = (t^-, t^+)$, the d angles $\{t(c)/d, t(c)/d + 1/d, \cdots, t(c)/d + (d-1)/d\}$ lie in the d intervals $\mathcal{I}/d, (\mathcal{I}/d + 1/d), \cdots, (\mathcal{I}/d + (d-1)/d)$ respectively and no element of $\mathcal{P}$ belongs to $\bigcup_{j=0}^{d-1} (\mathcal{I}/d + j/d)$. First note that the rays $\mathcal{R}_{t^-}^c$ and $\mathcal{R}_{t^+}^c$ indeed land at $J(f_c)$ as $t(c) \notin$ the finite sets $\{t^\pm, dt^\pm, d^2 t^\pm, \cdots\}$. Each $\mathcal{A}_j$ is contained a unique C_r. Therefore, for each $n \geq 0$, the angles $d^n t^-$ and $d^n t^+$ belong to the same L_i. So t^- and t^+ have the same $t(c)$-itinerary; which implies that the two characteristic rays $\mathcal{R}_{t^-}^c$ and $\mathcal{R}_{t^+}^c$ land at the same point of $J(f_c)$. $\qquad \square$

It is now easy to prove the realization theorem for formal orbit portraits.

Theorem 3 (Realizing Orbit Portraits). *Let $\mathcal{P}(\mathcal{O}) = \{\mathcal{A}_1, \mathcal{A}_2, \cdots, \mathcal{A}_p\}$ be a formal orbit portrait for some $d \geq 2$. Then there exists some $c \in \mathbb{C} \setminus \mathcal{M}_d$, such that $f(z) = z^d + c$ has a repelling periodic orbit with associated orbit portrait $\mathcal{P}(\mathcal{O})$.*

Proof. Let $\mathcal{P} = \{\mathcal{A}_1, \mathcal{A}_2, \cdots, \mathcal{A}_p\}$ be a formal orbit portrait with characteristic arc $\mathcal{I}_\mathcal{P} = (t^-, t^+)$ such that $\{t^-, t^+\} \subset \mathcal{A}_1$. Choose c outside the Multibrot set $\mathcal{M}_d$ with $t(c) \in (t^-, t^+)$. Then, the two dynamical rays $\mathcal{R}_{t^-}^c$ and $\mathcal{R}_{t^+}^c$ land at the same point $z \in J(f_c)$. Let $\mathcal{P}' = \{\mathcal{A}'_1, \mathcal{A}'_2, \cdots, \mathcal{A}'_{p'}\}$ be the orbit portrait associated with $\mathcal{O}(z)$ such that $\mathcal{A}'_1$ is the set of angles of the rays landing at z.

First let's assume that $|\mathcal{A}_j| = 2$ and the first return map (multiplication by d^p) of $\mathcal{A}_j$ fixes each ray. In this case, $\{t^-, t^+\} = \mathcal{A}_1$ and $\mathcal{A}_1 \subset \mathcal{A}'_1$. By Lemma 3, we have $\mathcal{P}' = \mathcal{P}$.

On the other hand, if multiplication by d acts transitively on $\mathcal{P}$, there exists $l \in \mathbb{N}$ such that $d^{lp} t^- = t^+$. It follows that $f_c^{\circ lp}(z) = z$. Since t^- and t^+ are adjacent angles in $\mathcal{A}_1$, it follows that multiplication by d^{lp} acts transitively on $\mathcal{A}_1$ and all the rays in $\mathcal{A}_1$ land at z. Evidently, $\mathcal{A}_1 \subset \mathcal{A}'_1$ and once again Lemma 3 implies that $\mathcal{P}' = \mathcal{P}$. $\qquad \square$

Corollary 4 (Characteristic Angles Determine Formal Orbit Portraits). *Let $\mathcal{P}$ be a non-trivial formal orbit portrait with characteristic angles t^- and t^+. Then a formal orbit portrait $\mathcal{P}' = \{\mathcal{A}'_1, \cdots, \mathcal{A}'_p\}$ equals $\mathcal{P}$ if and only if some $\mathcal{A}'_i$ contains t^- and t^+.*

Proof. Follows from the proof of the previous theorem. $\qquad \square$

Remark. Conversely, it is easy to show that for some c not in $\mathcal{M}_d$, the unicritical polynomial f_c can admit the orbit portrait $\mathcal{P} = \{\mathcal{A}_1, \mathcal{A}_2, \cdots, \mathcal{A}_p\}$ only if

$t(c) \in (t^-, t^+)$, where (t^-, t^+) is the characteristic arc of $\mathcal{P}$. Indeed, the characteristic arc must be a critical value arc for some $\mathcal{A}_j$ and in the dynamical plane, the corresponding critical value sector bounded by the two rays $\mathcal{R}^c_{t^-}$ and $\mathcal{R}^c_{t^+}$ together with their common landing point contains the critical value c. Therefore the external angle $t(c)$ of c will lie in the interval (t^-, t^+).

Lemma 8 (Characteristic Point). *Every periodic orbit with a non-trivial orbit portrait has a unique point z, called the* characteristic *point of the orbit, with the following property: two dynamical rays landing at z separate the critical value from the critical point, from all points of the orbit other than z, and from all other dynamical rays landing at the orbit of z. The external angles of these two rays are exactly the characteristic angles of the orbit portrait associated with the orbit of z.*

Proof. The characteristic rays land at a common point z of the orbit and divide $\mathbb{C}$ into two open complementary components. By definition, one of the domains, say U_1 contains exactly the external angles from the characteristic arc $\mathcal{I}_\mathcal{P}$; let the other component be U_0. Then clearly the closure $\overline{U_0}$ must contain all dynamical rays with angles from the portrait, and hence the entire orbit of z. The critical value must be contained in U_1 (or U_1 would have a pre-image bounded by rays from the portrait, yielding a complementary arc of the portrait which was shorter than the characteristic arc). Also, the critical point must be contained in U_0. Indeed, if 0 belonged to U_1, then U_1 would be a critical sector and $\mathcal{I}_\mathcal{P}$ would be a critical arc of $\mathcal{P}$. But a critical arc has length greater than $(1 - 1/d)$ and $\mathcal{I}_\mathcal{P}$ being the characteristic arc, has the smallest length amongst all complementary arcs of $\mathcal{P}$. Clearly, the smallest complementary arc of $\mathcal{P}$ cannot have length greater than $(1 - 1/d)$. This proves the lemma. $\qquad\square$

2.2. *Stability of Orbit Portraits*

Lemma 9 (Landing of Dynamical Rays). *For every map f_c and every periodic angle t of some period n, the dynamical ray $\mathcal{R}^c_t$ lands at a repelling periodic point of period dividing n, except in the following circumstances:*

- *$c \in \mathcal{M}_d$ and $\mathcal{R}^c_t$ lands at a parabolic orbit;*
- *$c \notin \mathcal{M}_d$ is on the parameter ray at some angle $d^k t$, for $k \in \mathbb{N}$.*

Conversely, every repelling or parabolic periodic point is the landing point of at least one periodic dynamical ray for every $c \in \mathcal{M}_d$.

Proof. This is well known. For the case $c \in \mathcal{M}_d$, see [16, Theorem 18.10]; if $c \notin \mathcal{M}_d$, see [9, Appendix A]. For the converse, see [16, Theorem 18.11]. $\qquad\square$

We should note that for every periodic point z_0 of f_{c_0} with multiplier $\lambda(c_0, z_0) \neq 1$, the orbit of z_0 remains locally stable under perturbation of the parameter by the Implicit Function Theorem. With some restrictions this is even true for the associated portrait.

Lemma 10 (Stability of Portraits). *Let c_0 be a parameter such that $f_{c_0} = z^d + c$ has a repelling periodic point z_0 so that the rays at angles $A := \{t_1, \cdots, t_v\}$ ($v \geq 2$) land at z_0. Then there exist a neighborhood U of c_0 and a unique holomorphic function $z \colon U \to \mathbb{C}$ with $z(c_0) = z_0$ so that for every $c \in U$ the point $z(c)$ is a repelling periodic point for f_c and all rays with angles in A land at $z(c)$.*

Let n be the common period of the rays in A. Let U be an open, path connected neighborhood of c_0 which is disjoint from all parabolic parameters of ray period n and from all parameter rays at angles in $\tilde{A} := \bigcup_{j \geq 0} d^j A.$, then for all parameters $c \in U$, exactly the rays at angles in A (and no more) co-land. The same is true at any parabolic parameter of ray period n on the boundary of U.

Proof. The point z_0 can be continued analytically as a repelling periodic point of f_{c_0} in a neighborhood of c_0. By [9, Lemma B.1], this orbit will keep its periodic dynamical rays in a neighborhood of c_0; it might possibly gain extra rays.

Since U doesn't intersect any dynamical ray of period n, for all $c \in U$, the dynamical rays in A indeed land. Let U' be the subset of U where the rays in A continue to land together. For any $c \in U'$, the common landing point of the rays in A must be repelling as U doesn't contain any parabolic point of ray period n. By local stability of co-landing rays at repelling periodic points, we conclude that U' is open in U.

Let c be a limit point of U' in U. The external rays at angles in A do land in the dynamical plane of f_c; we claim that they all co-land. Otherwise, at least two rays at angles θ_1 and θ_2 (say) in A land at two different repelling periodic points in the dynamical plane of f_c and by implicit function theorem, these two rays would continue to land at different points for all nearly parameters. This contradicts the fact that c lies on the boundary of U'. Hence, U' is closed in U. Therefore, $U' = U$; i.e. the rays in A co-land throughout U.

Finally, it follows from the maximality property (Lemma 3) of non-trivial orbit portraits that no dynamical ray at angle $\theta \notin A$ can co-land along with the rays at angles in A. $\qquad\square$

The local stability of external rays landing at repelling periodic points is true for arbitrary polynomials (not necessarily unicritical) as well. However, there is a danger that such an orbit gains periodic rays: if for a parameter c_0 some periodic ray lands at a parabolic orbit, then under small perturbations all continuations of the parabolic orbit may lose the periodic ray, and this ray can land at a different repelling orbit for all sufficiently small (non-zero) perturbations. This happens for general cubic or biquadratic polynomials (compare [19, §6]).

2.3. *Wakes*

We begin a basic lemma which states that the set of parabolic parameters are isolated.

Lemma 11 (Parabolic Parameters Are Countable). *For $n \in \mathbb{N}$ the number of parameters with parabolic orbits of ray period n is finite.*

Proof. Let $Q(c, z) := f_c^{on}(z) - z$, considering it as a polynomial in z whose coefficients are polynomials in c. If f_c has a parabolic orbit of ray period n, then we have: $f_c^{on}(z) = z$ and $\frac{d}{dz} f_c^{on}(z) = 1$.

In other terms, this reads: $Q(c, z) = 0$ and $\frac{d}{dz} Q(c, z) = 0$. Therefore, for such a parameter c, $Q(c, z)$ has a multiple root in z forcing its discriminant to vanish. The discriminant of $Q(c, z)$ (viewed as a polynomial in z) is simply a polynomial in c. Therefore, there are only finitely many such values of c, which finishes the proof. □

The previous lemma is false if "ray period n" is replaced by "orbit period n": the boundary of every hyperbolic component of period n contains a dense set of parabolic parameters with orbit period n.

As a consequence of the previous lemma and the stability of orbit portraits near repelling periodic points, we deduce the fundamental fact that every parameter ray at a periodic angle lands.

Lemma 12 (Periodic Parameter Rays Land). *Let θ be an angle of exact period n under multiplication by d. Then the parameter ray $\mathcal{R}_\theta$ lands at a parabolic parameter c_0 with parabolic orbit of exact ray period n such that the dynamical ray $\mathcal{R}_\theta^{c_0}$ lands at a point of the parabolic orbit.*

Proof. We follow the method of Goldberg and Milnor (see [9, Theorem C.7]). Let $c \in \mathcal{M}_d$ be a limit point of the parameter ray $\mathcal{R}_\theta$. If the dynamical ray $\mathcal{R}_\theta^c$ landed at a repelling periodic point, then it would continue to do so in a neighborhood U of c by Lemma [9, Lemma B.1]. But for any parameter on the parameter ray $\mathcal{R}_\theta$, this is impossible since for such a parameter, the dynamical ray at angle θ bounces off a pre-critical point and fails to land. Therefore, by Lemma 9, the dynamical ray $\mathcal{R}_\theta^c$ must land at a parabolic periodic point and c is one of the finitely many parabolic parameters with a parabolic orbit of ray period n (Lemma 11). Since the set of limit points of any ray is connected, the claim follows. □

Theorem 4 (Parameter Rays Landing at a Common Point). *Let $\mathcal{P}$ be a non-trivial portrait with characteristic angles t^- and t^+. Then the parameter rays $\mathcal{R}_{t^-}$ and $\mathcal{R}_{t^+}$ land at a common parabolic parameter.*

Proof. For $\mathcal{P} = \{A_1, \cdots, A_p\}$, let n be the common period of the angles in $A_1 \cup \ldots \cup A_p$ and let F_n be the set of all parabolic parameters of ray period n. Consider

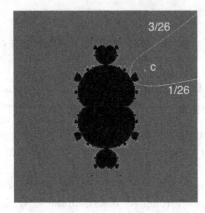 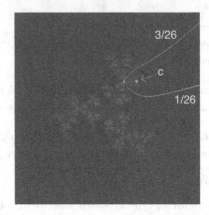

Fig. 4.   Left : The $\mathcal{P}$-wake of $\mathcal{M}_3$ for the portrait $\mathcal{P}$ with the characteristic interval $(1/26, 3/26)$. Right : The Julia set of $z \mapsto z^3 + c$ for a parameter c in the $\mathcal{P}$-wake.

the connected components U_i of $\mathbb{C} \setminus (\bigcup_{t \in A_1 \cup \cdots \cup A_p} \mathcal{R}_t \cup F_n)$. Since $\bigcup A_i$ and F_n are finite (Lemma 11) there are only finitely many components and by Proposition 12 they are open. By Lemma 10, throughout every component U_i the same rays with angles in $A_1 \cup \ldots \cup A_p$ land at common points.

Let U_1 be the component which contains all parameters c outside $\mathcal{M}_d$ with external angle $t(c) \in (t^-, t^+)$ (there is such a component as (t^-, t^+) does not contain any other angle of $\mathcal{P}$). U_1 must have the two parameter rays $\mathcal{R}_{t^+}$ and $\mathcal{R}_{t^-}$ on its boundary. By Theorem 3 and Lemma 10, each $c \in U_1 \setminus \mathcal{M}_d$ has a repelling periodic orbit admitting the portrait $\mathcal{P}$. If the two parameter rays at angles t^+ and t^- do not co-land, then U_1 would contain parameters c outside $\mathcal{M}_d$ with $t(c) \notin (t^-, t^+)$. It follows form the remark at the end of Theorem 3 that such a parameter can never admit the orbit portrait $\mathcal{P}$, a contradiction. Hence, the parameter rays $\mathcal{R}_{t^+}$ and $\mathcal{R}_{t^-}$ at the characteristic angles must land at the same point of $\mathcal{M}_d$. $\qquad\square$

Definition 4 (Wake). For an essential portrait $\mathcal{P}$ with characteristic angles t^- and t^+, the *wake* of the portrait (or simply the $\mathcal{P}$-wake of $\mathcal{M}_d$) is defined to be the component of $\mathbb{C} \setminus (\mathcal{R}_{t^-} \cup \mathcal{R}_{t^+} \cup \{c_0\})$ not containing 0, where c_0 is the common landing point of $\mathcal{R}_{t^-}$ and $\mathcal{R}_{t^-}$. The wake associated with the orbit portrait $\mathcal{P}$ is denoted by $\mathcal{W}_{\mathcal{P}}$.

The wake is well-defined because the parameter 0 can never be on the boundary of the partition.

Lemma 13 (Portrait Realized Only in Wake). *For every non-trivial portrait $\mathcal{P}$, the associated wake $\mathcal{W}_{\mathcal{P}}$ is exactly the locus of parameters c for which the map f_c has a repelling orbit with portrait $\mathcal{P}$.*

Proof. The proof is similar to [15, Theorem 3.1], so we only give a sketch omitting the details.

We use the notations of the previous theorem. If c belongs to $\mathcal{W}_{\mathcal{P}} \setminus \mathcal{M}_d$, then $t(c) \in (t^-, t^+)$. By the proof of Theorem 3, such an f_c has a repelling periodic orbit with associated orbit portrait $\mathcal{P}$. Since $\mathcal{W}_{\mathcal{P}} \setminus F_n$ is disjoint from all parabolic parameters of ray period n and from all parameter rays at angles in $A_1 \sqcup \cdots \sqcup A_p$, it follows from Lemma 10 that every c in $\mathcal{W}_{\mathcal{P}} \setminus F_n$ has a repelling cycle admitting the portrait $\mathcal{P}$. By the same reasoning (using the fact that for some c not in $\mathcal{M}_d$, f_c can admit the orbit portrait $\mathcal{P}$ only if $t(c) \in (t^-, t^+)$), no parameter in $\mathbb{C} \setminus (\mathcal{W}_{\mathcal{P}} \cup F_n)$ has a repelling cycle admitting the portrait $\mathcal{P}$. If for some c_0 in $F_n \setminus \mathcal{W}_{\mathcal{P}}$, f_{c_0} has a repelling cycle with associated portrait $\mathcal{P}$, then by Lemma 10, each nearby map f_c would have a repelling cycle with associated portrait $\mathcal{P}$. But this contradicts the fact that F_n is finite and no parameter in $\mathbb{C} \setminus (\mathcal{W}_{\mathcal{P}} \cup F_n)$ has a repelling cycle admitting the portrait $\mathcal{P}$. The treatment of the parabolic parameters $F_n \cap \mathcal{W}_{\mathcal{P}}$ is slightly more subtle: one can show that the landing points of the dynamical rays $\mathcal{R}_{t^-}^c$ and $\mathcal{R}_{t^+}^c$ depend holomorphically throughout $\mathcal{W}_{\mathcal{P}}$; since these two rays land at a common repelling periodic point for each c in $\mathcal{W}_{\mathcal{P}} \setminus F_n$, it follows from holomorphicity that $\mathcal{R}_{t^-}^c$ and $\mathcal{R}_{t^+}^c$ land at a common repelling periodic point for any c in $F_n \cap \mathcal{W}_{\mathcal{P}}$. Now it follows from the proof of Theorem 3 that for every c in $F_n \cap \mathcal{W}_{\mathcal{P}}$, f_c has a repelling cycle admitting the portrait $\mathcal{P}$. $\qquad\square$

3. Parameter and Dynamical Rays at the Same Angle

Definition 5 (Characteristic Point of a Parabolic Orbit). Let $\mathcal{O}$ be a parabolic orbit for some f_c. The unique point on this orbit which lies on the boundary of the bounded Fatou component containing the critical value is defined as the *Characteristic point of $\mathcal{O}$.*

It is easy to see that if $\mathcal{O}$ has a non-trivial orbit portrait, this definition coincides with the one in 8.

Consider the *parameter* ray $\mathcal{R}_\theta$ at a periodic angle θ with landing point c. We show in this section that the corresponding *dynamical* ray $\mathcal{R}_\theta^c$ lands at the characteristic point of the parabolic orbit (Theorem 5). This proves the Structure Theorem 1 in the primitive case (Corollary 5) and "half" of the Structure Theorem in the non-primitive case (Lemma 20).

The main tool is the concept of orbit separation (Subsection 3.3), which in turn is based on Hubbard trees as introduced in the Orsay Notes [6].

3.1. *Hubbard Trees*

Recall the following fact for a parameter c with super-attracting orbit: for every Fatou component U of K_c there is a unique periodic or pre-periodic point z_U of the super-attracting orbit and a Riemann map $\phi_U : U \to \mathbb{D}$ with $\phi_U(z_U) = 0$ that extends to a homeomorphism from $\overline{U}$ onto $\overline{\mathbb{D}}$. Such a map ϕ_U is unique except for rotation around 0. The point z_U is called the *center of U* and for any $\theta \in \mathbb{R}/\mathbb{Z}$ the pre-image $\{\phi_U^{-1}(r^{2\pi i\theta}) : r \in [0,1)\}$ is an *internal ray of U* with well-defined landing point. Moreover, since hyperbolic filled-in Julia sets are connected and locally connected, hence arcwise connected [16, Lemmas 17.17 and 17.18], every two different points $z, z' \in K_c$ are connected by an arc in K_c with endpoints z and z' (an arc is an injective path).

Definition 6 (Regular Arc). Let $c \in \mathcal{M}_d$ be a parameter with super-attracting orbit and $z, z' \in K_c$. A closed arc $[z, z']$ in K_c is a *regular arc* if

- $[z, z']$ has the endpoints z and z',
- for every Fatou component U of f_c, the intersection $[z, z'] \cap \overline{U}$ is contained in the union of at most two internal rays of U together with their landing points.

We do not distinguish regular arcs which differ by re-parametrization.

Lemma 14 (Regular Arcs). *Let c be a parameter with a super-attracting orbit. Then any two points $z, z' \in K_c$ are connected by a unique regular arc in K_c.*

Proof. For the existence of a regular arc $[z, z']$, take any arc in K_c connecting z to z'. For any bounded Fatou component U of K_c, it suffices to assume that $[z, z'] \cap \overline{U}$ is connected (in fact, this is automatic). It is then easy to modify the arc within U so as to satisfy the conditions on regular arcs.

For uniqueness, assume that there are two different regular arcs $[z, z']$ and $[z, z']'$ for different points $z, z' \in K_c$. If $[z, z'] \neq [z, z']'$, then $\mathbb{C} \setminus ([z, z'] \cup [z, z']')$ has a bounded component V (say) and ∂V is a simple closed curve formed by parts of the two regular arcs. Since the complement of K_c is connected, $V \subset K_c$, so V is contained within one bounded Fatou component U of K_c. But now ∂V must be contained in finitely many internal rays of U together with their landing points such that two distinct external rays land at the same point: an impossibility.   □

Definition 7 (Hubbard Tree). For $c \in \mathcal{M}_d$ with super-attracting orbit $\mathcal{O}$, the *Hubbard tree* of f_c is defined as $\bigcup_{(z,z') \in \mathcal{O} \times \mathcal{O}} [z, z']$.

It follows from Lemma 14 that every super-attracting map f_c has a unique Hubbard tree. It is easy to show that this is indeed a finite tree in the topological sense: it has finitely many branch points and no loops.

Definition 8 (Branch Points and Endpoints). Let c be a parameter with a super-attracting orbit $\mathcal{O}$ and associated Hubbard tree Γ. For $z \in \Gamma$ the components of $\Gamma \setminus \{z\}$ are called *branches of Γ at z*.

If the number of branches with respect to z is at least three then z is called a *branch point* of Γ. If the number is one, then z is called an *endpoint* of Γ.

Fig. 5. The Julia set of a map $z \mapsto z^3 + c$ with a 3-periodic super-attracting orbit. The points on the critical orbit are connected by the Hubbard tree. The tree has a branch point and the critical value is an end-point of the tree.

Lemma 15 (The Hubbard Tree). *Let $c \neq 0$ be a parameter with a super-attracting orbit $\mathcal{O}$ and associated Hubbard tree Γ. Then Γ intersects the boundary of the Fatou component containing the critical value in exactly one point, which is periodic and the boundary of any other bounded Fatou component in most d points which are periodic or pre-periodic. In particular, the critical value is an endpoint of the Hubbard tree.*

Proof. Let U_0 be the Fatou component containing the critical point and $U_1, \cdots, U_{n-1}$ the other bounded periodic components such that $f(U_i) = U_{i+1}$ with $U_n := U_0$.

We denote the number of points of intersection of Γ and ∂U_i by a_i. For two different points $z, z' \in \mathcal{O}$ we consider their regular arc $[z, z']$ and pick $z^* \in \partial U_l \cap [z, z']$. Since Γ is unique and invariant, $f_c(z^*) \in \partial U_{l+1} \cap \Gamma$. This means that the set of intersection points of Γ with the boundary of periodic Fatou components is forward invariant. Since $f_c \colon \overline{U_l} \to \overline{U_{l+1}}$ is a one-to-one map for $l \in \{1, 2, \ldots, n-1\}$ and a d-to-one map for $l = 0$, we have $0 < a_0/d \leq a_1 \leq a_2 \leq \cdots \leq a_{n-1} \leq a_0$.

We show that there is an $l^* \in \{1, 2, \cdots, n-1\}$ with $a_{l^*} = 1$. Indeed, let $c_l := f_c^{\circ l}(0)$ be an endpoint of Γ. If $a_l > 1$, it's easy to check that there are at least two ways to connect c_l to the other points on the critical orbit by regular arcs: a contradiction. This along with the previous inequality yields the result.

$\square$

Lemma 16 (Properties of Branch Points). *Let c be a parameter with a super-attracting orbit and associated Hubbard tree Γ. Consider a $z \in \Gamma$ such that Γ has m branches at z. Then:*

- *If $z \neq 0$ then Γ has at least m branches at $f_c(z)$.*

- If $z = 0$ then Γ has exactly one branch at $f_c(z) = c$.
- If z is a branch point then it is periodic or pre-periodic and lies on a repelling or the super-attracting orbit.
- Every point on the critical orbit has at least one and at most d branches.

Proof. The first statement follows from forward invariance of the Hubbard tree, and the second was shown in Lemma 15.

This implies the third statement because Γ has only finitely many branch points, and all periodic orbits other than the unique super-attracting one are repelling.

The last claim is a restatement of Lemma 15. □

3.2. Hyperbolic Components

Definition 9 (Roots, Co-Roots and Centers). A *hyperbolic component* H *with period* n of $\mathcal{M}_d$ is a connected component of $\{c \in \mathcal{M}_d : f_c$ has an attracting orbit with exact period $n\}$.

A *root* of H is a parameter on ∂H with an essential parabolic orbit of exact ray period n (so that the parabolic orbit disconnects the Julia set). Similarly a *co-root* of H is a parameter on ∂H with a non-essential parabolic orbit of exact ray period n (so that the parabolic orbit does not disconnect the Julia set). A *center* of H is a parameter in H which has a super-attracting orbit of exact period n.

Since our maps f_c have only one critical point, we can see exactly as in the quadratic case [4, Chapter VIII, Theorem 1.4] that every hyperbolic component H with period n is a connected component of the interior of $\mathcal{M}_d$ and that there is a (non-unique) holomorphic map $z \colon H \to \mathbb{C}$ such that $z(c)$ is attracting for all $c \in H$ and has exact orbit period n.

Moreover, there is a holomorphic *multiplier map* $\mu \colon H \to \mathbb{D}$ so that the unique attracting orbit of f_c with $c \in H$ has multiplier $\mu(c)$. Since $|\mu(c)| \to 1$ as $c \to \partial H$, the multiplier map $\mu \colon H \to \mathbb{D}$ is a proper holomorphic map, hence extends surjectively from $\overline{H}$ to $\overline{\mathbb{D}}$. It follows that every hyperbolic component has (at least) one center and at least one root or co-root (in fact, exactly one center and one root when the period of the hyperbolic component is different from one); see Theorems 7 and Corollary 8. Moreover, we will see in Corollary 7 that the number of co-roots of a hyperbolic component of period different from one is exactly $d - 2$. In particular in the quadratic case there are no co-roots. For the hyperbolic component of period one, there are $d - 1$ co-roots and no root.

Lemma 17 (In the Neighborhood of Parabolic Parameters). *Let c_0 be a parabolic parameter with exact parabolic orbit period k, exact ray period n and let z_0 be a point of the parabolic orbit. Then:*

- *If the parabolic orbit portrait is non-primitive ($k \neq n$), then c_0 lies on the boundary of hyperbolic components with period k and n. Moreover, there exists* a neighborhood U of c_0 and a holomorphic function $z_1 \colon U \to \mathbb{C}$ such that $z_1(c)$ is a point of exact period k and $z_1(c_0) = z_0$, and with the following property: for every $c \in U \setminus \{c_0\}$ there is an orbit $\mathcal{O}(c)$ with exact period n that merges into the parabolic orbit $\mathcal{O}(c_0)$ as $c \to c_0$ and for which the multiplier map $c \mapsto \lambda(c, \mathcal{O}(c))$ is holomorphic on U.

- If the parabolic orbit portrait is primitive ($k = n$) then c_0 is a root or co-root of a hyperbolic component with period n. Furthermore, there are a two-sheeted cover $\pi \colon U' \to U$ of a neighborhood U of c_0 with the only ramification point $\pi(c_0') = c_0$ and a holomorphic function $z \colon U' \to \mathbb{C}$ such that $z(c')$ is a point of exact period n and $z(c_0') = z_0$.

Proof. The proof for the quadratic case generalizes directly to $d \geq 2$ (see [15, Lemmas 6.1 and 6.2] and [23, Lemma 5.1]). However, for completeness and because our organization differs from the one in [15] and [23], we include the proof here.

The non-primitive case: The multiplier of the parabolic orbit is a root of unity different from 1, so by the implicit function theorem there is a neighborhood U of c_0 and a holomorphic function $z_1 \colon U \to \mathbb{C}$ such that $z_1(c)$ is a point with exact orbit period k and $z_1(c_0) = z_0$. This implies that the multiplier $\lambda(c, z_1(c))$ is a holomorphic function in c on U. By the Open Mapping Theorem and $|\lambda(c_0, z_1(c_0))| = 1$ it follows that every neighborhood of c_0 contains parameters c such that $z_1(c)$ is attracting. Thus, c_0 lies on the boundary of a hyperbolic component with period k.

We now show that it lies also on the boundary of a hyperbolic component with period n: since $\lambda(f_{c_0}^{\circ n}, z_0) = 1$, we obtain $f_{c_0}^{\circ n}(z) = z + a(z - z_0)^{q+1} + O\left((z - z_0)^{q+2}\right)$ for an integer $q \geq 1$, $a \in \mathbb{C}$, as the Taylor expansion of $f_{c_0}^{\circ n}$ near z_0. It follows from the Leau-Fatou flower theorem (see [16, Theorem 10.5]) that z_0 has q attracting petals and that $f_{c_0}^{\circ n}$ is the first iterate of f_{c_0} which fixes them and the (at least q) dynamical rays landing at z_0. These rays are permuted transitively by the first return map $f_{c_0}^{\circ k}$ (Lemma 6), and hence the q attracting petals are also permuted transitively. Since n is the least integer with $\lambda(f_{c_0}^{\circ n}, z_0) = 1$, it follows that $q = n/k$ and $\lambda(f_{c_0}^{\circ k}, z_0)$ is an exact q-th root of 1. It follows that there are neighborhoods U of c_0 and V of z_0 such that for every $c \in U$, $f_c^{\circ n}(z)$ has exactly $q + 1$ fixed points in V, counted with multiplicities. We are interested in the exact periods of these points with respect to f_c for $c \in U \setminus \{c_0\}$. By the above discussion exactly one of them has exact period k and no one has a lower period. Since $\lambda(f_{c_0}^{\circ l \cdot k}, z_0) \neq 1$ for $l = 2, \ldots, q - 1$, it follows that the iterates $f_c^{\circ l \cdot k}$ have for $c \in U$ exactly one fixed point in V.

Therefore, q points in V have exact period n, and these lie on a single orbit. We thus have for every $c \in U \setminus \{c_0\}$ an orbit $\mathcal{O}(c)$ with exact period n and well-defined multiplier such that q points of $\mathcal{O}(c)$ each coalesce at one point of the parabolic orbit $\mathcal{O}(c_0)$ as $c \to c_0$. The multiplier $c \mapsto \lambda(c, \mathcal{O}(c))$ defines a holomorphic function on U (One cannot, in general, follow the individual points of the period n orbit holomorphically. What one can rather follow holomorphically are symmetric functions of the points on the periodic orbit; the multiplier is indeed a symmetric function of the periodic points and it extends holomorphically to c by Riemann's removable singularity theorem). As before it follows by the Open Mapping Theorem that c_0 lies on the boundary of a hyperbolic component with period n.

The primitive case: In the primitive case we have again $\lambda(f_{c_0}^{on}, z_0) = 1$ and therefore the Taylor expansion $f_{c_0}^{on}(z) = z + a(z - z_0)^{q+1} + O((z - z_0)^{q+2})$ near z_0 for an integer $q \geq 1$, $a \in \mathbb{C}$. Since each of the q petals must absorb a critical orbit and f_{c_0} has only one, we see $q = 1$ and the multiplicity of z_0 as a root of $f_{c_0}^{on}$ is exactly 2. Therefore, it splits into two simple fixed points of $f_{c_0}^{on}$ when c_0 is perturbed. These two fixed points have exact period n under the original map f_{c_0}.

As c traverses a small loop around c_0, these two fixed points are interchanged. But they are at their original positions after two loops around c_0. Hence, on a two-sheeted cover U' (let the projection map be π so that π is branched only over c_0) of a neighborhood of c_0, the two fixed points can be defined as the values of two holomorphic functions $z_1(c'), z_2(c')$ with corresponding holomorphic multipliers $\lambda_1(c'), \lambda_2(c')$ (initially, the functions $z_i(c')$ are defined on a two-sheeted cover of a punctured neighborhood of c_0, but the puncture can then be filled in). Since we have $\lambda_i(\pi^{-1}(c_0)) = +1$, the Open Mapping Theorem implies that there is a parameter $c_1' \in U'$ with $c_1 = \pi(c_1') \in \mathcal{M}_d$ such that $|\lambda_1(c_1')| < 1$. This implies that there is a point c_1 arbitrarily close to c_0 with an attracting periodic orbit of exact period n. Thus, c_0 lies on the boundary of a hyperbolic component of period n. This proves the lemma. $\qquad\square$

3.3. *Orbit Separation Lemmas*

Now we establish a couple of orbit separation lemmas which will be useful in the sequel. They show the existence of pairs of dynamical rays landing at a common repelling periodic or pre-periodic point which separate certain points within the filled-in Julia set, in the following sense: two points $z, z' \in K_c$ are *separated* in the dynamical plane if there are two dynamical rays $\mathcal{R}_\theta^c, \mathcal{R}_{\theta'}^c$ landing at a common repelling point z_0 such that z and z' are in different components of $\mathbb{C} \setminus (\mathcal{R}_\theta^c \cup \mathcal{R}_{\theta'}^c \cup \{z_0\})$. It is convenient to call these two rays together with their landing point the *ray pair* at angles (θ, θ'). One useful feature is that such a co-landing ray pair may be stable even small perturbation when the landing behavior of other dynamical rays is not.

Lemma 18 (Orbit Separation in Super-attracting Case). *Suppose that f_c has a super-attractive orbit of some period n. Then for every two repelling periodic points z and z' on the same orbit with non-trivial orbit portrait, there exists a repelling periodic or pre-periodic point w on a different orbit so that two periodic dynamical rays landing at w separate z from z'.*

Proof. By Lemma 8, the characteristic point on the orbit of z is on the Hubbard tree Γ, hence the entire orbit of z is. Let Γ' be the union of regular arcs between the points on the orbit of z; clearly $\Gamma' \subset \Gamma$. Let $\Gamma'' \subset \Gamma$ be the component of $\Gamma \setminus \{f_c^{-1}(z)\}$ containing the critical point. Then any regular arc in $\Gamma' \setminus \Gamma''$ has its

f_c-image in Γ', so any regular arc in Γ has its f_c-image in $\Gamma' \cup [z, c]$ (where c is the critical value).

Suppose first that z is the characteristic point of its orbit; then z is an endpoint of the tree Γ' (similarly as in Lemma 15). We may suppose that (z, z') does not contain any point on the orbit of z. If (z, z') contains a branch point w of Γ', then all points on the forward orbit of w have at least as many branches in Γ' as w, except when the orbit runs through z; in the latter case, the orbit may lose one branch (the branch to the critical value c), but z is in fact an endpoint of Γ' and thus cannot be on the orbit of w. Therefore, this branch point w must be (pre)periodic. If w is a repelling (pre)periodic point, then it must be the landing point of at least two rational dynamical rays and we are done. If w lies on the critical orbit (let U be the periodic Fatou component containing w), then there must be at least one repelling (pre)periodic point w' other than z and z' on $[z, z'] \cap \partial U$ separating z from z'. As before, w' must be the landing point of at least two rational dynamical rays. Thus the lemma is proved if (z, z') contains a branch point of Γ'.

We now assume that (z, z') contains no branch point of the tree, and no point on the orbit of z. Let n be the period of z, and let $k \in \{1, 2, \cdots, n-1\}$ be so that $f^{\circ k}(z') = z$. If $f^{\circ k}$ is injective on $[z, z']$, then $[z, z']$ must contain a (repelling) fixed point of $f^{\circ k}$ and the lemma is proved in this case.

Otherwise, there is a minimal $k' < k < n$ such that $f_c^{\circ k'}([z, z']) \ni 0$. Then there is a point $z'' \in [z, z']$ such that $f^{\circ(k'+1)} \colon [z'', z'] \to \left[c, f^{\circ(k'+1)}(z')\right] \supset [z', z]$ is a homeomorphism, and again there is a point $w \in (z, z')$ which is fixed under $f^{\circ(k'+1)}$ (necessarily repelling), so this case is done as well.

In order to treat the case that z is not the characteristic point of its orbit, it suffices to iterate f_c until it brings z to the characteristic point. $\qquad\square$

We will need the concept of parabolic trees, which are defined in analogy with Hubbard trees for post-critically finite polynomials. Our definition will follow [10, Section 5]. The proofs of the basic properties of the tree can be found in [23, Lemma 3.5, Lemma 3.6].

Definition 10 (Parabolic Tree). If c lies on a parabolic root arc of period k, we define a *loose parabolic tree* of f_c as a minimal tree within the filled-in Julia set that connects the parabolic orbit and the critical orbit, so that it intersects the critical value Fatou component along a simple $f_c^{\circ k}$-invariant curve connecting the critical value to the characteristic parabolic point, and it intersects any other Fatou component along a simple curve that is an iterated pre-image of the curve in the critical value Fatou component.

Since the filled-in Julia set of a parabolic polynomial is locally connected and hence path connected, any loose parabolic tree connecting the parabolic orbit is uniquely defined up to homotopies within bounded Fatou components. It is easy to see that any loose parabolic tree intersects the Julia set in a Cantor set, and these points of intersection are the same for any loose tree (note that for simple parabolics, any two periodic Fatou components have disjoint closures).

By construction, the forward image of a loose parabolic tree is again a loose parabolic tree. A simple standard argument (analogous to the post-critically finite case) shows that the boundary of the critical value Fatou component intersects the tree at exactly one point (the characteristic parabolic point), and the boundary of any other bounded Fatou component meets the tree in at most d points, which are iterated pre-images of the characteristic parabolic point [23, Lemma 3.5]. The critical value is an endpoint of any loose parabolic tree. All branch points of a loose parabolic tree are either in bounded Fatou components or repelling (pre-)periodic points; in particular, no parabolic point (of odd period) is a branch point.

Following [23, §3], we now define a preferred parabolic tree as follows: Let U be the critical value Fatou component of f_c, and let w be the characteristic parabolic periodic point. First we want to connect the critical value c in U to w by a simple curve which is forward invariant under the dynamics. We will use Fatou coordinates for the attracting petal of the dynamics [16, §10]. In these coordinates, the dynamics is simply addition by $+1$, and our curve will just be the pre-image under the Fatou coordinate of a horizontal straight line in a right half-plane connecting the images of the critical orbit. Since any bounded Fatou component eventually maps onto the critical value Fatou component, we now require the parabolic tree in any other bounded Fatou component to be a pre-image of this chosen curve. With this choice, we have specified a preferred tree which is invariant under the dynamics. We will refer to this tree as *the* parabolic tree.

Lemma 19 (Orbit Separation Lemma For Two Parabolic Points).

Suppose that f_c has a parabolic orbit. Then for any two parabolic periodic points $z \neq z'$, there exists a ray pair landing at a repelling periodic or pre-periodic point which separates z' from z.

Proof. It suffices to prove the lemma when z is the characteristic point of the parabolic cycle (otherwise we can iterate f_c until z satisfies this condition). We may assume that the part of the parabolic tree between z and z' neither traverses a periodic Fatou component except at its ends, nor does it traverse another parabolic periodic point. If there is a branch point of the tree between z and z', this branch point must be the landing point of two rational rays separating the orbit and we are done.

Otherwise, we argue as in the previous lemma: let n be the period of z, and let $k \in \{1, 2, \cdots, n-1\}$ be so that $f^{\circ k}(z') = z$. Put $f^{\circ k}(z) = z''$, then $f^{\circ k}$ maps $[z', z]$ onto $[z, z''] \supseteq [z, z']$. Since $[z', z]$ contains no branch point of the tree, $[z, z'']$ cannot branch off as well; which implies that there exists $z^* \in (z', z)$ fixed by $f^{\circ k}$. Clearly, z^* is a repelling periodic point disconnecting the Julia set and hence is the landing point of two rational dynamical rays separating the orbit. This completes the proof of the lemma. $\qquad\square$

3.4. *Results*

Now we can show that at least certain rays land pairwise and give a complete description of the periodic rays landing at primitive parabolic parameters.

Theorem 5 (A Necessary Condition). *If a parameter ray $\mathcal{R}_\theta$ at a periodic angle lands at a parameter c_0, then the dynamical ray $\mathcal{R}_\theta^{c_0}$ lands at the characteristic point of the parabolic orbit of c_0.*

Proof. The landing point c_0 of $\mathcal{R}_\theta$ is necessarily parabolic by Theorem 12, and the dynamical ray $\mathcal{R}_\theta^{c_0}$ lands at a point of the parabolic orbit of c_0. Without restriction we assume that the exact parabolic orbit period is at least 2.

Let, z be the characteristic point on the parabolic cycle. We assume that the dynamical ray $\mathcal{R}_\theta^{c_0}$ lands at some point z' of the parabolic orbit where $z' \neq z$. Then the Orbit Separation Lemma 19 shows that there is a rational ray pair at angles θ_1, θ_2 landing at some common (pre)periodic repelling point w separating z from z'. In particular, the dynamical ray $\mathcal{R}_\theta^{c_0}$ and the critical value c_0 belong to two different regions of the partition $\mathbb{C} \setminus (\mathcal{R}_{\theta_1}^{c_0} \cup \mathcal{R}_{\theta_2}^{c_0})$. For all parameters c close to c_0, the two parameter rays at angles θ_1 and θ_2 continue to land together [9, Lemma B.1] and the partition is stable [23, Lemma 2.2] in following sense: the dynamical ray $\mathcal{R}_\theta^c$ and the critical value c belong to different regions of this partition in the dynamical plane of f_c. But c_0 is a limit point of the parameter ray $\mathcal{R}_\theta$ and there are parameters c arbitrarily close to c_0 such that in the dynamical plane of f_c, the critical value c lies on the dynamical ray $\mathcal{R}_\theta^c$; a contradiction. This proves that the landing point of the dynamical ray $\mathcal{R}_\theta^{c_0}$ must be the characteristic point z on the parabolic orbit. □

The next result shows that any parabolic parameter with a non-trivial orbit portrait is the landing point of at least two rational parameter rays.

Lemma 20 (At Least the Characteristic Rays Land at a Parameter).
Every parabolic parameter with a non-trivial portrait is the landing point of the parameter rays at the characteristic angles of the parabolic orbit portrait.

Proof. Let c be a parabolic parameter and denote its parabolic orbit portrait by $\mathcal{P} = \{\mathcal{A}_1, \cdots, \mathcal{A}_p\}$. Let t^- and t^+ be the characteristic angles and label the elements of $\mathcal{P}$ cyclically so that $t^-, t^+ \in \mathcal{A}_1$.

By Theorem 4, the two parameter rays $\mathcal{R}_{t^-}$ and $\mathcal{R}_{t^+}$ land at a common parabolic parameter c' (say) with associated wake W and let $\mathcal{P}' = \{\mathcal{A}'_1, \cdots, \mathcal{A}'_q\}$ be the orbit portrait of its parabolic orbit. By Theorem 5, the dynamical rays $\mathcal{R}_{t^-}^{c'}$ and $\mathcal{R}_{t^+}^{c'}$ land at the same point of the parabolic orbit. There is thus an element in $\mathcal{P}'$ which contains both t^- and t^+; call this element $\mathcal{A}'_1$. It follows from Lemma 4 that $\mathcal{P} = \mathcal{P}'$. Therefore, the parabolic parameters c and c' have the same parabolic orbit portrait $\mathcal{P}$.

By [15, Theorem 4.1] (the quadratic case easily generalizes to unicritical polynomials of any degree), both c and c' are limit points of parameters with a repelling periodic orbit with associated orbit portrait $\mathcal{P}$. Now Lemma 13 tells us that $c, c' \in \partial W$. This clearly implies that $c = c'$ and c is the landing point of the parameter rays at the characteristic angles of the parabolic orbit portrait. $\quad\square$

Lemma 21 (Parabolic Parameters with Trivial Portrait). *Let c_0 be a parabolic parameter with a trivial orbit portrait. Then at least one parameter ray lands at c_0.*

Proof. We continue with the proof of the Lemma 17. Let f_{c_0} have a parabolic orbit of period n and z_0 be a representative point of the parabolic orbit. Then z_0 is a fixed point of multiplicity 2 for the map $f_{c_0}^{\circ n}$ and there exists a two-sheeted cover U' (with a projection map π so that π is branched only over c_0) of a neighborhood of c_0 such that the two simple fixed points of the perturbed maps can be defined as the values of two holomorphic functions $z_1(c'), z_2(c')$ with corresponding holomorphic multipliers $\lambda_1(c'), \lambda_2(c')$. Since we have $\lambda_i(\pi^{-1}(c_0)) = +1$, the Open Mapping Theorem implies that there is a parameter $c_1' \in U'$ such that $c_1 = \pi(c_1') \in \mathcal{M}_d$ and $|\lambda_1(c_1')| < 1$ and therefore $|\lambda_2(c_1')| > 1$ (there can be at most one non-repelling orbit). Since $z_2(c_1')$ is repelling, it is the landing point of at least one periodic dynamical ray (Lemma 9); so the corresponding orbit has a portrait $\mathcal{P} \neq \emptyset$. Similarly, there is another parameter $c_2' \in U'$ such that $c_2 = \pi(c_2') \in \mathcal{M}_d$ and $|\lambda_2(c_2')| < 1$; so no periodic dynamical ray lands at $z_2(c_2')$. We may assume that U' is small enough so that it does not contain parabolic parameters other than c_0 of the same ray period n. Then Lemma 10 implies that any path connecting c_1 and c_2 must cross a parameter ray at an angle of period n, which lands at c_0. $\quad\square$

Now we can prove the Structure Theorem 1 in the primitive case, in particular Statement 2:

Corollary 5 (Parameter Rays Landing at Primitive Parameters). *Let c be a primitive parabolic parameter. Then c is the landing point of the parameter rays at precisely those angles θ such that the corresponding dynamical rays $\mathcal{R}_\theta^c$ land at the characteristic parabolic point of f_c. The number of such rays is exactly two if the parabolic orbit is essential, and exactly one otherwise.*

Proof. We know by Lemma 20 and Lemma 21: at least two parameter rays at periodic angles land at c if f_c has a non-trivial parabolic orbit portrait, and at least one parameter ray at a periodic angle lands at c if the associated parabolic portrait is trivial.

By Theorem 5, only those parameter rays $\mathcal{R}_\theta$ can land at c for which the dynamical ray $\mathcal{R}_\theta^c$ lands at the characteristic point of the parabolic orbit and there are only two (resp. one) such rays available in the primitive case. $\quad\square$

The state of affairs is now as follows: every parameter ray at a periodic angle lands at a parabolic parameter. Conversely, given a primitive parabolic parameter c, we have shown that it is the landing point of *exactly* two or one parameter ray(s) at periodic angle(s) depending on whether the parabolic orbit portrait is non-trivial or not. On the other hand, if c is a non-primitive (satellite) parabolic parameter, then it is the landing point of *at least* two parameter rays at the characteristic angles of the associated parabolic orbit portrait. Thus it remains to show that a parabolic parameter of satellite type is the landing point of *at most* two parameter rays at periodic angles. The following two sections will be devoted to the proof of this statement.

4. Roots and Co-Roots of Hyperbolic Components

The goal of this section is to show that every hyperbolic component of period different from one has exactly one root and exactly $d - 2$ co-roots.

Theorem 6 (Continuous Dependence of Landing Points on Parameters). *Let z_0 be a repelling or parabolic periodic point of f_{c_0}. For a dynamical ray $\mathcal{R}_\theta^{c_0}$ landing at z_0, let $\Omega(\theta) := \{c \in \mathbb{C} : \mathcal{R}_\theta^c$ lands $\}$. Then there is a continuous $z \colon \Omega(\theta) \to \mathbb{C}$ such that $z(c)$ is the landing point of $\mathcal{R}_\theta^c$.*

Remark. Although the landing point of the dynamical ray depends continuously on the parameter c (if the ray lands), the portrait may be destroyed. This is certainly always the case whenever the orbits of the landing points have different periods. For example it may occur that $z(c)$ splits into several periodic points while perturbing away from a parabolic parameter, among which the rays of the parabolic point are distributed.

Proof. The proof is analogous to the one in the quadratic case, see [23, Proposition 5.1]. □

Corollary 6 (Stability of Portraits in a Hyperbolic Component). *Let H be a hyperbolic component and E be the set of all roots and co-roots of H. If $c_0 \in H \cup E$ and z_0 be a repelling or parabolic periodic point of f_{c_0}, then there is a continuous map $z \colon H \cup E \to \mathbb{C}$ such that $z(c_0) = z_0$ and the portrait of the orbit of $z(c)$ is the same for all $c \in H \cup E$.*

Proof. Note that for all $c \in H \cup E$, all periodic dynamical rays land. Hence, their landing points depend continuously on the parameter by Theorem 6. In order to prove the corollary, it is enough to show the following: if the dynamical rays, say $\mathcal{R}_\theta^{c_0}$ and $\mathcal{R}_{\theta'}^{c_0}$ land at z_0 for some $c_0 \in H \cup E$, then $\mathcal{R}_\theta^c$ and $\mathcal{R}_{\theta'}^c$ land together for all $c \in H \cup E$.

Let $z(c)$ and $z'(c)$ be two continuous functions on $H \cup E$ such that $z(c_0) = z'(c_0) = z_0$ and $z(c)$ (resp. $z'(c)$) is the landing point of $\mathcal{R}_\theta^c$ (resp. $\mathcal{R}_{\theta'}^c$).

If z_0 is a repelling point, then $\mathcal{R}_\theta^c$ and $\mathcal{R}_{\theta'}^c$ continue to land together at repelling periodic points for all c close to c_0. By Lemma 10, they co-land throughout H; i.e.

$z(c) = z'(c)$ for all c in H. By continuity, $z(c) = z'(c)$ for all c in E and we are done.

Now suppose that z_0 is a parabolic point. Let k be the orbit period and n the ray period of z_0. By Lemma 17, points of a k-periodic and an n-periodic orbit coalesce at z_0 and no further orbits are involved. Since one of them is attracting (namely the n-periodic orbit in the non-primitive case), $z(c)$ and $z'(c)$ always have period k. Since there is only one orbit of period k available, it follows that $z(c) = z'(c)$ for c close to c_0 with $c \in H$. As above, Lemma 10 implies that $z(c) = z'(c)$ for $c \in H$. By continuity, the same holds at the roots and co-roots. Therefore, $\mathcal{R}_\theta^c$ and $\mathcal{R}_{\theta'}^c$ land together for all $c \in H \cup E$. $\hfill \square$

Definition 11 (Multiplier Map of a Hyperbolic Component). Let H be a hyperbolic component and for $c \in H$ let $\lambda(c, \mathcal{O})$ be the multiplier of the unique attracting orbit of f_c. Then the map $\lambda_H \colon H \to \mathbb{D}$, $c \mapsto \lambda(c, \mathcal{O})$ is called the *multiplier map of H*.

The multiplier map λ_H is well-defined: for hyperbolic parameters there is a unique attracting orbit and the absolute value of the multiplier of an attracting orbit is less than one. Precisely as in the quadratic case we see that the multiplier map λ_H of a hyperbolic component H is a proper holomorphic map and has a continuous extension $\lambda_{\overline{H}}$ from $\overline{H}$ onto $\overline{\mathbb{D}}$.

Lemma 22. *Let H be a hyperbolic component of period n with center c_0 and set of roots and co-roots E (a subset of ∂H). For any $c \in H \cup E$, there are exactly $(d-1)$ points on the boundary of the characteristic Fatou component U_1 of f_c which are fixed by the first return map of U_1. All these periodic points are repelling or parabolic.*

Proof. Note that for all $c \in H \cup E$, the Julia set $J(f_c)$ is locally connected. Since U_1 is simply connected, there exists a Riemann map $\phi \colon U_1 \to \mathbb{D}$ which extends to a homeomorphism of the closures (this is implied by the local connectivity of $J(f_c)$ and the fact that the filled-in Julia set of f_c is full). Then ϕ conjugates the first return map $f_c^{\circ n}$ of U_1 to a proper degree d holomorphic self-map of $\mathbb{D}$, hence a Blaschke product of degree d, say $\mathcal{B}_c$. As conjugate dynamical systems have the same number of fixed points, the number of points on the boundary of U_1 which are fixed by the first return map of U_1 is equal to the number of fixed points of $\mathcal{B}_c$ on $\partial \mathbb{D}$.

In the (super-)attracting case, there is exactly one fixed point of $\mathcal{B}_c$ in $\mathbb{D}$ and by reflection, exactly one fixed point in $\mathbb{C} \setminus \overline{\mathbb{D}}$. Since a rational map of degree d has $(d+1)$ fixed points counted with multiplicity, there must be $(d-1)$ fixed points (counted with multiplicity) on $\partial \mathbb{D}$. Since these fixed points are never parabolic, each of them has multiplicity 1; i.e. there are exactly $(d-1)$ fixed points of $\mathcal{B}_c$ on $\partial \mathbb{D}$. In the parabolic case, there are no fixed points of $\mathcal{B}_c$ in $\mathbb{D} \cup (\mathbb{C} \setminus \overline{\mathbb{D}})$. So all the $(d+1)$ fixed points lie on $\partial \mathbb{D}$. $\mathcal{B}_c$ has a parabolic fixed point on $\partial \mathbb{D}$ and the Julia set is all of $\partial \mathbb{D}$. Clearly, there are two attracting petals; i.e. the parabolic fixed point has multiplicity 3 and there are exactly $(d-2)$ simple fixed points, all distinct. Hence, the total number of distinct fixed points of $\mathcal{B}_c$ on $\partial \mathbb{D}$ is again $(d-1)$.

Therefore, U_1 has exactly $(d-1)$ points on its boundary which are fixed by its first return map. Since there can be only one non-repelling periodic orbit of c, these periodic points are either all repelling or exactly one of them is parabolic. $\qquad\square$

Lemma 23 (On the Boundary of the Characteristic Fatou Component).
Let H be a hyperbolic component of period n with center c_0 and set of roots and co-roots $E (\subset \partial H)$. There are $(d-1)$ continuous functions $z^{(1)}, \cdots, z^{(d-1)}$ on $H \cup E$ such that for any $c \in H \cup E$, $\{z^{(1)}(c), \cdots, z^{(d-1)}(c)\}$ are precisely the $(d-1)$ points on the boundary of the characteristic Fatou component U_1 which are fixed by the first return map of U_1. At exactly one of them, more than one dynamical rays land. Moreover, at every $c \in E$, one of the $z^{(i)}(c)$'s is the characteristic point of the parabolic orbit.

Proof. The super-attracting point c_0 in the dynamical plane of f_{c_0} can be holomorphically followed throughout H yielding an analytic function $z^*\colon H \to \mathbb{C}$ with $z^*(c_0) = c_0$ and $z^*(c)$ periodic of period n for all c in H. z^* can be extended to a continuous function on $H \cup E$ and since the multiplier map $\lambda_{\overline{H}}$ is proper holomorphic on $\overline{H}$, $z^*(c)$ must have multiplier 1 for every $c \in E$. Also, $z^*(c)$ lies in the closure of the critical value Fatou component for every $c \in H \cup E$, so it must be the characteristic point of the parabolic orbit for $c \in E$.

Let the $(d-1)$ points on the boundary of the characteristic Fatou component U_1 of f_{c_0} which are fixed by the first return map of U_1 be $z_0^{(1)}, \ldots, z_0^{(d-1)}$. Since the $z_0^{(i)}$'s are repelling, by Corollary 10 and 6 there are continuous functions $z^{(i)}$ on $H \cup E$ with $z^{(i)}(c_0) = z_0^{(i)}$ for $1 \le i \le d-1$ and the $z^{(i)}(c)$'s are repelling for $c \in H$ such that for any fixed i, the portrait of the orbit of $z^{(i)}(c)$ remains constant for all $c \in H \cup E$. Also, each $z^{(i)}(c)$ lies on the boundary of the characteristic Fatou component of f_c and is fixed by the first return map of the component. Since $z^*(c)$ has period n and lies on the boundary of the characteristic Fatou component of c for $c \in E$, it must be one of the points $z^{(i)}(c)$'s.

Finally we show that exactly one of the $z^{(i)}(c)$'s is the landing point of more than one rays. We first prove the super-attracting case following the proof of [20, Lemma 3.4]: without restriction we assume $n > 1$ (If $n = 1$, the only parameter with a super-attracting fixed point is 0 and the dynamical rays at angles 0 and 1, which we consider as two different rays in this case, land trivially at a common point in the dynamical plane of z^d). One of the $z_0^{(i)}$'s, say $z_0^{(1)}$, lies on the Hubbard tree Γ of c_0 and it is the only point of $\Gamma \cap \overline{U_1}$ by Lemma 15. Therefore, $z_0^{(1)}$ disconnects ∂K_c and is the landing point of at least two dynamical rays. If there is a second $z_0^{(i)}$ $(i \ne 1)$ which disconnects the filled-in Julia set of c_0, then there must be a periodic or pre-periodic Fatou component U' which is separated from the characteristic Fatou component by $z_0^{(i)}$. Let γ be an injective curve in the filled-in Julia set which connects the critical value to the center of U' (that is the unique point of U' which maps onto the critical orbit); γ becomes unique if we require that it maps onto the Hubbard tree by the time that U' lands on a periodic component. From then on, all forward iterates of γ will be on the Hubbard tree, so they will not meet $z_0^{(i)}$; this is in contradiction to periodicity of $z_0^{(i)}$ and shows that exactly one of the $z_0^{(i)}$s can disconnect the filled-in Julia set of c_0; i.e. exactly one of them (say $z_0^{(1)}$) is the landing point of more than one rays. By Corollary 6, only $z^{(1)}(c)$ must be the landing point of more than one dynamical rays for each $c \in H \cup E$. $\qquad\square$

Theorem 7 (Mapping Degree of $\lambda_{\overline{H}}$). *The multiplier map $\lambda_{\overline{H}}$ has mapping degree $d - 1$ and every hyperbolic component H of period greater than one has exactly one root.*

Proof. Let H be a hyperbolic component with period n. The mapping degree of $\lambda_{\overline{H}}$ is at least $d - 1$: for a center c_0 of H there is a holomorphic function $z(c)$ on H such that $z(c_0) = 0$ and we can locally write $\lambda_{\overline{H}}(c) = d^n \, f_c^{\circ(n-1)} \, (z(c))^{d-1}$ $\cdots f_c \, (z(c))^{d-1} \, z(c)^{d-1}$. Since $z(c)$ has the only zero at c_0, it follows that $\lambda_{\overline{H}}(c) = d^n \, (c - c_0)^{d-1} \, g(c)^{d-1}$ for some holomorphic function g that does not vanish in a neighborhood of c_0, and $\lambda_{\overline{H}}$ has mapping degree at least $d - 1$.

Note that the map $\lambda_{\overline{H}}$ takes the value 1 precisely at the roots and co-roots of H. Therefore, the mapping degree of $\lambda_{\overline{H}}$ is bounded above by the total number of roots and co-roots. By the previous Lemma 23, exactly one of the $z^{(i)}$'s becomes the characteristic point of the parabolic orbit at each root or co-root. Fix $i \in \{1, 2, \cdots, d - 1\}$; we show that there exists a unique $c \in E$ for which $z^{(i)}(c)$ is the characteristic point of the parabolic orbit. Suppose there was another c' with the same property. Let the orbit portrait of the orbit of $z^{(i)}$ be $\mathcal{P}_i$ which remains constant throughout $H \cup E$ and let θ be a characteristic angle of $\mathcal{P}_i$ (if $\mathcal{P}_i$ is trivial, then θ is the angle of the only dynamical ray landing at the characteristic point $z^{(i)}$). By Lemma 20 and Corollary 5, the parameter ray at angle θ lands both at c and at c'; a contradiction which proves our claim. Therefore, there are only $d-1$ candidates for parabolic parameters with ray period n on the boundary of H implying that the total number of roots and co-roots of H is at most $d - 1$. Thus, the mapping degree of $\lambda_{\overline{H}}$ is at most $d - 1$ and hence precisely $d - 1$.

Moreover, this shows that all candidates for characteristic points are realized. Since portraits are stable for all parameters in $H \cup E$ (Corollary 6) we obtain by Lemma 23 that exactly one parameter in E has a parabolic orbit with a nontrivial orbit portrait (when the period is different from one). This proves that each hyperbolic component of period different from one has exactly one root. $\qquad\square$

Corollary 7 (Number of Co-Roots). *Every hyperbolic component of period greater than one has exactly $d - 2$ co-roots.*

The previous two statements, Theorem 7 and Corollary 7, prove the last assertion of the Structure Theorem.

Corollary 8. *Every hyperbolic component of $\mathcal{M}_d$ has exactly one center.*

Proof. By Theorem 7 and its proof, H has at least one center c_0 (i.e. $\lambda_{\overline{H}}(c_0) = 0$) such that the local mapping degree of $\lambda_{\overline{H}}$ at c_0 is d 1. But, the mapping degree of $\lambda_{\overline{H}}$ is $d - 1$, and hence, c_0 is the unique parameter where $\lambda_{\overline{H}}$ vanishes. Therefore, H has a unique center. $\qquad\square$

So far we have showed that *at least d* parameter rays at n-periodic angles land on the boundary of every hyperbolic component of the same period n. For the proof of the Structure Theorem it remains to show that *at most d* parameter rays land at every hyperbolic component. For this purpose we need to connect the landing points of the various parameter rays at periodic angles landing on the boundary of a common hyperbolic component by *internal rays*.

Definition 12 (Internal Rays of a Hyperbolic Component). An $\quad$ *internal ray* of a hyperbolic component H is an arc $c\colon [0,1] \to \overline{H}$ starting at the center such that there is an angle θ with $\lambda_{\overline{H}}(c) = [0,1] \cdot e^{2\pi i\theta}$.

Remark. Since λ_H is a $(d-1)$-to-one map, an internal ray of H with a given angle is not uniquely defined. In fact, a hyperbolic component has $(d-1)$ internal rays with any given angle θ.

5. Kneading Sequences

In this section we complete the description of the landing properties of parameter rays at periodic angles by an induction proof on the ray period, Theorem 10. We use a similar strategy as in [23, Section 3]. However, contrary to the quadratic case, we need for $d > 2$ some knowledge on the hyperbolic components, which we accumulated in the previous sections.

Definition 13 (Itineraries and Kneading Sequences). Fix $d \geq 2$. For an angle $\theta \in \mathbb{R}/\mathbb{Z}$ label the components of $\mathbb{R}/\mathbb{Z} \setminus \{\theta/d, (\theta+1)/d, \cdots, \{\theta + (d-1)\}/d\}$ in the following manner:

$$L_\theta(\eta) := \begin{cases} m & \text{if } \eta \in \left(\{\theta + (m-1)\}/d, (\theta + m)/d\right) \\ (m_1, m_2) & \text{if } \eta = \{\theta + (m_2 - 1)\}/d = (\theta + m_1)/d \end{cases}$$

for some $m, m_1, m_2 \in \{0, 1, \cdots, d-1\}$.

The infinite sequence $I_\theta(\eta) := \{L_\theta(\eta), L_\theta(d\eta), L_\theta(d^2\eta), \ldots\}$ is called the θ-*itinerary* of η under the d-tupling map. We call the special itinerary $K(\theta) := I_\theta(\theta) = \{L_\theta(\theta), L_\theta(d\theta), L_\theta(d^2\theta), \ldots\}$ the *kneading sequence* of θ.

The symbols $(0,1), (1,2), \cdots, (d-2, d-1), (d-1, 0)$ are called *boundary symbols*. Sometimes we replace them by an asterisk $(*)$.

It is convenient to write $K(\theta_1) = K(\theta_2)$ for angles θ_1, θ_2 if both angles have matching boundary symbols at the same entries and all other symbols coincide.

In the following theorem we define a partition of initial kneading sequences and do most of the proof of the induction step of Theorem 10.

Theorem 8 (The Induction Step). *Let $n \geq 2$ be an integer. Suppose that the root of every hyperbolic component with period $n - 1$ or lower is the landing point of exactly two parameter rays. Then any two parameter rays with angles θ_1, θ_2 of exact ray period n can land at the same parameter only if $K(\theta_1) = K(\theta_2)$.*

Proof. We claim that there is a partition P_{n-1} of $\mathbb{C}$ such that every parameter ray with exact ray period n together with its landing point is completely contained in an open component of P_{n-1}. Moreover, for any two parameter rays $\mathcal{R}_{\theta_1}, \mathcal{R}_{\theta_2}$ which are both in the same open component of P_{n-1} the kneading sequences of θ_1 and θ_2 coincide in the first $n - 1$ entries. Then the theorem follows because any parameter ray at an angle of exact period n has a kneading sequence of period n and the n-th entry of the kneading sequence is $(*)$.

We construct such a partition: let Θ_k be the set of all angles with exact period k and Λ_k the set of multiplier maps of the k-periodic hyperbolic components. We define $P_{n-1} := \bigcup\limits_{k=1}^{n-1} \left(\bigcup\limits_{\theta \in \Theta_k} \mathcal{R}_\theta \cup \bigcup\limits_{\lambda_{\overline{H}} \in \Lambda_k} \lambda_{\overline{H}}^{-1} ([0,1]) \right)$ and assert that P_{n-1} is a partition with the required properties. By construction P_{n-1} (together with the components of $\mathbb{C} \setminus P_{n-1}$) is a partition of $\mathbb{C}$. Parameter rays with exact ray period k land at a parameter which has a parabolic orbit with exact ray period k (Lemma 12). For a hyperbolic component H the inverse image $\lambda_{\overline{H}}^{-1} ([0,1])$ is the set of all internal rays with angle 0. Each of these $d - 1$ internal rays lands at a root or co-root of H and conversely the root and every co-root of H is a landing point of one of these internal rays. It follows that every parameter ray of period n together with its landing point is contained in one of the open components of $\mathbb{C} \setminus P_{n-1}$.

Now assume that two parameter rays $\mathcal{R}_{\theta_1}, \mathcal{R}_{\theta_2}$ are both contained in the same open component of $\mathbb{C} \setminus P_{n-1}$. We know that every hyperbolic component has $d - 2$ co-roots and exactly one root. Also by assumption, exactly two (resp. one) parameter ray(s) land at every root (resp. co-root) of hyperbolic components of period k, for $k \in \{1, \cdots, n - 1\}$. It follows that on the boundary of every hyperbolic component of period k ($k \in \{1, \cdots, n - 1\}$), exactly d parameter rays of period k land. Thus, for every $k \in \{1, \cdots, n - 1\}$ the number of angles which are in $\Theta_k \cap (\theta_1, \theta_2)$ is $m \cdot d$ for some $m \in \mathbb{N} \cup \{0\}$. This yields that the k-th entry ($k \in \{1, \cdots, n - 1\}$) of $K(\theta)$ is incremented $m \cdot d$ times as θ travels from θ_1 to θ_2. Therefore, θ_1 and θ_2 have the same kneading sequences. $\qquad \square$

Theorem 9 (Different Kneading Sequences). *Consider the angles of periodic rays landing at the characteristic point of the parabolic orbit in the dynamical plane* *of c_0, where c_0 is a root. Then all these angles have pairwise different kneading sequences, except for possibly the two characteristic angles.*

Proof. We introduce notations: let z_1 be the characteristic point and $\mathcal{R}^{c_0}_{\theta_1}, \cdots, \mathcal{R}^{c_0}_{\theta_s}$ the dynamical rays landing at z_1. We call the associated orbit portrait $\mathcal{P}$. For $s = 2$ (where s is the number of dynamical rays landing at z_1) there is nothing to prove, so we assume $s \geq 3$ and are automatically in the non-primitive case (Lemma 6). Denote the exact period of the angles $\theta_1, \cdots, \theta_s$ by n and the orbit period of the parabolic orbit by $k = n/s$. Without restriction we assume $k \geq 2$. For $i \in \{1, \cdots, s\}$ denote the d inverse images of θ_i with respect to the d-tupling map by $\theta_i^{(l)} := (\theta_i + l)/d \in \mathbb{S}^1$ and the landing point of $\theta_i^{(l)}$ by $z_0^{(l)}$, $l \in \{0, \cdots, d-1\}$. The points $z_0^{(0)}, \ldots, z_0^{(d-1)}$ do not depend on the choice of a specific angle θ_i ($i \in \{1, \cdots, s\}$); in fact they are just the pre-images of z^1 under f_{c_0}. Finally, let H be the hyperbolic component of period n with $c_0 \in \partial H$ and c_1 a center of H.

By Corollary 6 and Lemma 23 there are continuous functions $z^{(0)}, \cdots, z^{(d-1)}$ on $H \cup \{c_0\}$ such that $z^{(l)}(c_0) = z_0^{(l)}$ and at $z^{(l)}(c)$ land the dynamical rays at the same angles as at $z_0^{(l)}$ for all $i \in \{0, 1, \cdots, d-1\}$ and $c \in H \cup \{c_0\}$. The points $z^{(l)}(c_1)$ lie on the boundary of the critical Fatou component U_0 of f_{c_1}. Let Γ be the Hubbard tree of c_1, U_1 the characteristic Fatou component and γ be the regular arc which connects the unique intersection point $\Gamma \cap \partial U_1$ with the critical value c_1. Then γ has d inverse images and each of them lies in $\overline{U_0}$ and connects the critical point with one of the points $z^{(l)}(c_1)$. Therefore, $\mathbb{C} \backslash P_{\theta_i}$ (where, $P_{\theta_i} := f_{c_1}^{-1}(\gamma) \cup \bigcup_{l=0}^{d-1} \mathcal{R}^{c_1}_{\theta_i^{(l)}}$) has precisely d open components for each $i \in \{1, 2, \cdots, s\}$. We label the boundary P_{θ_i} by (∗) and the component containing the critical value c_1 by L_1. The subsequent components are labelled (as L_i) in anti-clockwise direction. By construction the branch of $\Gamma \backslash \overline{U_0}$ on which the critical value lies is contained in the component with label L_1. Since f_{c_1} is orientation preserving this implies that the label of any branch of $\Gamma \backslash \overline{U_0}$ does not depend on P_{θ_i}. Now it's easy to see that for a fixed $j \neq k - 1 (mod\ k)$, all the dynamical rays at angles $\{d^j \cdot \theta_i : i = 1, 2, \cdots, s\}$ have the same label with respect to the corresponding partition P_{θ_i}. Thus the kneading sequences of the θ_i's can differ only in the $(mk-1)$-th position, for some $m \in \mathbb{N}$.

The kneading sequences of all θ_i except for two angles are pairwise different at an $(mk-1)$-th position: the $(mk-1)$-th entry of the kneading sequence of θ_i is just the label of the ray $\mathcal{R}^{c_1}_{d^{(mk-1)} \cdot \theta_i}$ with respect to the partition P_{θ_i}. Let the dynamical rays landing at the parabolic point on the boundary of the critical Fatou component be $\mathbb{S} = \{\mathcal{R}^{c_1}_{\theta_1^{(l)}}, \mathcal{R}^{c_1}_{\theta_2^{(l)}}, \cdots, \mathcal{R}^{c_1}_{\theta_s^{(l)}}\}$, for some $l \in \{0, 1, \cdots, d-1\}$. Now, $K(\theta_i) = K(\theta_j)$ only if the number of rays (amongst $\mathbb{S}$) which lie in a given component with label L_r with respect to the partition P_{θ_i} is equal to the number of such rays that lie in the corresponding component with label L_r with respect to the partition P_{θ_j}. However, if at least two rays at angles in $\mathbb{S}$ have different labels with respect to P_{θ_i}, then the number of rays with the smaller label is different with respect to P_{θ_i} and P_{θ_j} for $i \neq j$. Therefore, all these dynamical rays must have the same label. This is

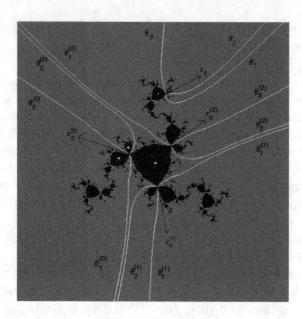

Fig. 6. The Julia set of a polynomial $z \mapsto z^3 + c$, $(c \approx 0.2253 + 0.9414i)$ with a 6-periodic super-attracting orbit (marked in green). With the notation of the proof, the dynamical rays landing at z_1 have angles $\theta_1 = 92/728, \theta_2 = 100/728$ and $\theta_3 = 172/728$. The rays $\theta_i^{(l)}$ landing at the three pre-images of z_1 are also drawn in.

only possible if none of the rays at angles in $\mathbb{S}$ lies in the connected component of $\mathbb{C} \setminus (\mathcal{R}_{\theta_i^{(l)}}^{c_1} \cup \mathcal{R}_{\theta_j^{(l)}}^{c_1} \cup \{z^{(l)}(c_1)\})$ containing U_0, i.e. if θ_i and θ_j are the characteristic angles. $\qquad\qquad\qquad\qquad\qquad\qquad\qquad\square$

We finish the proof of the Structure Theorem in the periodic case:

Theorem 10 (Precisely Two Parameter Rays Land at Every Root).
Every root c_0 is the landing point of exactly two parameter rays. The angles of the parameter rays landing at c_0 are the characteristic angles of the parabolic orbit of c_0.

Proof by induction on the ray period n of c_0. For $n = 1$, there are $d - 1$ co-roots; but there is a unique co-root which is the landing point of the parameter rays at angles 0 and 1 that we consider as two rays in this case. Assume that the roots of all hyperbolic components with period $n - 1$ or lower are the landing points of exactly two parameter rays. We obtain by Theorem 8 that only the parameter rays at n-periodic angles with same kneading sequences can land at the root c_0 of any hyperbolic component with period n. Note that a parameter ray with a given angle can land at c_0 only if the dynamical ray at the same angle lands at the characteristic point z_0 of the parabolic orbit (Theorem 5) and that the angles of all the dynamical rays landing at z_0, except possibly for the characteristic angles t^- and t^+, have

, different kneading sequences. Therefore, the only candidates for landing at c_0 are the two parameter rays at angles t^-, t^+. By Corollary 20, we know that they indeed land at c_0. This finishes the induction. $\square$

6. Pre-periodic Parameter Rays

In this section, we record the landing properties of the parameter rays of the multi-brot sets at pre-periodic angles. The generalization of the following results from the quadratic case is straight-forward in this case and does not require any new technique. We refer the readers to [23, Section 4] for a more comprehensive account on the combinatorics of the parameter rays (of the Mandelbrot set) at pre-periodic angles.

Definition 14 (Misiurewicz Point). A parameter c for which the critical orbit is strictly pre-periodic is called a *Misiurewicz point*.

Theorem 11 (Pre-periodic Parameter Rays Land). *Every parameter ray at a pre-periodic angle θ lands at a Misiurewicz point c_0. The corresponding dynamical ray $\mathcal{R}_\theta^{c_0}$ lands at the critical value c_0.*

Proof. See the proof of the pre-periodic case in [23, Theorem 1.1]. $\square$

Theorem 12 (Every Misiurewicz Point is a Landing Point). *Every Misiu-rewicz parameter is the landing point of a parameter ray at a pre-periodic angle.*

Proof. See the proof of the pre-periodic case in [23, Theorem 1.1]. $\square$

Theorem 13 (Number of Rays at a Misiurewicz Point). *Suppose that a pre-periodic angle θ has pre-period l and period n. Then the kneading sequence $K(\theta)$ has the same pre-period l, and its period k divides n. If $n/k > 1$, then the total number of parameter rays at pre-periodic angles landing at the same point as the ray at angle θ is n/k; if $n/k = 1$, then the number of parameter rays is 1 or 2.*

Proof. See [23, Lemma 4.4]. $\square$

References

[1] A.A. KAHN, J. KAHN, M. LYUBICH & WEIXIAO SHEN (2009): Combinatorial rigidity for unicritical polynomials, Annals of Mathematics, 170, pp. 783-797.

[2] B. BIELEFELD, Y. FISHER & J. HUBBARD (1992): The classification of critically preperiodic polynomials as dynamical systems, Journal of the American Mathematical Society, 5(4), pp. 721-762.

[3] A. CHÉRITAT (2014): Near Parabolic Renormalization for Unisingular Holomorphic Maps, URL: http://arxiv.org/pdf/1404.4735.pdf.

[4] L. CARLESON & T. W. GAMELIN (1993): Complex Dynamics, Springer, Berlin.

[5] A. DOUADY & J. H. HUBBARD (1982): Itération des polynômes quadratiques complexes, C. R. Acad. Sci. Paris Ser. I Math., 294, pp. 123-126.

[6] A. DOUADY & J. H. HUBBARD (1984-1985): Étude dynamique des polynômes complexes I, II, Publications Mathématiques d'Orsay, Université de Paris-Sud, Département de Mathématiques, Orsay.

[7] D. EBERLEIN (1999): Rational Parameter Rays of Multibrot Sets, Zentrum Mathematik, Technische Universität München.

[8] L. R. GOLDBERG (1992): Fixed points of polynomial maps I: Rotation subsets of the circles, Ann. Scient. École Norm. Sup., 4e série", 25, pp. 679-685.

[9] L. R. GOLDBERG & JOHN MILNOR (1993): Fixed points of polynomial maps II: Fixed point portraits, Ann. Scient. École Norm. Sup., 4e série, 26, pp. 51-98.

[10] J. HUBBARD & D. SCHLEICHER (2014): Multicorns are not Path Connected, pp. 73-102, in Frontiers in Complex Dynamics: In Celebration of John Milnor's 80th Birthday, Princeton University Press

[11] H. INOU & S. MUKHERJEE (2014): Non-landing parameter rays of the multicorns, to appear in Inventiones Mathematicae,
URL: http://link.springer.com/article/10.1007/s00222-015-0627-3.

[12] J. KAHN & M. LYUBICH (2005): Local connectivity of Julia sets for unicritical polynomials, URL: http://arxiv.org/pdf/math/0505194v1.pdf.

[13] P. LAVAURS (1989): Systèmes dynamiques holomorphes: explosion de points périodiques paraboliques, Université de Paris-Sud Centre d'Orsay.

[14] J. MILNOR (1999): Periodic Orbits, External Rays and the Mandelbrot Set, Institute for Mathematical Sciences, SUNY, Stony Brook NY, Stony Brook IMS Preprint, 3, URL: http://www.math.sunysb.edu/preprints/.

[15] J. MILNOR (2000): Periodic Orbits, External Rays and the Mandelbrot Set, Astérisque, 261, pp. 277-333.

[16] J. MILNOR (2006): Dynamics in one complex variable, 3rd edition, Princeton University Press, New Jersey.

[17] J. MILNOR (2014): Arithmetic of Unicritical Polynomial Maps, pp. 15-23, in Frontiers in Complex Dynamics: In Celebration of John Milnor's 80th Birthday, Princeton University Press.

[18] S. MUKHERJEE (2015): Orbit Portraits of Unicritical Antiholomorphic polynomials, Conformal Geometry and Dynamics of the AMS, 19, pp. 35-50.

[19] S. MUKHERJEE, S. NAKANE & DIERK SCHLEICHER (2015): On Multicorns and Unicorns II: Bifurcations in Spaces of Antiholomorphic Polynomials, to appear in *Ergodic Theory and Dynamical Systems*,
URL: http://journals.cambridge.org/action/displayAbstract?fromPage=online&aid=10048271&fulltextType=RA&fileId= S0143385715000656.

[20] SHIZUO NAKANE AND DIERK SCHLEICHER (2003): On Multicorns and Unicorns I: Antiholomorphic Dynamics, Hyperbolic Components and Real Cubic Polynomials, International Journal of Bifurcation and Chaos, 13, pp. 2825-2844.

[21] C. L. PETERSEN & G. RYD (2000): Convergence of rational rays in parameter spaces, pp. 161-172, In T. LEI (Ed.): The Mandelbrot Set, Theme and Variations, Cambridge University Press, London Mathematical Society Lecture Note Series, 274.

[22] A. POIRIER (1993): On Postcritically Finite Polynomials, Part 1: Critical Portraits, URL: http://arxiv.org/abs/math/9305207.

[23] D. SCHLEICHER (2000): Rational Parameter Rays of the Mandelbrot Set, Astérisque, 261, pp. 405-443.

Chapter 4

The Matovich-Pearson Equations Revisited

Thomas Hagen

Department of Mathematical Sciences, The University of Memphis, Memphis, TN 38152, USA
E-mail: thagen@memphis.edu

The Matovich-Pearson equations are a well-known asymptotic regime of the axisymmetric Navier-Stokes equations with moving boundary. They arise as the slender body approximation of the full equations for a highly viscous, Newtonian fluid and are used to describe the dynamics and evolution of thin, viscous fluid filaments, e.g. in the context of fiber spinning. While these equations can be systematically derived, additional boundary conditions have to be imposed to make the resulting initial-boundary value problem well-posed.

In this work we shall extend existing results and techniques for the Matovich-Pearson equations to the case of a prescribed pulling force at the downstream boundary. We will prove a local well-posedness result and show that global existence of solutions holds true. Even though this latter result relies on earlier techniques developed for the classical initial-boundary value problem of the Matovich-Pearson equations with prescribed take-up velocity, it requires additional arguments to overcome the absence of *a priori* bounds on the filament velocity. Global existence is, at its core, based on the fundamental long-term behavior of viscous fluids in extension when other factors are omitted: Viscous fluid filaments do not break.

Contents

1. Introduction

In industrial applications such as fiber spinning, film casting and film blowing a polymer melt is extruded through a spinneret or die to form a synthetic fiber or film. The highly viscous liquid melt is withdrawn from a reservoir, axially stretched, solidified at a fixed distance from the outflow point and finally wound up on a spool, roll or similar device. Flows of this kind occur in the textile and chemical industry to manufacture fibers and thin sheets, both for everyday use and more specialized needs. Production output and product quality can be considerably influenced by the flow dynamics, including detrimental flow instabilities, film necking, filament break-up or film rupture. The potential use of these products is broad and ranges from packaging material to protective clothing and medicinal equipment. A schematic of the fiber spinning process is given in Fig. 1.

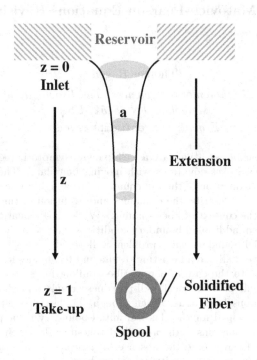

Fig. 1.   Fiber spinning

   The model equations we will study arise as the thin filament/slender body approximation of the axisymmetric Navier-Stokes (or Stokes) equations with moving boundary. For the purely viscous case discussed here, these equations were first used by Kase and Matsuo [13] in their stability study of fiber spinning. Matovich and Pearson [14, 17] gave a formal, yet systematic derivation of the governing equations from the axisymmetric (Navier-) Stokes equations by exploiting the smallness of the aspect ratio of radial vs. axial length scale. By proceeding in this way, the two-dimensional flow is reduced to one spatial dimension and the moving boundary condition is eliminated from the resulting model. At the same time, however, many qualitative features of the flow are preserved so that the predicted flow behavior (such as the onset of instabilities) agrees reasonably well with experimental findings if suitable boundary conditions are imposed to close the model formulation. Other features such as fiber break-up are, however, lost in the classical Matovich-Pearson equations (without surface tension). Hence a thorough understanding of the model's limitations is essential. However, the failure (or strong inhibition) of fluid filaments to break is expected when the fluid is highly viscous (such as honey). An excellent account about fluid fibers and films in engineering applications was given by Yarin [23]. The study of fiber break-up due to surface tension was initiated in the seminal works by Eggers, Papageorgiou and others, see e.g. [1, 2, 15, 16]. A fairly recent overview of these and related developments appeared in [3]. Analytical work on fiber break-up is due to Renardy [18, 19, 21]. An initial "no break-up" result in a simplified flow scenario was given in [12].

In their original (classical) form the Matovich-Pearson equations assume a highly viscous, Newtonian fluid of constant viscosity with negligible inertia, gravity and surface tension. All equations are stated in dimensionless form. We denote time by t, the axial variable by z, the fiber cross-sectional area by $a = a(t, z)$, and the axial velocity by $v = v(t, z)$. Then the governing equations consist of conservation of mass

$$a_t + (v\,a)_z = 0 \tag{1}$$

and conservation of linear momentum

$$(a\,v_z)_z = 0. \tag{2}$$

The equations are stated on the normalized domain $0 \leq z \leq 1$, $t \geq 0$. Here, the inlet/spinneret and take-up point are assumed at $z = 0$ and $z = 1$, respectively. To close the formulation of the problem, we impose the initial condition

$$a(0, z) = \alpha(z), \quad 0 \leq z \leq 1 \tag{3}$$

and the boundary conditions

$$a(t, 0) = a_0(t), \quad v(t, 0) = v_0(t), \quad t \geq 0 \tag{4}$$

together with the force condition at the take-up point

$$a(t, 1)\,v_z(t, 1) = f(t), \quad t \geq 0. \tag{5}$$

Here, f denotes a prescribed non-negative pulling force (modulo division by the constant tensile viscosity). The flow situation is depicted in Fig. 2.

It was pointed out by Renardy [20] that the boundary conditions for the Matovich-Pearson equations are an idealization of the physical reality. There is no real theoretical basis to decide which conditions are physically most appropriate and enforceable in actual spinning applications. In fact, instead of the force boundary condition (5), many authors have considered the velocity boundary condition at the take-up point

$$v(t, 1) = v_1(t), \quad t \geq 0. \tag{6}$$

It is this condition which is most commonly used in the literature since the resulting dynamical behavior of solutions appears to agree well with actual observations, in particular with respect to instabilities such as "draw resonance." The alternative

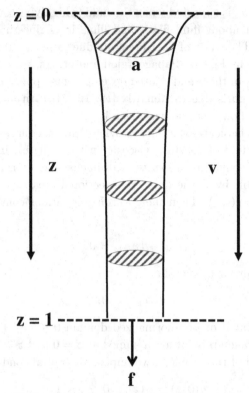

Fig. 2.   The flow domain and the flow variables

downstream boundary condition (5) was considered by Renardy [20] in the context of linear stability of stationary solutions. It has also been discussed in the
engineering literature, see e.g. [17, 22].

    In this work we address the solvability of the initial-boundary value problem
given by Eqs. (1)–(5). We will obtain a local-in-time existence and uniqueness
result of smooth solutions. The existence/uniqueness proof is classical and follows
a fixed-point strategy, similar to the one devised in [6, 9, 10]. In the second part of
our work we consider the question of global-in-time existence of solutions. While
it was proved by Feireisl, Laurençot and Mikelić [5] and independently by Hagen
and Renardy [11] that the model equations with the velocity boundary condition
(6) admit global solutions, the analogous result for the force boundary condition
(5) has so far been left open. Our discussion will be based on a fundamental
dynamical result for solutions of the governing equations: highly viscous fluid fibers
do not break in finite time (when surface tension and inertia are neglected). Hence
solutions are expected to exist for all times. Even though the global existence proof
we present borrows ideas from our result in [11], a new technical problem arises
here: no a priori velocity bounds are available. Hence a new argument is needed to
overcome this difficulty.

Mathematically rigorous work on the Matovich-Pearson equations and related equations of film casting is generally sparse apart from the publications mentioned above, a recent collection of works on mathematical issues in fiber spinning [4] and some older articles. This is rather unsatisfactory given the importance of these asymptotic regimes of the Navier-Stokes equations. Dynamical behavior of fibers and films and challenging physical phenomena such as instabilities still await a more complete mathematical examination and rigorous explanation. Detailed studies on the linearization of the governing equations, both for fibers and films, have become available (see e.g. [7, 8]) and are expected to serve as a basis for further work.

2. Local Well-Posedness

In this section we will prove that the governing equations (1)–(5) have a unique, local-in-time solution which is smooth enough to be interpreted in the classical sense. Our approach is based on a fixed-point argument and resembles related results for viscoelastic and non-isothermal regimes of the Matovich-Pearson equations with velocity boundary condition (6), see e.g. [6, 9, 10].

Using the momentum balance (2) and the force boundary condition (5), we first recast the governing equations in the reduced form

$$a_t + v\, a_z = -f \tag{7}$$

with

$$v(t,z) = v_0(t) + f(t) \int_0^z \frac{1}{a(t,x)}\, dx \tag{8}$$

and the initial and boundary conditions

$$a(0,z) = \alpha(z), \quad 0 \le z \le 1 \quad \text{and} \quad a(t,0) = a_0(t), \quad t \ge 0. \tag{9}$$

We will always assume that both α, a_0 and v_0 are strictly positive and that f is non-negative. In addition, it suffices to assume that a_0, v_0 and f are given on a time interval $[0, T]$ for some $T > 0$. Since we are only interested in smooth solutions, we impose the compatibility conditions on the boundary and initial data

$$\alpha(0) = a_0(0), \tag{10}$$
$$a_0'(0) + v_0(0)\, \alpha'(0) = -f(0). \tag{11}$$

These conditions are necessary to ensure a continuously differentiable solution a. We will tacitly assume throughout that the data α, a_0, v_0 and f satisfy these two conditions. In [5] less regular solutions were constructed for the Matovich-Pearson equations with the velocity boundary condition (6). An approach similar to that one seems also possible here.

Let us introduce some abbreviations that will prove useful later on. Let $t^* > 0$, $1 \le p \le \infty$ and $m, n \in \mathbb{N}_0$. We will write

(1) $\|\cdot\|_p$ for the norm on the Lebesgue space $L^p(0,1)$,

(2) $\|\cdot\|_{H^1}$ for the norm on the Sobolev space $H^1(0, t^*)$,

(3) $\|\cdot\|_{m,n}$ for the norm on the Sobolev space $W^{m,\infty}(0, t^*; H^n(0, 1))$.

It is implicitly understood that t^* must be clear from the context.

The notion of *boundary regularity* as discussed in [9] plays a prominent role in the existence theory of quasilinear, first-order transport equations as encountered here. Hence we will briefly introduce some of the key results needed in later developments.

Definition 1. For $t_0 > 0$ the space $\mathbb{BR}(0, t_0)$ of boundary-regular functions consists of all functions $g = g(t, x)$ on $[0, t_0] \times [0, 1]$ such that

$$g \in W^{1,\infty}(0, t_0; H^1(0, 1)) \cap L^\infty(0, t_0; H^2(0, 1)), \tag{12}$$

$$\partial_x g(\cdot, 0), \partial_x g(\cdot, 1) \in H^1(0, t_0). \tag{13}$$

The space $\mathbb{BR}(0, t_0)$ is endowed with the norm

$$\mathcal{E}(g) = \left(\|g\|_{0,2}^2 + \|g\|_{1,1}^2 + \|\partial_x g(\cdot, 0)\|_{H^1}^2 + \|\partial_x g(\cdot, 1)\|_{H^1}^2 \right)^{\frac{1}{2}}. \tag{14}$$

Note that any function $g \in W^{1,\infty}(0, t_0; H^1(0, 1)) \cap L^\infty(0, t_0; H^2(0, 1))$ has the property that $g(t, \cdot) \in H^2(0, 1)$ for *every* $t \in [0, t_0]$. We will tacitly make use of this property later on.

The central result highlighting the importance of boundary regularity is the following:

Theorem 1 (Hagen and Renardy [9]). *Suppose that for $t^* > 0$*

$$p, q \in \mathbb{BR}(0, t^*), \quad p > 0, \tag{15}$$

$$u_\alpha \in H^2(0, 1), \quad u_0 \in H^2(0, t^*), \tag{16}$$

$$u_\alpha(0) = u_0(0), \quad u_0'(0) + p(0, 0) u_\alpha'(0) = q(0, 0). \tag{17}$$

Then the initial-boundary value problem on $[0, t^*] \times [0, 1]$

$$u_t + p u_z = q, \tag{18}$$

$$u(0, z) = u_\alpha(z) \text{ for } 0 \leq z \leq 1 \quad and \quad u(t, 0) = u_0(t) \text{ for } 0 \leq t \leq t^* \tag{19}$$

has a unique solution u in $\mathbb{BR}(0, t^*) \cap C^1([0, t^*]; H^1(0, 1)) \cap C([0, t^*]; H^2(0, 1))$.

The proof is based on regularization techniques and is given in detail in [9].

The main goal of this section is the following local existence and uniqueness result for smooth solutions of the governing equations.

Theorem 2. *For $\alpha \in H^2(0, 1)$, $a_0 \in H^2(0, T)$, $v_0 \in C^1([0, T])$ and $f \in C^1([0, T])$, there exists $t^* \in (0, T]$ such that the initial-boundary value problem (7)–(9) has a unique solution (a, v) with the following properties:*

$$a \text{ is strictly positive for } 0 \leq t \leq t^*, \ 0 \leq z \leq 1, \tag{20}$$

$$a \in C\left([0, t^*]; H^2(0, 1)\right) \cap C^1\left([0, t^*]; H^1(0, 1)\right), \tag{21}$$

$$v \in C\left([0, t^*]; H^3(0, 1)\right) \cap C^1\left([0, t^*]; H^2(0, 1)\right). \tag{22}$$

Note that this result implies that the fiber cross-sectional area a is continuously differentiable. For the proof of Theorem 2 we proceed in several steps.

Definition 2. For $L > 0$ and $t_0 \in (0, T]$, let $\mathcal{S}(t_0, L)$ be the set of functions $b \in \mathbb{BR}(0, t_0)$ such that

$$\mathcal{E}(b) \leq L \quad \text{and} \quad b(0, z) - \alpha(z) \text{ for } 0 \leq z \leq 1. \tag{23}$$

Lemma 1. *If* $L > \mathcal{E}(\alpha)$, *the set* $\mathcal{S}(t_0, L)$ *is nonempty for every* $t_0 \in (0, T]$. *Moreover, if* $t_0 \in (0, T]$ *is sufficiently small, the elements of* $\mathcal{S}(t_0, L)$ *are strictly positive.*

Proof. The initial datum α is contained in $\mathcal{S}(t_0, L)$ if L is larger than $\mathcal{E}(\alpha)$. If $b \in \mathcal{S}(t_0, L)$, then

$$|b(t, z) - \alpha(z)| \leq \int_0^t |b_t(\tau, z)| \, d\tau \leq C L t_0 \tag{24}$$

where C is a constant independent of t, t_0, z and L. Hence the claim follows. $\square$

From now on we tacitly assume that the conclusions of Lemma 1 apply to t_0 and L.

Lemma 2. $\mathcal{S}(t_0, L)$ *is a complete metric space in the metric* $d(b_1, b_2) = \|b_1 - b_2\|_{0,0}$ *for* b_1, $b_2 \in \mathcal{S}(t_0, L)$.

The proof is based on an elementary comparison between weak* convergence in $W^{1,\infty}(0, t_0; H^1(0, 1)) \cap L^\infty(0, t_0; H^2(0, 1))$ and strong convergence in $L^2((0, t_0) \times (0, 1))$. We refer to [6, 9, 10] for the proof of analogous results in a similar context. The idea to consider a metric space consisting of a set bounded in a strong norm, but with a metric given through a very weak norm is classical and a well-known tactic in the analysis of nonlinear hyperbolic problems.

Theorem 3. *For* $0 < t^* \leq t_0$, *let* Σ *be the map that assigns to each* $b \in \mathcal{S}(t^*, L)$ *the solution* u *of the initial-boundary value problem on* $[0, t^*] \times [0, 1]$

$$u_t + w \, u_z = -f \quad \text{with} \quad w(t, z) = v_0(t) + f(t) \int_0^z \frac{1}{b(t, x)} \, dx, \tag{25}$$

$$u(0, z) = \alpha(z) \text{ for } 0 \leq z \leq 1 \quad \text{and} \quad u(t, 0) = a_0(t) \text{ for } 0 \leq t \leq t^*. \tag{26}$$

If $L > 0$ *is sufficiently large and* $t^* > 0$ *sufficiently small,* Σ *is a contraction on* $\mathcal{S}(t^*, L)$ *with respect to the metric* d.

Proof. For the proof we make use of Theorem 1. First note that Theorem 1 ensures that the map Σ is well-defined due to the regularity and compatibility of the data in (25), (26). Then we invoke fairly technical energy estimates from [9] for the auxiliary problem (18), (19) to conclude that for sufficiently large $L > 0$ and sufficiently small $t^* > 0$, the image of $\mathcal{S}(t^*, L)$ under Σ is contained in $\mathcal{S}(t^*, L)$. Similar arguments were given in [6, 9, 10]. Finally, it is left to show that Σ is contractive in the metric d if t^* is small enough. To this end, let $u = \Sigma(b)$ and $\tilde{u} = \Sigma(\tilde{b})$. We also define w

and $\tilde{w}$ via (25) for b and $\tilde{b}$, respectively. As we subtract the equations for u and $\tilde{u}$ and multiply by $u - \tilde{u}$, we obtain

$$\frac{1}{2}\frac{\partial}{\partial t}(u - \tilde{u})^2 + \frac{1}{2}w\frac{\partial}{\partial z}(u - \tilde{u})^2 + \tilde{u}_z(w - \tilde{w})(u - \tilde{u}) = 0. \tag{27}$$

Now we integrate each term in z from 0 to 1. This gives

$$\int_0^1 \frac{1}{2}\frac{\partial}{\partial t}(u - \tilde{u})^2\,dz = \frac{1}{2}\frac{d}{dt}\int_0^1 (u - \tilde{u})^2\,dz, \tag{28}$$

$$\int_0^1 \frac{1}{2}w\frac{\partial}{\partial z}(u - \tilde{u})^2\,dz = \frac{1}{2}\left(w(u - \tilde{u})^2\right)\Big|_{z=1} - \frac{1}{2}\int_0^1 w_z(u - \tilde{u})^2\,dz, \tag{29}$$

$$\int_0^1 \tilde{u}_z(w - \tilde{w})(u - \tilde{u})\,dz = f\int_0^1 \tilde{u}_z\int_0^z \frac{\tilde{b} - b}{b\tilde{b}}\,dx\,(u - \tilde{u})\,dz. \tag{30}$$

Next we find positive constants C_1, C_2 and C_3 that depend on α, L and t_0 ($\geq t^*$) such that on $[0, t^*] \times [0, 1]$

$$|w_z| \leq C_1, \quad |\tilde{u}_z| \leq C_2, \quad \left|\frac{1}{b\tilde{b}}\right| \leq C_3. \tag{31}$$

Hence using the Hölder and Young inequalities, we obtain

$$\left|\int_0^1 w_z(u - \tilde{u})^2\,dz\right| \leq C_1\int_0^1 (u - \tilde{u})^2\,dz, \tag{32}$$

$$\left|\int_0^1 \tilde{u}_z\int_0^z \frac{\tilde{b} - b}{b\tilde{b}}\,dx\,(u - \tilde{u})\,dz\right| \leq C_2 C_3\int_0^1 |b - \tilde{b}|\,dz\int_0^1 |u - \tilde{u}|\,dz$$

$$\leq \frac{C_2 C_3}{2}\left(\int_0^1 (b - \tilde{b})^2\,dz + \int_0^1 (u - \tilde{u})^2\,dz\right). \tag{33}$$

These estimates prove that there exists a constant $C > 0$ which depends on α, f, L and t_0 such that

$$\frac{d}{dt}\int_0^1 (u - \tilde{u})^2\,dz \leq C\left(\int_0^1 (u - \tilde{u})^2\,dz + \int_0^1 (b - \tilde{b})^2\,dz\right). \tag{34}$$

This differential inequality gives for $0 \leq t \leq t^*$

$$\|u(t, \cdot) - \tilde{u}(t, \cdot)\|_2^2 \leq C\,t^*\,e^{C t^*}\,\|b - \tilde{b}\|_{0,0}^2. \tag{35}$$

Hence if $0 < t^* \leq t_0$ is chosen such that $C\,t^*\,e^{C t^*} < 1$, we have the contraction property

$$d(u, \tilde{u}) \leq \kappa\,d(b, \tilde{b}) \quad \text{with } \kappa = \left(C\,t^*\,e^{C t^*}\right)^{1/2} < 1. \tag{36}$$

This implies the claim.   □

Theorem 3 proves the existence and uniqueness of a fixed point a for the map Σ for sufficiently small time t^*. This fixed point a is strictly positive on $[0, t^*] \times [0, 1]$ and satisfies a problem of the form (18), (19), namely problem (7), (9) with v given by (8) and time restricted to $[0, t^*]$. Hence $a \in \mathbb{BR}(0, t^*) \cap C^1([0, t^*]; H^1(0, 1)) \cap$

$C([0, t^*]; H^2(0, 1))$, and consequently $v \in C^1([0, t^*]; H^2(0, 1)) \cap C([0, t^*]; H^3(0, 1))$. This concludes the proof of our local-in-time existence and uniqueness result, Theorem 2.

3. No Breakup and Global Existence

Our objective is to extend local-in-time solutions to global solutions. A local solution (a, v) will exist as long as

- a stays positive (i.e. the filament does not break), and
- the solution retains its smoothness.

Let us assume for the moment that $a = a(t, z)$ is a strictly positive, continuously differentiable solution of Eqs. (7)–(9) with velocity v on $[0, T) \times [0, 1]$ for some $T > 0$. The full regularity afforded by Theorem 2 is not required at this point. We also assume that the boundary data a_0, v_0 and f are given on the closed interval $[0, T]$. Our first goal is to prove that the area a and the velocity v can be extended *continuously* into $t = T$.

By Eq. (8), the velocity v is strictly positive, while f is non-negative. Consequently, by integration along characteristics we find a constant $A > 0$ such that

$$0 < a(t, z) \leq A \quad \text{for } 0 \leq t < T, 0 \leq z \leq 1. \tag{37}$$

In contrast to the classical case of the Matovich-Pearson equations with the boundary condition (6), here it is not clear that the velocity is bounded. In fact, as time t approaches T, the velocity migh very well become unbounded if the cross-sectional area a were to become zero at some point z. Similarly, the cross-sectional area a might vanish at some point if the velocity blew up. This dilemma makes the problem at hand interesting and different from the classical case discussed in [11]. While we can borrow some of the techniques presented there, a new argument will be necessary.

To set the stage, let us introduce the *Lagrangian variable* $Z = Z(t, z)$, first studied in [11]:

$$Z(t, z) = \int_0^z a(t, x) \, dx - \int_0^t a_0(s) \, v_0(s) \, ds, \quad 0 \leq t < T, \quad 0 \leq z \leq 1. \tag{38}$$

For later use we abbreviate

$$\mathbb{S} = \left(-\int_0^T a_0(s) \, v_0(s) \, ds, \int_0^1 \alpha(x) \, dx \right]. \tag{39}$$

We make some elementary observations:

Lemma 3 (Hagen and Renardy [11]). *The Lagrangian variable Z has the following properties:*

(1) $Z \in C^1([0, T) \times [0, 1])$,

(2) Z *satisfies* $Z_t + v Z_z = 0$ *for* $0 \leq t < T$, $0 \leq z \leq 1$,
(3) $Z_z > 0$ *and* $Z_t < 0$ *for* $0 \leq t < T$, $0 \leq z \leq 1$,
(4) *Image* $Z = \mathbb{S}$.

Since Z is monotone in t and bounded, we can define the extension Z^* of Z to all of $0 \leq t \leq T$, $0 \leq z \leq 1$ by setting

$$Z^*(t, z) = \begin{cases} Z(t, z) & \text{if } 0 \leq t < T, 0 \leq z \leq 1, \\ \lim_{s \nearrow T} Z(s, z) & \text{if } t = T, 0 \leq z \leq 1. \end{cases} \tag{40}$$

While Z^* is uniquely defined for $t = T$, there is no guarantee that $Z^*(T, \cdot)$ is continuous. This problem is owed to the fact that v might not be bounded.

The central result for the Lagrangian variable Z is the following:

Theorem 4 (Hagen and Renardy [11]). *Let h be a continuously differentiable solution of the advection equation*

$$h_t + v \, h_z = 0 \tag{41}$$

for $0 \leq t < T$ and $0 \leq z \leq 1$ with v given by Eq. (8). Then there exists a continuously differentiable function Φ on $\mathbb{S}$ such that

$$h(t, z) = \Phi(Z(t, z)) \quad \text{for } 0 \leq t < T, 0 \leq z \leq 1. \tag{42}$$

The proof of this result relies on the basic observation that $\begin{pmatrix} h_t \\ h_z \end{pmatrix}$ and $\begin{pmatrix} Z_t \\ Z_z \end{pmatrix}$ are parallel at each point of their common domain. Hence it can be shown that h and Z are *functionally dependent* as expressed in (42). For proof details we refer to [11].

Let us define

$$F(t) = \int_0^t f(s) \, ds. \tag{43}$$

Then using Eq. (7), we can write

$$(a + F)_t + v \, (a + F)_z = 0. \tag{44}$$

Consequently, by Theorem 4 we have:

Corollary 1. *There exists a continuously differentiable function Φ on $\mathbb{S}$ such that*

$$a(t, z) = \Phi(Z(t, z)) - F(t) \quad \text{for } 0 \leq t < T, 0 \leq z \leq 1. \tag{45}$$

Since $Z^*(T, z) \in \mathbb{S}$ for $0 < z \leq 1$, it is natural to extend a into $t = T$ by defining

$$a^*(t, z) = \begin{cases} a(t, z) & \text{if } 0 \leq t < T, 0 \leq z \leq 1 \\ \Phi(Z^*(T, z)) - F(T) & \text{if } t = T, 0 < z \leq 1 \\ a_0(T) & \text{if } t = T, z = 0. \end{cases} \tag{46}$$

We can now prove the following key result.

Theorem 5. *The extension a^* of the cross-sectional area a is continuous on all of $0 \leq t \leq T$, $0 \leq z \leq 1$.*

Proof. By definition of a^* (and Z^*), a^* is bounded for $0 \leq t \leq T$, $0 \leq z \leq 1$, and $a^*(T, \cdot)$ is measurable. Hence the function

$$X(t,z) = \int_0^z a^*(t,x)\,dx \tag{47}$$

is well-defined and continuous for $0 \leq t < T$, $0 \leq z \leq 1$ and $t = T$, $0 \leq z \leq 1$. By the Dominated Convergence Theorem it follows that

$$\lim_{t \nearrow T} X(t,z) = X(T,z) \quad \text{for each } 0 \leq z \leq 1. \tag{48}$$

Hence X is continuous on all of $0 \leq t \leq T$, $0 \leq z \leq 1$. Now we observe that by definition of Z and Z^*,

$$Z^*(t,z) = X(t,z) - \int_0^t a_0(s)\,v_0(s)\,ds \quad \text{for } 0 \leq t \leq T, 0 \leq z \leq 1. \tag{49}$$

By definition (46), we obtain that a^* continuously extends a to all of $0 \leq t \leq T$, $0 \leq z \leq 1$ with the possible exception of the point $t = T$, $z = 0$. There, however, setting $a^*(T,0) = a_0(T)$ as in (46) makes a^* continuous everywhere. $\square$

Using the definition of Z in (38) and the corresponding equation for Z^*, we conclude:

Corollary 2. *The extension Z^* of the Lagrangian variable Z is continuously differentiable with respect to z for $0 \leq t \leq T$, $0 \leq z \leq 1$ such that $Z_z^* = a^*$.*

The extension result above is the key to invoke the following crucial no breakup result in [11].

Theorem 6 (Hagen and Renardy [11]). *The continuous extension a^* of the cross-sectional area a is strictly positive for $0 \leq t \leq T$, $0 \leq z \leq 1$.*

This result proves that the cross-sectional area a has a positive lower bound. Hence we can extend the velocity v using Eq. (8).

Corollary 3. *The velocity v has a continuous extension v^* to all of $[0,T] \times [0,1]$.*

Next, as we observed in [11], we have for $0 \leq t < T$

$$a_0'(t) + v_0(t)\,a_z(t,0) + f(t) = 0 \tag{50}$$

and

$$a_z(t,0) = \Phi'(Z(t,0))\,a_0(t). \tag{51}$$

Thus

$$\Phi'(Z(t,0)) = -\frac{a_0'(t) + f(t)}{a_0(t)\,v_0(t)}. \tag{52}$$

Hence since the only point $Y \in \bar{\mathbb{S}} \setminus \mathbb{S}$ is

$$Y = -\int_0^T a_0(s)\,v_0(s)\,ds, \tag{53}$$

we can extend Φ' into Y by setting

$$
\Omega(Z) = \begin{cases} \Phi'(Z) & \text{if } Z \in \mathbb{S}, \\ -\lim_{t \nearrow T} \dfrac{a_0'(t) + f(t)}{a_0(t)\, v_0(t)} & \text{if } Z = Y. \end{cases} \tag{54}
$$

Consequently, Φ can be extended to a continuously differentiable function Φ^* on $\bar{\mathbb{S}}$ by setting

$$
\Phi^*(Z) = \Phi(0) + \int_0^Z \Omega(R)\, dR. \tag{55}
$$

Hence, in light of (46), we can draw the following conclusion:

Lemma 4. *The extensions a^* and v^* are continuously differentiable for $0 \le t \le T$, $0 \le z \le 1$ and*

$$
a_t^* + v^*\, a_z^* = -f \qquad \text{on } [0,T] \times [0,1]. \tag{56}
$$

Finally, we are ready to prove global existence of local solutions. See [11] for the corresponding case with the velocity boundary condition (6). Our proof proceeds along similar lines with appropriate modifications.

Theorem 7. *For $\alpha \in H^2(0,1)$, $a_0 \in H^2(0,T)$, $v_0 \in C^1([0,T])$ and $f \in C^1([0,T])$, let (a,v) be the local-in-time solution of the initial-boundary value problem (7)–(9) obtained in Theorem 2 on $[0,t^*] \times [0,1]$ for some $0 < t^* \le T$. Then (a,v) extends to a unique solution on $[0,T] \times [0,1]$ with the following properties:*

$$
a \text{ is strictly positive for } 0 \le t \le T,\ 0 \le z \le 1, \tag{57}
$$
$$
a \in C\left([0,T]; H^2(0,1)\right) \cap C^1\left([0,T]; H^1(0,1)\right), \tag{58}
$$
$$
v \in C\left([0,T]; H^3(0,1)\right) \cap C^1\left([0,T]; H^2(0,1)\right). \tag{59}
$$

Proof. Let $\mathbb{T}$ be the set of all times $t_0 \in (0,T]$ with the property that a local solution in the sense of Theorem 2 exists on $[0,t_0] \times [0,1]$. Clearly, $\mathbb{T}$ is nonempty. We can extend any local solution on $[0,t^*] \times [0,1]$ with $0 < t^* < T$ to $[0,t^*+\epsilon] \times [0,1]$ by invoking Theorem 2 to solve the problem for a small time span ϵ with initial data $a(t^*,\cdot) \in H^2(0,1)$ and appropriately shifted boundary data. Pasting this solution to the given solution produces the desired result. Hence the set $\mathbb{T}$ is open in $(0,T]$. Now we show that the set $\mathbb{T}$ is closed in $(0,T]$. To this end, let $\tau = \sup \mathbb{T}$. Then the problem has a unique solution (a,v) on $[0,t_0] \times [0,1]$ for each $0 < t_0 < \tau$. Consequently, by uniqueness and smoothness according to Theorem 2, (a,v) must be continuously differentiable on $[0,\tau) \times [0,1]$. By Lemma 4, (a,v) extends to a pair (a^*,v^*) of continuously differentiable functions on $[0,\tau] \times [0,1]$ such that

$$
a_t^* + v^*\, a_z^* = -f \qquad \text{on } [0,\tau] \times [0,1]. \tag{60}
$$

By Theorem 6, a^* is strictly positive. In light of Eq. (8) with a^* and v^* taking the place of a and v, v^* belongs to $\mathbb{BR}(0,\tau)$, and so does f. Hence by Theorem 1,

a^* belongs to $C\left([0,\tau];H^2(0,1)\right)\cap C^1\left([0,\tau];H^1(0,1)\right)$. In return, v^* belongs to $C\left([0,\tau];H^3(0,1)\right)\cap C^1\left([0,\tau];H^2(0,1)\right)$. Consequently, τ belongs to $\mathbb{T}$. Since $\mathbb{T}$ is both open and closed in $(0,T]$, the claim follows. □

Acknowledgement

A mathematician's career does not happen in a vacuum. In our early years we encounter fellow students and teachers, and if we are really lucky, a rare master of his art who opens our eyes to the beauty and power of our discipline outside the dire reality of endless repetition and mindless exercise. Professor Armin Leutbecher is such a person. I will always be indebted to him for his generosity and kindness. This work is dedicated to him on the occasion of his 80th birthday.

References

[1] J. Eggers, Universal pinching of 3d axisymmetric free surface flows, *Phys. Rev. Lett.* **71**, 3458–3460 (1993).

[2] J. Eggers, Nonlinear dynamics and breakup of free-surface flows, *Rev. Mod. Phys.* **69**, 865–929 (1997).

[3] J. Eggers and E. Villermaux, Physics of liquid jets, *Rep. Prog. Phys.* **71**, 036601 (2008).

[4] A. Fasano (Ed.), *Mathematical Models in the Manufacturing of Glass* (Springer, Berlin, 2011).

[5] E. Feireisl, Ph. Laurençot and A. Mikelić, Global-in-time solutions for the isothermal Matovich-Pearson equations, *Nonlinearity.* **24**, 277–292 (2011).

[6] T. Hagen, Jeffreys fluids in forced elongation, *J. Math. Anal. Appl.* **288**, 634–645 (2003).

[7] T. Hagen, Linear theory of nonisothermal forced elongation, *J. Evol. Equ.* **5**, 417–440 (2005).

[8] T. Hagen, On the membrane approximation in isothermal film casting, *Zeitsch. Angew. Math. Phys.* **65**, 729–745 (2014).

[9] T. Hagen and M. Renardy, On the equations of fiber spinning in non-isothermal viscous flow. In eds. J. Escher and G. Simonett, *Topics in Nonlinear Analysis. The Herbert Amann Anniversary Volume* (Birkhäuser, Basel, 1999).

[10] T. Hagen and M. Renardy, Non-adiabatic elongational flows of viscoelastic melts, *Z. Angew. Math. Phys.* **51**, 845–866 (2000).

[11] T. Hagen and M. Renardy, Non-failure of filaments and global existence for the equations of fiber spinning, *IMA J. Appl. Math.* **76**, 834–846 (2011).

[12] O. Hassager, M. I. Kolte and M. Renardy, Failure and nonfailure of fluid filaments in extension, *J. Non-Newtonian Fluid Mech.* **76**, 137–152 (1998).

[13] S. Kase and T. Matsuo, Studies on melt spinning. I. Fundamental equations on the dynamics of melt spinning, *J. Appl. Polym. Sci. A.* **3**, 2541–2554 (1965).

[14] M. A. Matovich and J. R. A. Pearson, Spinning a molten threadline – steady-state isothermal viscous flows, *Ind. Eng. Chem. Fundam.* **8**, 512–520 (1969).

[15] D. T. Papageorgiou, On the breakup of viscous liquid threads, *Phys. Fluids* **7**, 1529–1544 (1995).

[16] D. T. Papageorgiou, Analytical description of the breakup of liquid jets, *J. Fluid Mech.* **301**, 109–132 (1995).

[17] J. R. A. Pearson and M. A. Matovich, Spinning a molten threadline - stability, *Ind. Eng. Chem. Fundam.* **8**, 605–609 (1969).

[18] M. Renardy, Finite time breakup of viscous filaments, *Z. Angew. Math. Phys.* **52**, 881–887 (2001).

[19] M. Renardy, A comment on self-similar breakup for inertialess Newtonian liquid jets, *IMA J. Applied Math.* **70**, 353–358 (2005).

[20] M. Renardy, Draw resonance revisited, *SIAM J. Appl. Math.* **66**, 1261–1269 (2006).

[21] M. Renardy, "Finite time breakup of viscous filaments," ZAMP 52 (2001), 881-887 (Corrigendum), *Z. Angew. Math. Phys.* **58**, 904–905 (2007).

[22] W. W. Schultz, S. H. Davis, Effects of boundary conditions on the stability of slender viscous fibers, *J. Applied Mech.* **51**, 1–4 (1984).

[23] A. L. Yarin, *Free Liquid Jets and Films: Hydrodynamics and Rheology* (Wiley, New York, 1993).

Chapter 5

Diffeomorphisms with Stable Manifolds as Basin Boundaries

Sandra Hayes & Christian Wolf[a]

The City College of New York
E-mail: {shayes|cwolf}@gc.cuny.edu

If a diffeomorphism of $\mathbb{R}^2$ has exactly 2 periodic points and one is attracting while the other is a saddle, it is of interest to find out when the boundary of the basin of attraction is the stable manifold of the saddle. Only in a few special cases of Hénon maps (see for example [5]) has this been proved, although in the standard literature on Hénon maps (see for example the books [1, 6]) it is often said to be true based on computer experiments. In this paper we consider a general family of diffeomorphisms

$$F(x,y) = (g(x) + h(y), h(x)), \tag{1}$$

where $g(x)$ is a unimodal C^2-map which has the same dynamical properties as the logistic map $P(x) = \mu x(1-x)$, e.g. $g(0) = g(1) = 0$, and $h(x)$ is a C^2-map with $h(0) = 0$ which is a small perturbation of a linear map. We prove that for certain maps of the form (1) there are exactly two periodic points, namely an attracting fixed point and a saddle fixed point and the boundary of the basin of attraction is the stable manifold of the saddle. The basin boundary also has the same regularity as F, in contrast to the frequently observed fractal nature of basin boundaries. To establish these results we describe all possibilities for the forward and backward orbits of every point in the plane.

Dedicated to Armin Leutbecher on the occasion of his birthday

Contents

[a]The work of Christian Wolf was partially supported by a grant from the PSC-CUNY (TRADA-45-356) and by a grant from the Simons Foundation (#209846).

99

1. Introduction

A widely observed phenomenon in deterministic dynamical systems is the unpredictability of its trajectories. More precisely, one notion of chaotic behavior of a dynamical system is that it has sensitive dependence on initial conditions which roughly means the divergence of trajectories over time even when they originate at points arbitrarily close to each other. This feature is observed in various applications in science and engineering such as weather prediction or the turbulent behavior of fluid flows. There are other features of chaotic behavior such as the exponential growth rate of periodic orbits or positive entropy (see [7] and the references therein). Another frequently observed phenomenon is that attracting periodic points have fractal basin boundaries [1, 6, 9].

A typical one-dimensional real chaotic map is the logistic map

$$f_\mu(x) = \mu x(1 - x) \tag{2}$$

for certain values of the parameter μ. However, higher dimensional systems are necessary to characterize most of the real world applications. Prototypes of two-dimensional real chaotic systems are given by the family of Hénon maps

$$F_{\delta,\mu}(x, y) = (\mu x(1 - x) + \delta y, \delta x), \tag{3}$$

according to [4]. These maps have a wide range of applications from control theory [8] to astronomy and astrophysics and even to physiology and geology (see [3]). As the parameters δ and μ of a Hénon map change, the dynamics changes and the system can suddenly lose its chaotic behavior: the number of periodic points can dwindle down to just finitely many, and the attracting ones can have smooth basin boundaries. In this paper we will first study those parameters of Hénon maps for which chaos ceases, stability prevails, and the long term behavior of trajectories is predictable. Because it is the shape of the maps and not the algebraic form which is important, these results will be shown to hold for a more general class of maps of the form (1).

2. Notation and results

For certain μ and δ we will give a complete description of all possible orbits of the Hénon family $F_{\delta,\mu}$ under forward as well as backward iteration. In particular, for some parameters (δ, μ) for which $F_{\delta,\mu}$ has exactly two periodic points, namely an attracting fixed point and a saddle fixed point, we show that the boundary of the basin of attraction is the stable manifold of the saddle.

Theorem 1. *For certain parameters (δ, μ) the following holds.*

   (i) The fixed points α and the origin $(0,0)$ are the only periodic points of $F = F_{\delta,\mu}$.

   (ii) The boundary $\partial W^s(\alpha)$ of the basin of α is the stable manifold $W^s(0,0)$ of the origin.

(iii) The filled Julia set K of points with bounded forward and backward orbits for F is $\{\alpha, (0,0)\} \cup (W^s(\alpha) \cap W^u(0,0))$.

The origin is also an unstable one-sided flip saddle point meaning one eigenvalue is negative and only one connected component of the punctured unstable manifold $W^u(0,0) \setminus (0,0)$ of $(0,0)$ meets K.

An immediate consequence of this theorem is the following.

Corollary 1. *For each (δ, μ) as in Theorem 1 the map $F_{\mu,\delta}$ is a Morse-Smale diffeomorphism.*

Note that $F_{\delta,\mu}$ is conjugate to $F_{-\delta,\mu}$ via $G(x,y) = (x,-y)$. Therefore, for the area decreasing case $0 \neq |\delta| < 1$ we only need to consider $0 < \delta < 1$. The area increasing case will have a similar description due to the conjugacy of F^{-1}. Thus, for the non-area preserving case we only have to consider parameters for which $F_{\mu,\delta}$ has two distinct fixed points, namely we consider parameters in

$$H_{1,1} = \{(\delta, \mu) : 0 < \delta < 1 \quad and \quad 1 - \delta^2 < \mu < 3(1 - \delta^2)\} \tag{4}$$

Next we consider the general case of diffeomorphisms of the form (1). We assume that $g : \mathbb{R} \to \mathbb{R}$ is a C^2-unimodal map with $g(0) = 0, g(1) = 0, g([0,1]) \subset [0,1)$, $g'(0) > 1$ and $g'(1) < -1$. We also assume that there exists $\gamma < 0$ such that $g''(x) < \gamma$ for all $x \in \mathbb{R} \setminus (0,1)$. Finally, we assume that there exists an attracting fixed point $x = x_g \in (0,1)$ of g with $W^s(x) = (0,1)$. For $\delta \in \mathbb{R}$ we denote by L_δ the linear map $x \mapsto \delta x$. Moreover, if $f : \mathbb{R} \to \mathbb{R}$ is a C^2-map, we write $||f||_2 = \sup_{x \in \mathbb{R}} \{|f(x)|, |f'(x)|, |f''(x)|\}$. Note that $||.||_2$ may be infinite and is, in particular not a norm. For maps g with the properties above we have the following result.

Theorem 2. *There exists $\delta(g) > 0$ such that for all $0 < \delta < \delta(g)$ there is $\epsilon > 0$ such that for any C^2-map $h : \mathbb{R} \to \mathbb{R}$ with $h(0) = 0$ and $||h - L_\delta||_2 < \epsilon$ the following holds.*

(i) The map $F(x,y) = (g(x) + h(y), h(x))$ is a C^2-diffeomorphism of $\mathbb{R}^2$.

(ii) The map F has precisely two periodic points, both of which are fixed points. The origin $(0,0)$ is a saddle point and the other fixed point α is attracting.

(iii) The boundary $\partial W^s(\alpha)$ of the basin of α is the stable manifold $W^s(0,0)$ of the origin.

(iv) The set K of points with bounded forward and backward orbits for F is $\{\alpha, (0,0)\} \cup (W^s(\alpha) \cap W^u(0,0))$.

3. The dynamics of $F_{\mu,\delta}$ for small δ

The purpose of this section is to prove Theorem 1. We construct a partitioning of $\mathbb{R}^2$ and use it to investigate the forward and backward orbits of every point. A key ingredient of the proof is that the dynamics of the maps $F_{\delta,\mu}$ is controlled by the set K of points with bounded (forward and backward) orbits. Unless stated otherwise we use the maximum norm in $\mathbb{R}^2$. Points with backward (resp. forward) orbits escaping to infinity under this norm will simply be denoted by $W^u(\infty)$ (resp. $W^s(\infty)$). We begin by noting some elementary properties of the eigenvalues associated with $F_{\delta,\mu}$. Throughout this section we use as a standing assumption that $(\delta, \mu) \in H_{1,1}$ as defined in (4).

(i) The eigenvalues of the Jacobian matrix

$$DF(x,y) = \begin{pmatrix} \mu - 2\mu x & \delta \\ \delta & 0 \end{pmatrix} \tag{5}$$

are given by

$$\lambda_1 = \lambda_1(x, \delta, \mu) = \frac{(1 - 2x)\mu - \sqrt{\mu^2(2x - 1)^2 + 4\delta^2}}{2}, \tag{6}$$

and

$$\lambda_2 = \lambda_2(x, \delta, \mu) \frac{(1 - 2x)\mu + \sqrt{\mu^2(2x - 1)^2 + 4\delta^2}}{2}. \tag{7}$$

(ii) We have that $\lambda_1 < 0$ and $\lambda_2 > 0$ for every x and μ when $\delta \neq 0$.

(iii) For the origin $(0,0)$, $\lambda_2 > 1$ if and only if $\delta^2 > 1 - \mu$ and $\lambda_1 > -1$ if and only if $\delta^2 < 1 + \mu$.

(iv) Let $\mu > 1$ and $\delta \neq 0$. Then the origin is a flip saddle fixed point for F if and only if $\delta^2 < 1 + \mu$.

(v) If $1 < \mu < 3$, then the fixed point $\alpha = (x_\alpha, y_\alpha) = (x_\mu + \frac{\delta^2}{\mu}, \delta x_\alpha) = (1 - \frac{1}{\mu} + \frac{\delta^2}{\mu}, \delta x_\alpha)$ is attracting if and only if $\mu < 3(1 - \delta^2)$.

(vi) The inverse of F is given by the formula $F^{-1}(x, y) =$

$$(\delta^{-1}y, \delta^{-1}[x - P_\mu(\delta^{-1}y)]) = (\delta^{-1}y, \delta^{-1}[x - \mu\delta^{-1}y(1 - \delta^{-1}y)]), \tag{8}$$

where $P_\mu(x) = \mu x(1 - x)$.

Our goal is to partition the region above the line $y = 2\delta$ into 3 parts. Similarly for the region below $y = -2\delta$. We define

$$W_\delta = \{(x,y) : |y| \geq 2\delta, |y| \geq \delta|x|\}.$$

Lemma 1. *Let $1 < \mu < 3$ and let $0 < \delta < \max\{\mu - 1, 1\}$. Then $W_\delta \subset W^u(\infty)$.*

Proof. Let $(x, y) \in W_\delta$ and $F^{-n}(x, y) = (x_{-n}, y_{-n})$. It will be shown that $|y_{-n}| \to \infty$ as $n \to \infty$.

$$\begin{aligned}
|y_{-1}| &= |\delta^{-1}[x - \mu\delta^{-1}y(1 - \delta^{-1}y)]| \\
&\geq \delta^{-1} [\mu|\delta^{-2}y^2| - |x| - \mu|\delta^{-1}y|] \\
&\geq \delta^{-1} |y| [2\mu\delta^{-1} - \delta^{-1} - \mu\delta^{-1}] \\
&\geq \delta^{-1} |y| [\delta^{-1}\mu - \delta^{-1}] \\
&\geq \delta^{-1} |y| [\delta^{-1}(\mu - 1)] > |y|,
\end{aligned} \tag{9}$$

since $\delta^{-1}(\mu - 1) > 1$ when $0 < \delta < \mu - 1$. Recall that $x_{-1} = \delta^{-1}y$. Therefore, (9) implies that $(x_{-1}, y_{-1}) \in W_\delta$. It now follows by induction (using (9)) that the sequence $(|y_{-n}|)_n$ is strictly increasing and $|y_{-n}| \to \infty$ as $n \to \infty$. $\square$

The remaining two parts of the region above $y = 2\delta$ will be treated separately. That part below $y = -\delta x$ for $x \leq -2$ is in $W^s(\infty)$ as follows from the next Lemma when δ is small enough.

Lemma 2. *Let $1 < \mu < 3$. Define $\beta = \beta(\delta) = \delta(\mu - 1)^{-1}$. Then for all $(x, y) \in \mathbb{R}^2$ with $x \leq \min\{-\beta y, 0\}$ and $(x, y) \neq (0, 0)$ we have $(x, y) \in W^s(\infty)$.*

Proof. Let $(x, y) \in \mathbb{R}^2$ with $x \leq \min\{-\beta y, 0\}$. First, we consider $y \geq 0$ in which case $x \leq -\beta y$. We will show that $x_n \to -\infty$ as $n \to \infty$ for $(x_n, y_n) = F^n(x, y)$. Since

$$x_1 = \mu x - \mu x^2 + \delta y \leq x - \mu x^2 + [\mu - 1 - \delta \beta^{-1}]x \leq x - \mu x^2 \qquad (10)$$

that follows by induction. The case $y < 0$ follows similarly with the simplification that $y_n < 0$ for all $n \in \mathbb{N}$. $\square$

It follows from Lemma 2 that every point (x, y) with $x \leq -2$ and $2\delta < y < -\delta x$ is in $W^s(\infty)$ when $\delta < \sqrt{\mu - 1}$. The region above $y = 2\delta$ and below $y = \delta x$ for $x \geq 2$ is in $W^s(\infty)$ as follows from the next Lemma.

Lemma 3. *Let $1 < \mu < 3$ and let $0 < \delta < \delta^* \overset{\text{def}}{=} \sqrt{\mu - 1}$. Let $x \geq 2, 0 \leq y \leq \delta x$. Then $(x, y) \in W^s(\infty)$.*

Proof. Let β be as in Lemma 2. Then

$$x_1 = \mu x (1 - x) + \delta y \leq -\mu x + \delta^2 x \leq -\mu x + (\mu - 1)x = -x. \qquad (11)$$

On the other hand,

$$-\beta y_1 = -\frac{\delta}{\mu - 1}\delta x \geq -x \geq x_1. \qquad (12)$$

Thus, the point (x_1, y_1) satisfies the conditions of Lemma 2 and (x_1, y_1) is in $W^s(\infty)$. $\square$

From Lemmas 1, 2 and 3 we conclude that above the line $y = 2\delta$ either the backward or the forward orbits escape to ∞. Therefore, K must lie below $y = 2\delta$.

Now we will partition the strip $-\infty < x < \infty, 0 < y < 2\delta$ and investigate the orbits in each partition. Note that the region below $y = 2\delta$ and above the x-axis for $x \geq 2$ is in $W^s(\infty)$ by Lemma 3. Next, we consider

$$A_\delta = \{(x, y) : 0 \leq x \leq 1, 0 \leq y \leq 2\delta\} \setminus \{(0, 0)\} \qquad (13)$$

for $\delta > 0$ and show that it is in the basin $W^s(\alpha)$ of attraction. We will show the existence of a polydisk with center α that is contained in $W^s(\alpha)$ and whose size is independent of δ and μ.

Proposition 1. *Let* $1 < \mu < 3$. *Then there exists* $r > 0$ *and* $\delta^* > 0$ *such that for all* $0 < \delta < \delta^*$ *we have* $\overline{P}(\alpha, r) = \{(x, y) \in \mathbb{R}^2 : |x - x_\alpha|, |y - y_\alpha| \leq r\} \subset W^s(\alpha)$.

Proof. Let $(x, y) = (x_0, y_0) = (x_\alpha + s_0, y_\alpha + t_0)$ and $(x_n, y_n) = F^n(x_0, y_0) = (x_\alpha + s_n, y_\alpha + t_n)$. Thus $(x, y) \in W^s(\alpha)$ if and only if $\|(s_n, t_n)\|_m \to 0$ as $n \to \infty$ where the maximum norm is used. Let $n \in \mathbb{N}$. Then

$$
\begin{aligned}
(x_{n+1}, y_{n+1}) &= F(x_\alpha + s_n, y_\alpha + t_n) \\
&= (\mu(x_\alpha + s_n)(1 - (x_\alpha + s_n)) + \delta(y_\alpha + t_n), \delta(x_\alpha + s_n)) \\
&= (\mu x_\alpha(1 - x_\alpha) + \delta y_\alpha + s_n(\mu - 2\mu x_\alpha - \mu s_s) + \delta t_n, \delta x_\alpha + \delta s_n) \\
&= (x_\alpha + s_n(2 - \mu - \frac{2\delta^2}{\mu} - \mu s_n) + \delta t_n, y_\alpha + \delta s_n).
\end{aligned}
$$

Hence,

$$
(x_{n+1}, y_{n+1}) - (x_\alpha, y_\alpha) = (s_n(2 - \mu - \frac{2\delta^2}{\mu} - \mu s_n) + \delta t_n, \delta s_n). \tag{14}
$$

Since $2 - \mu \in (-1, 1)$ there exists $0 < \gamma < 1$ and $r > 0$ such that $|2 - \mu - \mu s| < \gamma$ for all $|s| \leq r$. Moreover, an elementary continuity argument shows that there exists $\delta^* > 0$ such that for all $0 < \delta < \delta^*$ and all $|s| \leq r$ we have

$$
\left| 2 - \mu - \frac{2\delta^2}{\mu} - \mu s \right| < \gamma. \tag{15}
$$

Without loss of generality assume that $\delta^* < \frac{1-\gamma}{2}$. We conclude that if $0 < \delta < \delta^*$ and $(x_0, y_0) \in \overline{P}(\alpha, r)$ then $\|(s_n, t_n)\|_m$ converges geometrically to 0 and thus $(x_0, y_0) \in W^s(\alpha)$. $\square$

As a consequence of Proposition 1 we obtain the following.

Corollary 2. *Let* $1 < \mu < 3$ *and let* $h > 0$. *Then there exist* $r > 0$ *and* $\delta^* > 0$ *such that if* $0 < \delta < \delta^*$, $x_\mu - r \leq x \leq x_\mu + r$ *and* $0 \leq |y| \leq h$ *then* $(x, y) \in W^s(\alpha)$.

Proof. Note that the radius r in Proposition 1 only depends on δ^* and not on δ. Moreover, $x_\alpha \to x_\mu$ and $y_\alpha \to 0$ as $\delta \to 0$. Therefore the claim follows immediately from Proposition 1 and the fact that we can make δy as small as necessary by making δ^* small. $\square$

Note that in Corollary 2 the value of δ^* can be chosen to depend continuously on μ.

Proposition 2. *Let* $1 < \mu < 3$. *Then there exists* $\delta^* > 0$ *and such that if* $0 < \delta < \delta^*$ *then* $A_\delta \subset W^s(\alpha)$.

Proof. Let $\delta^*, h = 1$ and r be as in Corollary 2. Moreover, without loss of generality we may assume that $\delta^* < \frac{1}{4}$. Let $0 < \delta < \delta^*$. Obviously, $P_\mu([0, 1]) \subset [0, \frac{3}{4})$ which implies that $F(A_\delta) \subset A_\delta$. We have $P'_\mu(0) = \mu > 1$. Thus, by continuity there exists $\varrho_l > 0$ such that $P'_\mu(x) \geq \frac{\mu+1}{2} > 1$ for all $0 \leq x \leq \varrho_l$. Since x_μ is an attracting fixed point of P_μ we must have $\varrho_l < x_\mu$. Let $(x, y) \in A_\delta$ with $0 \leq x \leq \varrho_l$. If $x = 0$ then $x_1 > 0$ and therefore, without loss of generality we may assume that $x > 0$. It follows from the Mean Value Theorem that

$$x_1 = P_\mu(x) + \delta y \geq P_\mu(x) - 0 \geq \frac{\mu + 1}{2} x. \tag{16}$$

It now follows by induction that there exists $n \in \mathbb{N}$ such that $x_n \geq \varrho_l$. On the other hand, if $(x, y) \in A_\delta$ then

$$x_1 = P_\mu(x) + \delta y < \frac{3}{4} + 2\delta^2 < \frac{7}{8}. \tag{17}$$

We define $\varrho_u = \frac{1}{8}$. The above shows that in order to prove the claim it is sufficient to consider $(x, y) \in A_\delta$ with $\varrho_l \leq x \leq \varrho_u$. First, we consider the dynamics of P_μ on $[\varrho_l, \varrho_u]$. Since each $x \in [\varrho_l, \varrho_r]$ is contained in $W^s(x_\mu)$ (with respect to P_μ) and since $[\varrho_l, \varrho_r]$ is compact there exists $m \in \mathbb{N}$ such that $P_\mu^m(x) \in (x_\mu - r, x_\mu + r)$ for all $x \in [\varrho_l, \varrho_r]$. Note that

$$F^m(x, y) = (P_\mu^m(x) + P(x, y), Q(x, y)), \tag{18}$$

for some polynomials P and Q in two variables. Moreover, each of the coefficients of P and Q contains positive powers of δ. Thus, by making δ^* smaller if necessary, we can assure that if $(x, y) \in A_\delta$ with $\varrho_l \leq x \leq \varrho_u$ then $x_m \in (x_\mu - r, x_\mu + r)$ and the claim follows. $\quad\square$

Next we will investigate the negative part of the strip

$$S_\delta = \{(x, y) : -\infty < x < \infty, 0 < y < 2\delta\}, \tag{19}$$

i.e. $x \leq 0, 0 < y < 2\delta$, which contains a portion of $W^s(0, 0)$. In particular, we show that the local stable manifold of the orgin is contained in the basin of attraction of α.

Theorem 3. *Let $1 < \mu < 3$ and let $\beta = \delta(\mu - 1)^{-1}$. Then there exists $\delta^* > 0$ such that for all $0 < \delta < \delta^*$ the following holds:*

(i) *For all $0 \leq \overline{y} \leq 2\delta$ there exists a unique $-\beta\overline{y} \leq \overline{x} \leq 0$ such that $(\overline{x}, \overline{y}) \in W^s(0, 0)$.*

(ii) *If $0 < \overline{y} \leq 2\delta$ then $-\beta\overline{y} < \overline{x} < 0$;*

(iii) *Let $0 < \overline{y} \leq 2\delta$. If $\overline{x} < x \leq 0$ then $(x, \overline{y}) \in W^s(\alpha)$, and if $x < \overline{x}$ then $(x, y) \in W^s(\infty)$.*

Proof. Let δ^* be as in Proposition 2. We first prove the existence in (i). The uniqueness in (i) will follow from (iii). If $\overline{y} = 0$ then $\overline{x} = 0$. Assume now $0 < \overline{y} \leq 2\delta$. We define

$$\overline{x} = \inf\{x \leq 0 : (x, 0] \subset W^s(\alpha)\}. \tag{20}$$

By Proposition 2, $(0, \overline{y}) \in W^s(\alpha)$. Since $W^s(\alpha)$ is open we may conclude that $\overline{x} < 0$. It follows from the definition of $\overline{x}$ that $(\overline{x}, \overline{y}) \in \partial W^s(\alpha)$. Moreover, since

$W^s(\infty)$ is open we obtain $(\overline{x}, \overline{y}) \notin W^s(\infty)$. Combining this with Lemma 2 yields $\overline{x} > -\beta \overline{y}$. Thus, (ii) holds. For $n \in \mathbb{N}$ we write $(x_n, y_n) = F^n(\overline{x}, \overline{y})$. In particular,

$$(x_{2n}, y_{2n}) = (P_\mu(P_\mu(x_{2n-2}) + \delta y_{2n-2}) + \delta^2 x_{2n-2}, \delta(P_\mu(x_{2n-2}) + \delta y_{2n-2})). \quad (21)$$

We define $C = C(\delta) = \sup\{|P'_\mu(x)| : (x, y) \in C_\delta\}$. Clearly, $C(\delta) \to \mu$ as $\delta \to 0$. Since $\beta \to 0$ as $\delta \to 0$, we can assure (by making δ^* smaller if necessary) that

$$\delta(C\beta + \delta) < \frac{1}{2}. \quad (22)$$

We claim that $(x_{2n}, y_{2n}) \in C_\delta$ for all $n \in \mathbb{N}_0$. Since $(x_0, y_0) = (\overline{x}, \overline{y})$ the claim holds for $n = 0$. Assume now that the claim holds for $n - 1$. Then $y_{2n} \geq 0$. Otherwise (x_{2n}, y_{2n}) would be contained in the third quadrant and thus by Lemma 2 in $W^s(\infty)$. Applying the Mean Value theorem we obtain

$$|y_{2n}| = \delta |P_\mu(x_{2n-2}) + \delta y_{2n-2}| \leq \delta (|Cx_{2n-2}| + |\delta y_{2n-2}|)$$
$$\leq \delta(C\beta + \delta) |y_{2n-2}| < \frac{1}{2} |y_{2n-2}|. \quad (23)$$

Finally, if $x_{2n} < -\beta y_{2n}$ then by Lemma 2 we would have $(x_{2n}, y_{2n}) \in W^s(\infty)$ which is a contradiction to $(x_{2n}, y_{2n}) \in \partial W^s(\alpha)$ and the claim is proved. Since $-\beta y_{2n} \leq x_{2n} \leq 0$, equation (23) proves that $(\overline{x}, \overline{y}) \in W^s(0, 0)$.

It remains to prove (iii). Let $0 < y \leq 2\delta$. That $(\overline{x}, 0] \times \{\overline{y}\} \subset W^s(\alpha)$ is a direct consequence of the definition of $\overline{x}$. We now consider the case $x < \overline{x}$. If $x < -\beta y$ then $(x, y) \in W^s(\infty)$ by Lemma 2. Assume now $(x, y) \in C_\delta$. It follows from (21) (with $(\overline{x}, \overline{y})$ replaced by (x, y)) and Lemma 2 that if $y_{2n} < 0$ for some $n \in \mathbb{N}$ then $(x, y) \in W^s(\infty)$. It remains to consider the case $y_{2n} \geq 0$ for all $n \in \mathbb{N}$. Similar to the case of $(\overline{x}, \overline{y})$ it follows from (23) that $y_{2n} \to 0$ as $n \to \infty$. Note that $\inf\{|P'_\mu(x)| : (x, y) \in C_\delta\} = \mu > 1$. It now follows by induction by using (21) and the Mean Value Theorem that $x_{2n} < -\beta y_{2n}$ for some $n \in \mathbb{N}$. Therefore, Lemma 2 implies $(x, y) \in W^s(\infty)$ and the proof is complete. $\square$

As a result of Theorem 3, all points (x, y) with $x < 0$ and $0 < y < 2\delta$ have forward orbits which escape to ∞ or converge to the attracting fixed point α, depending on whether they lie to the left of $W^s(0, 0)$ or to the right. The only partition of the strip $-\infty < x < \infty, 0 < y < 2\delta$ not dealt with yet is the set of points (x, y) with $1 < x < 2, 0 < y < 2\delta$. It will be shown that those points have forward orbits converging to α or escaping to ∞ according to whether they lie to the left of $W^s(0, 0)$ or to the right.

Before we attack that proof, an interesting property of the map $\overline{y} \mapsto \overline{x}$ given in Theorem 3 can be derived.

Note that for small y values part (i) of Theorem 3 also follows from the Stable Manifold Theorem applied to the saddle point $(0, 0)$. For our purposes however, it is crucial to have a uniform estimate from below for the size of the local stable manifolds. On the other hand, the Stable Manifold Theorem implies the following:

Corollary 3. *Under the assumptions of Theorem 3 the map $\overline{y} \mapsto \overline{x}$ is real analytic.*

Proof. By the Stable Manifold Theorem there exists $\eta > 0$ such that

$$W^s_{\text{loc}}(0,0) \cap \{(x,y) : y \geq 0\} = \{(\overline{x},\overline{y}) : 0 \leq \overline{y} < \eta\}.$$

The statement now follows from the fact that

$$F^{2n}(\{(\overline{x},\overline{y}) : 0 \leq \overline{y} \leq 2\delta\}) \subset W^s_{\text{loc}}(0,0) \cap \{(x,y) : y \geq 0\}$$

for some $n \in \mathbb{N}$.   □

Finally, we can treat the last part of the strip S_δ, namely points (x,y) with $1 \leq x \leq 2, 0 < y < 2\delta$:

Theorem 4. *Let $1 < \mu < 3$. Then there exists $\delta^* > 0$ such that for all $0 < \delta < \delta^*$ the following holds:*

(i) *For all $0 \leq \overline{y} \leq 2\delta$ there exists a unique $1 < \overline{x} < 2$ such that $(\overline{x},\overline{y}) \in W^s(0,0)$. Moreover, $\overline{y} \mapsto \overline{x}$ is real-analytic.*

(ii) *Let $0 \leq \overline{y} \leq 2\delta$. If $1 \leq x < \overline{x}$ then $(x,\overline{y}) \in W^s(\alpha)$, and if $\overline{x} < x \leq 2$ then $(x,\overline{y}) \in W^s(\infty)$.*

Proof. The rectangle $B_\delta = (x,y) : 1 \leq 2, 0 \leq y \leq 2\delta$ for $\delta > 0$ has an image $F(B_\delta)$ which is a topological rectangle. Let $0 \leq \overline{y} \leq 2\delta$ be fixed. The curve $F(x,\overline{y})$, $1 \leq x \leq 2$, which is the image of a horizontal line, is a parabola in that topological rectangle opening to the left. It is parallel to the parabolas which are the images of $F(x,0)$ and $F(x,2\delta)$, $1 \leq x \leq 2$. Let $Q = W^s(0,0) \cap F(\overline{x},\overline{y})$ be the intersection of the parabola $F(x,\overline{y})$, $1 \leq x \leq 2$, and the local stable manifold of the origin. Then $Q' = F^{-1}(Q) = (\overline{x},\overline{y})$ is in $W^s(0,0)$. If $1 \leq x < \overline{x}$, then the x–coordinate of $Q_1 = F(x,\overline{y})$ is larger than that of Q, whereas its y-coordinate is smaller than that of Q, implying it is to the right of $W^s(0,0)$ and thus in the basin of attraction $W^s(\alpha)$ by Theorem 3. Similarly, if $\overline{x} < x \leq 2$, then $(x,\overline{y})$ is in $W^s(\infty)$, since the y-coordinate of Q_1 is larger than that of Q whereas the x-coordinate is smaller than that of Q.   □

Using the previous results, the filled Julia set K for the map $F_{\delta,\mu}$ must be below the line $y = 2\delta$. The behavior of all forward orbits of points in the strip S_δ can be described for small δ as follows.

Proposition 3. *Let $1 < \mu < 3$. There exists a $\delta^* > 0$ such that for all $0 < \delta < \delta^*$ the forward orbit under $F_{\delta,\mu}$ of every point (x,y) with $-\infty < x < \infty, 0 < y < 2\delta$ converges either to $(0,0)$, to α or to ∞.*

Proof. Theorem 3 describes the behavior of forward orbits for points with $x \leq 0$. For $0 < x \leq 2$, the statement follows from Proposition 2 and Theorem 4. Finally, Lemma 3 establishes the claim for $x > 2$.   □

To summarize, for all points in the upper half plane we know the fate of either the forward or the backward orbits. Now we will concentrate on the points below the x-axis. As a result of Lemma 2, the third quadrant is in $W^s(\infty)$. We now investigate the other possibilities.

Proposition 4. *Let* $1 < \mu < 3$. *Then*

(i) *If* $0 < \delta < \delta^* \stackrel{\text{def}}{=} \sqrt{3\mu(\mu-1)}$ *and* $(x,y) \in \mathbb{R}^2$ *with* $x \geq 2$ *and* $y \leq 0$ *then* $(x,y) \in W^s(\infty)$.

(ii) *If* $0 < \delta \leq 1$ *and* $(x,y) \in \mathbb{R}^2$ *with* $0 \leq x \leq 2$ *and* $y \leq \delta_0 = -\delta^{-1}\left(\frac{2\delta^2}{\mu-1}+1\right)$ *then* $(x,y) \in W^s(\infty)$.

Proof. (i) Suppose $x \geq 2$ and $y \leq 0$. Then $P_\mu(x) \leq -\mu x$. Hence,

$$x_1 = P_\mu(x) + \delta y \leq -\mu x < -\frac{\delta^2}{\mu-1}x = -\beta\delta x = -\beta y_1.$$

By Lemma 2, $(x_1, y_1) = F(x,y) \in W^s(\infty)$ which proves (i).

(ii) Let $(x,y) \in \mathbb{R}^2$ with $0 \leq x \leq 2$ and $y \leq -\delta^{-1}\left(\frac{2\delta^2}{\mu-1}+1\right)$. Then

$$x_1 = P_\mu(x) + \delta y \leq 1 + \delta y \leq -\frac{2\delta^2}{\mu-1} \leq -\beta\delta x = -\beta y_1.$$

Therefore, the same argument as in the proof of (i) shows $(x,y) \in W^s(\infty)$. □

In order to determine the behavior of the forward orbits for the remaining part of the fourth quadrant, we now investigate points (x,y) on the stable manifold of the origin satisfying $0 \leq x \leq 2$.

In the following we introduce some notation that will be used later. Let $Q = W^s(0,0) \cap F(\bar{x}, 0)$ be the intersection of the parabola $F(x,0), 1 < x < 2$, and the local stable manifold of the origin. Let W' be the part of $W^s(0,0)$ connecting Q to the origin and let $Q' = F^{-1}(Q) = (\bar{x}, 0)$. Then $C = F^{-1}(W')$ is a curve connecting Q' to the origin and is contained in $W^s(0,0)$. We claim that except for the two endpoints, the curve C is below the x-axis. To show the claim, we consider first $(x,y) \in W'$ with $y \leq \delta$. Because $x < 0$ and $F^{-1}(x,y) = \frac{1}{\delta}(y, x + \frac{\mu y(y-\delta)}{\delta^2})$, it follows that $F^{-1}(x,y)$ is below the x-axis. In particular, $P' = F^{-1}(P) = (1, \frac{x}{\delta}) \in C$ for $P = (x, \delta) \in W'$ is below the x-axis as well as on C. If there would exist a point (x,y) on C with $y > 0$ and $1 < x < \bar{x}$, then by the Intermediate Value Theorem there would have to be a point on the interval $(1, x)$ which is contained in $W^s(0,0)$. But this is a contradiction, since the entire interval $(0, \bar{x})$ is in $W^s(\alpha)$ by Proposition 2 and Theorem 3.

Lemma 4. *The region below* C *in the fourth quadrant is in* $W^s(\infty)$. *The region above* C *and below the interval* $(0, \bar{x})$ *on the x-axis is contained in* $W^s(\alpha)$.

Proof. Let $0 < x \leq \bar{x}$ be fixed with $Q' = (\bar{x}, 0)$ on $W^s(0,0)$. The vertical half-line $(x,y), y \leq 0$ is mapped by F to the horizontal half-line $(z, \delta x), z \leq P_\mu(x)$. Let (x, y_0) be the intersection of C and the vertical half-line. Then (x, y_0) and $F(x, y_0)$ are on $W^s(0,0)$. The horizontal half-line starts in the basin of attraction $W^s(\alpha)$.

All points (x, y) with $y_0 < y < 0$ are also in that basin by Theorem 3, since the first coordinate of $F(x, y) = (P_\mu(x) + \delta y, \delta x)$ is larger than that of $F(x, y_0)$. Similarly, if $y < y_0$, then all points (x, y) are in $W^s(\infty)$ by Theorem 3.   □

We are now able to provide a description of all forward orbits of points below the x-axis for $1 < \mu < 3$ and small δ.

Proposition 5. *There exists $\delta^* > 0$ such that for all $0 < \delta < \delta^*$ the forward orbit under $F_{\delta,\mu}$ of every point $(x, y) \in \mathbb{R}^2$ with $-\infty < x < \infty, y < 0$ converges either to $(0, 0)$, to α or to ∞.*

Proof. All points in the third quadrant escape to infinity under forward iteration by Lemma 2. If $0 < x < \overline{x}$, points (x, y) with $y < 0$ which are not on C are either in the basin of attraction of α or escape to infinity according to whether they are above C or below C by Lemma 4. Obviously, points on C converge to the origin. Theorems 3 and 4 show that points (x, y) for $\overline{x} < x < 2$ and $y < 0$ also escape to infinity.   □

It is now obvious that the forward orbit of every point in the real plane under $F_{\delta,\mu}$ converges either to $(0, 0)$, to α or to ∞ for $1 < \mu < 3$ and $\delta = \delta(\mu)$ small.

Lemma 5. *Let $1 < \mu < 3$. Then there exists $\delta^* > 0$ such that for all $0 < \delta < \delta^*$. the map $F_{\delta,\mu}$ satisfies properties (i) and (ii) of Theorem 1.*

Proof. (i) Let $1 < \mu < 3$ and let r be as in Proposition 1 with $r < 1$. If $0 < \delta < \delta^*$ for δ^* as in Propositions 3 and 5 and $(\delta, \mu') \in (\mathbb{R} \setminus \{0\}) \times \mathbb{R}$ with $\|(\delta, \mu') - (0, \mu)\| < r$. Then every forward orbit of $F_{\delta,\mu'}$ converges to either to 0, to α or to ∞. Consequently, there are no periodic points other than the two fixed points $(0, 0)$ and α.

(ii) It suffices to show that in a neighborhood U of the origin, the boundary $\partial W^s(\alpha)$ of the basin of α is the stable manifold $W^s(0, 0)$ of the origin. Without restriction every point in U converges under forward iteration either to 0, to α or to ∞. Let (x, y) be in U and on $W^s(0, 0)$. Since all points which are to the right of $W^s(0, 0)$ are in $W^s(\alpha)$ by Theorem 3, obviously (x, y) is in $\partial W^s(\alpha)$. On the other hand, let (x, y) be in U and on $\partial W^s(\alpha)$. If the point (x, y) were not on $W^s(0, 0)$, then it must be in the open set $W^s(\infty)$ contradicting the fact that it is on the boundary $\partial W^s(\alpha)$.   □

To prove part (iii) of Theorem 1, we now describe the backward orbits of $F_{\delta,\mu}$ for $1 < \mu < 3$ and small δ. The next lemma shows that the backward orbits of points on $W^s(0, 0)) \setminus \{0\}$ escape which will be proved by using the standard trapping regions V^+, V^- induced by a closed square V of side length r centered at the origin for an appropriate r (see [2], Lemma 2.1, and [4]). In particular, $W^u(\infty) \subset V^+$, $W^s(\infty) \subset V^-$.

Lemma 6. $W^s(0, 0)) \setminus \{(0, 0)\} \subset W^u(\infty)$.

Proof. Let $W_+^s(0,0)$ respectively $W_-^s(0,0)$ denote the two connected components of $W^s(0,0) \setminus \{0\}$ whose existence is guaranteed by the Stable Manifold Theorem. We use the immersed topology on these components. Note that this topology does not necessarily coincide with the relative topology induced by the topology of $\mathbb{R}^2$. Define $V_s^+ = W^s(0,0) \cap V^+$. Then $V_s^+ \subset W^u(\infty)$. Moreover, since $W^s(0,0)$ is the boundary of the basin of attraction of an attracting fixed point, and this basin is an unbounded set contained in $V \cup V^+$, we conclude that $V_s^+ \neq \emptyset$. By making the radius defining V larger if necessary we may assume that $V_s^+ \cap W_+^s(0,0)$ and $V_s^+ \cap W_-^s(0,0)$ are connected sets.

Let $F = F_{\delta,\mu}$. Since $F^{-2}(V_s^+) \subset V_s^+$, we conclude that $\bigcup_{n=1}^{\infty} F^{2n}(V_s^+)$ and $\bigcup_{n=1}^{\infty} F^{2n}(V_s^-)$ two increasing unions of connected curves. Moreover, these two unions are disjoint. Here we also use the fact that the stable eigenvalue of the saddle point 0 is negative and hence $W_+^s(0,0)$ and $W_-^s(0,0)$ are F^{2n}-invariant sets. Consider $p \in W^s(0,0)$ with $p \neq 0$. To prove the claim it suffices to show that p is contained in one of these two unions. Moreover, since $F^k(p) \in W_{\mathrm{loc}}^s(0,0)$ for some $k \in \mathbb{N}$ it is enough to consider the case $p \in W_{\mathrm{loc}}^s(0)$. Let now $n \in \mathbb{N}$ be such that $F^{2n}(V_s^+)$ contains points $p_1, p_2 \in W_{\mathrm{loc}}^s(0,0)$ on both sides of $(0,0)$ which are closer to $(0,0)$ than p. It now follows from the Stable Manifold Theorem that p must be contained in one of the two curves in $F^{2n}(V_s^+)$. This implies that $p \in W^u(\infty)$. □

Lemma 7. *Let $1 < \mu < 3$. There exists $\delta^* > 0$ such that for all $0 < \delta < \delta^*$ the map $F_{\delta,\mu}$ satisfies property (iii) of Theorem 1, that is $K = \{\alpha, (0,0)\} \cup (W^s(\alpha) \cap W^u(0,0))$.*

Proof. Obviously $W^s(\alpha) \cap W^u(0,0)$ is contained in K as well as are the fixed points α and $(0,0)$. Let $q \in K \setminus \{\alpha, (0,0)\}$. By Propositions 3 and 5 the forward orbit of q must converge to either $(0,0)$ or α, i.e. q is in $W^s(0,0)$ but is not the origin or q is in $W^s(\alpha)$ but is not α. However, Lemma 6 implies that only the second case is possible, since the backward orbit of q is bounded.

Now let q be in the basin of attraction $W^s(\alpha)$ as well as in K with $q \neq \alpha$. We will show that $q \in W^u(0,0)$. Let q_1 be an arbitrary accumulation point of the backwards orbit $(F^{-n}(q))$, where $F = F_{\delta,\mu}$. Since K is closed we conclude that $q_1 \in K$. Hence the forward orbit of q_1 must converge to either $(0,0)$, or α, again by Propositions 3 and 5.

If $q_1 \in W^s(0,0)$, then $q_1 = (0,0)$, because otherwise $q_1 \in W^u(\infty)$ by Lemma 6 which contradicts the fact that $q_1 \in K$. Since $q_1 = (0,0)$ is an arbitrary accumulation point of $F^{-n}(q)$, it follows that $q \in W^u(0,0)$ once we show that $q_1 \in W^s(\alpha)$ is not possible.

If $q_1 \in W^s(\alpha), q \neq \alpha$, then the backward orbit $(F^{-n}(q))_{n \geq 0}$ cannot have an accumulation point, because there is a sequence of mutually disjoint sets A_n with $F^{-1}(A_n) \subset A_{n+1}$ such that $W^s(\alpha) = \bigcup_{n \geq 0} A_n$. □

Remark 1. (i) It can be shown that the obtained δ^* can be derived as a continuous function of μ. Therefore, if $1 < \mu_0 < 3$ we can formulate Theorem 1 to be true for all parameters in the set $\{(\delta, \mu) : 0 < \delta < \delta^*, |\mu - \mu_0| < r\}$ for some $\delta^* > 0$ and some $r > 0$.

(ii) Computer experiments suggest that Theorem 1 might be true for all $(\delta, \mu) \in H_{1,1}$ (see (4) for the definition). However, our results are based on small perturbation techniques and do not apply to consider arbitrary parameters in $H_{1,1}$.

4. The general case

In this section we consider general maps of the form

$$F(x, y) = (g(x) + h(y), h(x)). \tag{24}$$

Here $g, h : \mathbb{R} \to \mathbb{R}$ are C^2-maps, where g is a unimodal map whose graph has the qualitative shape of the logistic function $x \mapsto \mu x(1 - x)$, $1 < \mu < 3$ and h is a small perturbation of the linear map L_δ. More precisely, we assume that $g : \mathbb{R} \to \mathbb{R}$ is a C^2-unimodal map with $g(0) = 0, g(1) = 0, g([0, 1]) \subset [0, 1), g'(0) > 1$ and $g'(1) < -1$. We also assume that there exists $\gamma < 0$ such that $g''(x) < \gamma$ for all $x \in \mathbb{R} \setminus (0, 1)$. Finally, we assume that there exists an attracting fixed point $x = x_g \in (0, 1)$ of g with $W^s(x) = (0, 1)$. For $\delta \in \mathbb{R}$ we denote by L_δ the linear map $x \mapsto \delta x$. We now discuss the proof of Theorem 2. Since the arguments for the general case are obvious adaptations of the case for Hénon maps, only a sketch will be outlined.

Proof of Theorem 2. (i) If $\delta > 0$ and $0 < \epsilon < \delta/2$ it follows that every C^2-map $h : \mathbb{R} \to \mathbb{R}$ with $||h - L_\delta||_2 < \epsilon$ is a C^2-diffeomorphism of $\mathbb{R}$. Therefore, it follows from (24) that F is a bijective C^2-map of $\mathbb{R}^2$. The statement that F is a C^2-diffeomorphism of $\mathbb{R}^2$ is now a consequence of the inverse mapping theorem.

(ii),(iii),(iv) First, we note that if $\delta > 0$ is small enough then the origin is a saddle fixed point of F and F has an attracting fixed point $\alpha = \alpha_F$ that is close to $(x_g, 0)$. Since $g((1, \infty)) \subset (-\infty, 0)$ and for all $x < 0$ we have that $g^n(x)$ converges to $-\infty$ at a geometric rate, it is straightforward to verify that analogous filtration properties to those in Lemmas 2 and 3 and Proposition 1 hold. Furthermore, since $W^s(x_g) = (0, 1)$, we can show that for small $\delta > 0$ the set A_δ (defined in (13)) is contained in the basin of attraction of α. The two key results (Theorems 3 and 4) which describe the iterates of F near the origin (respectively near the intersection of $F^{-1}(W_{loc}^s(0, 0))$ with the x-axis) are based on the geometric shape of the graph of g near zero and can be proven accordingly. Finally, the corresponding results to Lemmas 5, 6 and 7 can be proven using the qualitative shape of g, g' and g'' rather than the explicit formulas. These results show that Theorem 2 can be established by repeating the arguments of the proof of Theorem 1.

References

[1] K. Alligood, T. Sauer and J. Yorke, Chaos: An introduction to dynamical systems, Springer (2000).

[2] E. Bedford and J. Smillie, *Polynomial automorphisms of* $\mathbb{C}^2$ *. 2. Stable manifolds and recurrence*, J. Amer. Math. Soc. **4** (1991), 657–679.

[3] *Chaotic Modeling and Simulation, International Conference 2014.*

[4] S. Friedland and J. Milnor, *Dynamical properties of plane polynomial automorphisms*, Ergodic Theory and Dynamical Systems **9** (1989), no.1, 67–99.

[5] S. Hayes and C. Wolf, *Dynamics of a one-parameter family of Hénon maps*, Dynamical Systems: An International Journal **9** (2006), 399–407.

[6] R.C. Robinson, Dynamical Systems: Continuous and discrete, Pearson Prentice Hall (2004).

[7] Y. G. Sinai, *Chaos theory yesterday, today and tomorrow*, J. Stat. Phys. **138** (2010), no. 1-3, 2–7.

[8] J.C. Sprott, *High-dimensional dynamics in the delayed Hénon map*, Electronic Journal of Theoretical Physics **3** (2006), no. 12, 19–35.

[9] C. Wolf, *Hausdorff and topological dimension for polynomial automorphisms of* $\mathbb{C}^2$, Ergodic Theory Dynam. Systems **22** (2002), 1313–1327.

Chapter 6

A New Type of Functional Equations of Euler Products

Bernhard Heim[*]

Department of Mathematics and Sciences, German University of Technology in Oman, Halban Campus, Muscat, Sultanate of Oman
E-mail: bernhard.heim@gutech.edu.om

We investigate interplay between generating series and infinite products including partition numbers, functional equations and modular forms. We apply multiplicative Hecke operators on periodic functions and introduce a new type of functional equations characterizing infinite products of Euler type.

Contents

1. Introduction

The interplay between additive and multiplicative structures is one of the most fascinating topics in mathematics. Especially in algebra, combinatorics and number theory many beautiful and important results have been discovered and proven, some are still unproven (we refer to Ono [12] for a good source). For example, the famous Goldbach conjecture (1742), that every even number larger than 3 should be the sum of two prime numbers, is still open.

In this paper we focus on structures related to generating functions, introduced by Euler. We concentrate on the question: How does an additive structure (gener-

---
[*]The chapter was written during a research stay July-August 2015 at the RWTH Aachen: Graduiertenkolleg Experimentelle und konstruktive Algebra, supported by the DFG. The author thanks Prof. Dr. Nebe and Prof. Dr. Krieg for the invitation and for providing a stimulating research environment at the RWTH Aachen University.

ating functions) correspond to a multiplicative structure (infinite products)? This is closely related to the question, how can one deduce the divisor, the zeros and poles from the Fourier expansion of a meromorphic function.

We contribute to this very classical topic with a new viewpoint. We show that infinite products of Euler type can be determined by a new type of functional equations. Functional equations of a more symmetric flavor recently appeared in [5]. Borcherds lifts [1, 2, 9] (certain automorphic forms) on the orthogonal group $O(2, 2 + n)$ of signature $(2, 2 + n)$ are characterized by functional equations. In this paper we deal with the case $O(2, 1)$, which is a somehow degenerate case and not included. Moreover we work in a quite general setting.

It is interesting to note that these functional equations are related to properties of cyclotomic polynomials. It is well-known that roots of unity and cyclotomic polynomials $\Phi_m(X)$ play a fundamental role in algebra and number theory (see Leutbecher [10] page 218). Nevertheless is seems to us the first time in the literature that they are also closely related to infinite products of Euler type. Let p be a prime number. Then

$$\Phi_p(X) := \prod_{0 < a < p} \left(X - e^{2\pi i \frac{a}{p}} \right) = \frac{X^p - 1}{X - 1}. \tag{1}$$

(see Leutbecher [10], Beispiel 1, page 171). It is also possible to look at infinite products and functional equations on $O(2, 2)$. Then the analogue of the above property of the cyclotomic polynomials is the symmetry of the modular polynomial (modular equation, [3]).

Let $a, b \in \mathbb{Z}$. Then products of Euler type are given by

$$g_{a,b}(\tau) := \prod_{m=1}^{\infty} \left(1 - q^{bm} \right)^a = \sum_{m=0}^{\infty} a(n) q^n \qquad \left(q := e^{2\pi i \tau} \right). \tag{2}$$

In this paper we mainly study the cases $g_a := g_{a,1}$ and $\widehat{g}_a(\tau) := q^{\frac{a}{24}} g_{a,1}$. Note that $\widehat{g}_1$ is the Dedekind eta functions and $\widehat{g}_{24}$ the Ramanujan delta function.

Let us recall a result of Borcherds [1, 9] involving modular forms. The correspondence between the multiplicative and additive structure is mainly given by the special divisor of the form.

Let f be a weakly modular form for $\Gamma := \mathrm{SL}_2(\mathbb{Z})$ of integer weight k. Then Borcherds [1] proved that certain f are equal to an infinite product

$$q^h \prod_{m=1}^{\infty} (1 - q^m)^{c(m^2)}, \tag{3}$$

(not necessarily defined on the whole upper half space $\mathfrak{H} := \{\tau \in \mathbb{C} \mid \Im(\tau) > 0\}$. Here $h \in \mathbb{Z}$ and $c(n)$ are Fourier coefficients of a corresponding modular form of weight $\frac{1}{2}$. This is the case if and only if f has integral coefficient and all poles and zeros are located at infinity or Heegner points (see also [12] for a summary).

In this paper we present a method, which makes it possible to deduce from the Fourier coefficients of a generating series f, if f is equal to an infinite product of

Euler type. This involves functional equations $(*_n)$:

$$\prod_{a \cdot d = n} \prod_{b=0}^{d-1} f\left(\frac{a\tau + b}{d}\right) = \varepsilon_n f(\tau)^{\sigma(n)}, \qquad (*_n)$$

attached to multiplicative Hecke operators for every natural number n. Here $c_n \subset \mathbb{C}^\times$ only depends on n and f and $\sigma(n)$ denotes the sum of the divisors of n.

Employing well-known properties of Hecke operators shows that only $(*_p)$ for all prime numbers p is needed. If f is normalized the functional equations dictate that f is already determined by the first Fourier coefficient and that f is up to a power of q equal to powers of the Dedekind eta function. Conversely if we assume that f is a modular form and satisfies $(*_p)$ at least for one prime number, then f is already proportional to a power of the Dedekind eta function.

We begin with a classical example already discussed by Leibniz and J. Bernoulli (see Neher [11], Ono [12]). They had been interested in the number of partitions $P(n)$ of a natural number n. Partitions are nonincreasing sequences of positive integers whose sum is n.

For example the partitions of $3, 4, 5$ are

$$3 = 2 + 1 = 1 + 1 + 1$$
$$4 = 3 + 1 = 2 + 2 = 2 + 1 + 1 = 1 + 1 + 1 + 1$$
$$5 = 4 + 1 = 3 + 2 = 3 + 1 + 1 = 2 + 2 + 1$$
$$= 2 + 1 + 1 + 1 = 1 + 1 + 1 + 1 + 1$$

Hence $P(3) = 3, P(4) = 5, P(5) = 7$. It turns out that these numbers are growing very fast. Almost 90 years ago MacMahon, a major in the British Royal Artillery and master calculator, computed the values for n up to 200.

$$P(200) = 3972999029388$$

To calculate these numbers by employing the definition of $P(n)$ is obviously hopeless. The first general and structural result on partition functions was given by Euler in his "Introductio in analysin infinitorum". Euler and discovered and proved

Theorem 1 (Euler). *Let q be a complex number $|q| < 1$ then*

$$\prod_{n=1}^{\infty} (1 - q^n)^{-1} = \sum_{n=0}^{\infty} P(n) q^n. \qquad (4)$$

Here $P(0) := 1$.

With the support of Maple we have been able to calculate the partition numbers of 2000 in a short time by a view commands.

$$P(2000) = 4720819175619413888601432406799959512200344166 \qquad (> 4 \cdot 10^{45})$$

Another famous infinite product is related with the pentagonal numbers $1, 5, 7, 12, 22, \ldots$ given by $\omega(n) := \frac{1}{2}(3n^2 - n)$ $(n \in \mathbb{Z})$ studied already by Pythagoras. Euler discovered and proved that the infinite product given by

$$(1 - q)\left(1 - q^2\right)\left(1 - q^3\right)\left(1 - q^4\right)\cdots \tag{5}$$

is equal to

$$\sum_{m=-\infty}^{\infty} (-1)^m q^{\omega(m)+\omega(-m)} = 1 - q - q^2 + q^5 + q^7 - q^{12} \cdots \tag{6}$$

Hence it can be written as $1 + \sum_{n=1}^{\infty} a(n)q^n$ with interesting coefficients $a(n)$, where $a(1) = -1$. Please note that the product of the two infinite products just studied is equal to one. This fact easily be used to get recursion formulas for the partition numbers. We also refer to Remmert and Schumacher [13] completing the analytic aspects of the topic.

2. Results

In this section we present the main results. As far as we know it is the first time that infinite products of Euler type are characterized by functional equations of the type $(*_n)$.

We first state a uniqueness theorem and then give the solution of the functional equations $(*_n)$. Since the functional equations have an interpretation in terms of multiplicative Hecke operators, the characterization can be reduced to $(*_p)$ for all prime numbers p. This indicates already how to deal with more general Euler products of type $g_{a,b}$ (using for example the U_p or V_p operators at *bad* primes). Let f satisfy $(*_p)$ for all primes p, then f is almost equal to a power of the Dedekind eta function η. Conversely, assuming that f is a modular form, it is sufficient that f satisfies $(*_p)$ for only one prime to identify f as a $\eta(\tau)^m$, $m \in \mathbb{Z}$.

Theorem 2 (Uniqueness). *Let f be a holomorphic function on the complex upper half space $\mathfrak{H}$. Let f be periodic with Fourier expansion*

$$f(\tau) = q^{\frac{1}{l}} \sum_{m=m_0}^{\infty} a_m q^m \qquad (a_{m_0} \neq 0), \tag{7}$$

where $l \in \mathbb{N}$ and $m_0 \in \mathbb{Z}$. Let f satisfy the functional equations (8) for all natural numbers n.

$$\prod_{a \cdot d = n} \prod_{b=0}^{d-1} f\left(\frac{a\tau + b}{d}\right) = \varepsilon_n f(\tau)^{\sigma(n)}. \tag{8}$$

Here $\varepsilon_n \in \mathbb{C}^\times$ only depends on n and f and $\sigma(n)$ denotes the sum of the divisors of n. Then f is uniquely determined by a_{m_0} and a_{m_0+1}.

Theorem 3 (Existence). *Let f be a holomorphic function on the complex upper half space $\mathfrak{H}$. Let f be periodic with Fourier expansion*

$$f(\tau) = q^{\frac{1}{l}} \sum_{m=m_0}^{\infty} a_m q^m \qquad (a_{m_0} \neq 0),$$

where $l \in \mathbb{N}$ and $m_0 \in \mathbb{Z}$. Let $\frac{a_{m_0+1}}{a_{m_0}} \in \mathbb{Z}$. Then f satisfies the functional equations

$$\prod_{a \cdot d = n} \prod_{b=0}^{d-1} f\left(\frac{a\tau + b}{d}\right) = \varepsilon_n f(\tau)^{\sigma(n)} \tag{9}$$

for all natural numbers n if and only if

$$f(\tau) = q^{\frac{1}{l} + m_0} a_{m_0} \prod_{m=1}^{\infty} (1 - q^m)^{-\frac{a_{m_0+1}}{a_{m_0}}}. \tag{10}$$

Corollary 1 (All primes). *Let $f(\tau) = 1 + \sum_{m=1}^{\infty} a_m q^m$ be holomorphic on $\mathfrak{H}$. Let $a_1 \in \mathbb{Z}$, then*

$$f(p\tau) \prod_{b=0}^{p-1} f\left(\frac{\tau + b}{p}\right) = f(\tau)^{p+1} \tag{11}$$

holds for all prime numbers p if and only if

$$f(\tau) = \prod_{n=1}^{\infty} (1 - q^n)^{-a_1}. \tag{12}$$

Remarks.
(i) Note that the Theorem is false if we skip the functional equation for any prime number p_0. For example

$$f(\tau) = \prod_{n=1}^{\infty} (1 - q^n) \prod_{n=1}^{\infty} (1 - q^{p_0 n}) \tag{13}$$

does satisfy the assumptions of the theorem and $f(\tau)$ fulfills (11) for all prime numbers different from p_0.
(ii) Let $f(\tau) = 1 - q + \mathcal{O}(q^2)$ holomorphic satisfying the functional equations $(*_p)$ for all primes, then $f(\tau) = \prod_{n=1}^{\infty} (1 - q^n)$.

Although the remarks indicate that we need all prime numbers to recover the unique product expansion from the Fourier expansion as given in the theorem, by using the theory of modular forms we obtain the following results:

Theorem 4 (One prime). *Let f be a modular form for $\mathrm{SL}_2(\mathbb{Z})$ of half-integral or integral weight k and possible multiplier system. Let f satisfy the functional equation $(*_p)$ for one prime number p then f is proportional to η^{2k}.*

In the special case $f \in M_k(\Gamma)$ and k integral we have:

Corollary 2. *Let f be a modular form for $\mathrm{SL}_2(\mathbb{Z})$ of weight k with Fourier expansion $\sum_{n=n_0}^{\infty} a_n\, q^n$, where $a_{n_0} = 1$ and $a_{n_0+1} \in \mathbb{Z}$. Let f satisfy*

$$f(p\tau) \prod_{b=0}^{p-1} f\left(\frac{\tau+b}{p}\right) = \varepsilon_p\, f(\tau)^{p+1}$$

at least for one prime number p. Then f is equal to a power of the the discriminant function Δ:

$$\Delta(\tau)^{-a_{n_0+1}/24}.$$

Where $a_{n_0+1} = -2k$.

Finally we mention that it had been recently proved [6], that in the $O(2,2)$ split case, also one functional equation induced by a prime number is sufficient to determine the modular form. This is related with modular polynomials and the Andre-Oort conjecture.

3. Preliminaries

Basic properties of Modular Forms. We start with a general definition of a modular form, which includes modular forms with multipliers. For example it is well-known that the Dedekind eta function is a modular form of weight $1/2$ with respect to a multiplier system of the full modular group Γ. For more detailed information we refer to the two excellent books of Koecher, Krieg [7] and Koehler [8]. The group $\mathrm{GL}_2^+(\mathbb{R})$ acts on $\mathfrak{H}$ via $\gamma(\tau) := \frac{a\tau+b}{c\tau+d}$, where $\gamma = \left(\begin{smallmatrix} a & b \\ c & d \end{smallmatrix}\right)$. A fundamental domain of the action of Γ is given by

$$\mathcal{F} := \left\{\tau \in \mathfrak{H} \,\Big|\, -\frac{1}{2} < \Re(\tau) \le \frac{1}{2} \text{ and } |\tau| > 1\right\}$$

$$\bigcup \left\{\tau \in \mathfrak{H} \,\Big|\, 0 \le \Re(\tau) \le \frac{1}{2} \text{ and } |\tau| = 1\right\}.$$

Definition 1. Let $\widetilde{\Gamma}$ be a subgroup of $\mathrm{SL}_2(\mathbb{R})$ which is commensurable with the modular group Γ, and let k be a real number. A function $f : \mathfrak{H} \longrightarrow \mathbb{C}$ is called a (weakly) modular form of weight k and multiplier system v for $\widetilde{\Gamma}$ if f is holomorphic on $\mathfrak{H}$ and has the following two properties:

(i) The following equation holds for all $\gamma = \left(\begin{smallmatrix} a & b \\ c & d \end{smallmatrix}\right) \in \widetilde{\Gamma}$.

$$f(\gamma(\tau)) := f\left(\frac{a\tau+b}{c\tau+d}\right) = v(\gamma)\,(c\tau+d)^k\, f(\tau). \tag{14}$$

All $|v(\gamma)| = 1$.

(ii) The function f is (meromorphic) at all cusps. If f vanishes at all cusps, then f is a cusp form.

Let $z \in \mathbb{C}$ and $r \in \mathbb{R}$. Then we define z^r as given in [7], but not as in [8]. The reader not familiar with the concept of multipliers, commensurability, and cusps is adviced to consult [8], section 1.4. Since we are mainly dealing with $\widetilde{\Gamma} = \Gamma$ and (half-)integral weight, we also just state the facts on the attached Fourier expansions (for a comprehensive treatment on the topic, we refer again to [8], section 1.4).

One of the most important examples of this paper is given by the Dedekind eta function η (see also [8]):

$$\eta(\tau) := \widehat{g}_1(\tau) = q^{\frac{1}{24}} \prod_{n=1}^{\infty} (1 - q^n). \tag{15}$$

The normal convergence of the product implies that $\widehat{g}_a$ is a holomorphic function for all $a \in \mathbb{Z}$, which also implies that $\widehat{g}_a(\tau) \neq 0$. Note that

$$\eta(\tau) = \sum_{n=1}^{\infty} \left(\frac{12}{n}\right) e \left(\frac{n^2 \tau}{24}\right), \tag{16}$$

where $\left(\frac{*}{*}\right)$ denotes the Legendre-Jacobi-Kronecker symbol and $e(z) := e^{2\pi i z}$ $(z \in \mathbb{C})$.

Remark: It is well known that the Dedekind eta function satisfies a transformation law for all elements γ the full modular group Γ:

$$\eta(\gamma(\tau)) = v_\eta(\gamma) \, (c\tau + d)^{\frac{1}{2}} \, \eta(\tau). \tag{17}$$

Here v_η is the induced multiplier system. Actually it is known that v_η generates the group of multipliers of Γ of order 24. In particular one has

$$\eta(\tau + 1) = e^{\pi i / 12} \, \eta(\tau).$$

Please note that $\widehat{g}_a$ is in general a weakly modular forms of weight $\frac{a}{2}$. The multiplier system is trivial if $24|a$. Then $\widehat{g}_a(\tau + 1) = \widehat{g}_a(\tau)$ Let k be an nonnegative integer. Then we denote by $M_k(\Gamma)$ and $S_k(\Gamma)$ the space of modular forms and cusp forms of weight k. These are all modular forms with trivial multiplier with respect to Γ of weight k and with Fourier expansion

$$f(\tau) = \sum_{n=n_0}^{\infty} a_n q^n, \qquad q = e(\tau), \tau \in \mathfrak{H} \tag{18}$$

at infinity. Here $n_0 \geq 0$ and $n_0 \geq 1$ if f is a cusp form. Moreover $M_k^!(\Gamma)$ denotes the space of weakly modular forms (trivial multiplier system, but we allow poles at infinity).

Let $\gamma \in \mathrm{GL}_2^+(\mathbb{R})$ and $k \in \mathbb{Z}$ then we denote by $j(\gamma, \tau) := c\tau + d$ the usual cocycle and by

$$(f|_k \gamma)(\tau) := j(\gamma, \tau)^{-k} f(\gamma(\tau)). \tag{19}$$

the Petersson slash operator. It is well known that $M_k(\Gamma)$ and $S_k(\Gamma)$ are finite dimensional vector spaces over $\mathbb{C}$.

Before we give several further examples, we just recall some background facts, making this exposition more natural.

Let $k \geq 4$ even. Then $\dim M_k(\Gamma) = [\frac{k}{12}]$ for $k - 2$ divisible by 12 and $[\frac{k}{12}] + 1$ if not divisible by 12. Further we have $\dim S_k(\Gamma) = \dim M_k(\Gamma) - 1$. Here $[x]$ denotes the Gauss bracket. A meromorphic function f on $\mathfrak{H}$ is called meromorphic modular form with integer weight k on Γ if f satisfies

$$(f|_k\gamma)(\tau) = f(\tau) \tag{20}$$

for Γ and has at most poles at infinity, i.e. a Fourier expansion of type (18). Let $\varrho := \frac{1+\sqrt{-3}}{2}$ and let $\mathcal{F}^* := \{i\infty\} \cup \mathcal{F}$ be the compactified fundamental domain. To consider the zero/pole order $\mathrm{ord}_\omega f$ of f is well-defined for all $\omega \in \mathcal{F}^*$. Let further $\mathrm{ord}\,\omega := 2, 3, 1$ if $\omega = i, \varrho$, otherwise.

Let f be a meromorphic modular form of weight k. The famous valence formula (see for example Koecher and Krieg [7], page 169) states that

$$\sum_{\omega \in \mathcal{F}^*} \frac{1}{\mathrm{ord}\,\omega} \, \mathrm{ord}_\omega f = \frac{k}{12}. \tag{21}$$

The valence formula implies the dimension formula.

Examples.
Eisenstein series E_k:
Let $k \in \mathbb{N}_0$ be even. We denote by B_k the k-th Bernoulli number. Let $k \geq 2$ then

$$E_k(\tau) := 1 - \frac{B_k}{2k} \sum_{n=1}^{\infty} \sigma_{k-1}(n)\, q^n. \tag{22}$$

is a holomorphic function on $\mathfrak{H}$. Moreover $E_k \in M_k(\Gamma)$ for $k \geq 4$. Here $\sigma_s(n) := \sum_{d|n} d^s$.
Note that E_4 has exactly one zero at ϱ and E_6 at i in $\mathcal{F}^*$. The order for each zero is 1.

Ramanujan Delta Function Δ:
The dimension formula implies that there exists exactly one normalized cusp forms. This is given by

$$\Delta(\tau) := \frac{E_4(\tau)^3 - E_6(\tau)^2}{12^3} \tag{23}$$

$$= q - 24q + 252q^2 - \ldots.$$

It also follows from our observations that $\eta^{24} \in S_{12}(\Gamma)$. Hence

$$\Delta(\tau) = q \prod_{n=1}^{\infty} (1 - q^n)^{24} \quad (\tau \in \mathfrak{H}).$$

This implies that Δ has no zeros.

4. Hecke Theory for Periodic Functions

In this section we recall some standard notation and well-known properties of (additive) Hecke operators (for example, see [14]). Actually we generalize the theory slidely as given by Koecher and Krieg in chapter IV in [7]. At the same time we introduce multiplicative Hecke operators and indicate some useful properties used in this paper.

Let $V(\mathfrak{H})$ be the $\mathbb{C}$-vector space of all meromorphic functions f with $f(\tau) = f(\tau + 1)$ on the upper half space $\mathfrak{H}$. We assume that we have at most a pole at infinity. Hence for every $f \in V(\mathfrak{H})$ and $\Im(\tau)$ large enough one has

$$f(\tau) = \sum_{m \geq m_0} a_f(m) q^m \qquad (m_0 \in \mathbb{Z}). \tag{24}$$

Definition 2. Let k, n be integers and let n be positive. Let $f \in V(\mathfrak{H})$. Then we define the additive and multiplicative Hecke operators.

$$T_\Sigma^{(k)}(n)(f) := n^{k-1} \sum_{\substack{a \cdot d = n \\ b \pmod n}} f|_k \begin{pmatrix} a & b \\ 0 & d \end{pmatrix}, \tag{25}$$

$$T_\Pi(n)(f) := \prod_{\substack{a \cdot d = n \\ b \pmod n}} f|_0 \begin{pmatrix} a & b \\ 0 & d \end{pmatrix}. \tag{26}$$

We denote k the weight of the Hecke operator. Let $T_\Sigma^{(k)}(n)(f) = \lambda_f(n) f$ for $n \in \mathbb{N}$, then we call $0 \neq f$ Hecke eigenform of weight k with eigenvalues $\lambda_f(n)$.

Note that

$$T_\Sigma^{(k)}(n\,m)(f) = T_\Sigma^{(k)}(n) \left(T_\Sigma^{(k)}(m)(f) \right)$$

for all coprime n and m. Moreover let p a prime number. Then $T_\Sigma^{(k)}(p^n) \in \mathbb{Q}[T_\Sigma^{(k)}(p)]$ with suitable interpretation. All the additive Hecke operators build up a commutative algebra (see IV, section 2 [7] for more details). With some more effort one can deduce similar properties for the multiplicative Hecke operators.

$$T_\Pi(n\,m)(f) = T_\Pi(n) \left(T_\Pi(m)(f) \right) \tag{27}$$

for all coprime n and m. Moreover let p a prime number. Then the action of $T_\Pi(p^n)$ is deduced from $T_\Pi(p)$. For example

$$T_\Pi(p^2)(f) = T_\Pi(p) \left(T_\Pi(p)(f) \right) \cdot f^{-p}$$
$$T_\Pi(p^l)(f) = T_\Pi(p^{l-1}) \left(T_\Pi(p)(f) \right) \cdot \left(T_\Pi(p^{l-2}) f \right)^{-p} \quad (l \geq 2).$$

All the operators commute and everything is determined by the action of $T_\Pi(p)$. Hence it is easy to prove by induction that

$$T_\Pi(n)(f) = f^{\sigma(n)} \text{ for all } n \Leftrightarrow T_\Pi(p)(f) = f^{p+1} \text{ for all } p \text{ prime.} \tag{28}$$

The properties of additive Hecke operators acting on periodic functions are almost the same as the original Hecke operators, since most of the properties are already reflected in the underlying abstract Hecke algebra. In the following we note some of them.

Let f be a modular form, then (25) coincides with definition of the classical normalized Hecke operators (for example, see chapter two [12]). Let $g = T_\Sigma^{(k)}(n)(f)$ and let m_0 be the possible order of f. Then

$$a_g(m) = \sum_{d|(m,n)} d^{k-1} a_f\left(\frac{mn}{d^2}\right) \text{ for } m \geq \begin{cases} 1 & m_0 \geq 1 \\ 0 & m_0 = 0 . \\ nm_0 & m_0 < 0 \end{cases} \qquad (29)$$

Hence we obtain: $a_g(0) = \sigma_{k-1}(n)a_f(0)$ and $a_g(1) = a_f(n)$.

Corollary 3. *Let $0 \neq f \in V(\mathfrak{H})$. Then f is a Hecke eigenform of weight k if and only if for all $m, n \in \mathbb{N}$*

$$\lambda_f(n)a_f(m) = \sum_{d|(m,n)} d^{k-1} a_f\left(\frac{mn}{d^2}\right). \qquad (30)$$

We have $\lambda_f(n)a_f(1) = a_f(n)$. If $a_f(1) = 1$, we say f is normalized, which implies $\lambda_f(n) = a_f(n)$.

It is perhaps also interesting to see how the additive Hecke operators transform with respect to differentiation.

Lemma 1. *Let $f \in V(\mathfrak{H})$ then we have:*

$$n\left(T_\Sigma^{(k)}(n)(f)\right)' = T_\Sigma^{(k+2)}(n)(f'). \qquad (31)$$

In the following we indicate how the multiplicative and the additive structures can be related and transformed vice versa. Please note that we omit questions on convergence to directly see the algebraic structure. To complement them is straightforward.

The following result is very useful and gives more or less a tool to translate all formulas for additive Hecke operators into the corresponding formulas for multiplicative Hecke operators. It is also quite amusing how this is naturally related with logarithmic differentiation.

Proposition 1. *Let $F, G \in V(\mathfrak{H})$. Let F be of the form $1 + \mathcal{O}(q)$ and G of the form $q + \mathcal{O}(q^2)$ Suppose (formally) that*

$$F = \exp(G).$$

Then we have the following correspondence between multiplicative and additive Hecke operators

$$T_\Pi(n)(F) = \exp\left(nT_\Sigma^{(0)}(n)(G)\right). \qquad (32)$$

Moreover the logarithmic derivative on the left side leads to

$$(\ln (T_\Pi(n)(F)))' = T_\Sigma^{(2)}(n)(G').$$ (33)

The proof mainly depends on the following observation. Let $a, d \in \mathbb{N}$ and let $f \in V(\mathfrak{H})$ be an arbitrary element for a moment. Then we decompose the Hecke operator in the following way. Let

$$(T_{a,d}(f))(\tau) := \sum_{b \pmod d} f\left(\frac{a\tau + b}{d}\right).$$ (34)

The Fourier expansion of the action is given by

$$(T_{a,d}(f))(\tau) = \sum_{m \ge \frac{m_0}{d}} a_f(md)q^{ma}.$$

This implies that

$$\left(T_\Sigma^{(k)}(n)(f)\right) = n^{k-1} \sum_{ad=n} d^{-k} T_{a,d}(f).$$

Now let $F(\tau) = 1 + \sum_{m=1}^\infty a_F(m)q^m$ and $G(\tau) = \sum_{m=1}^\infty a_G(m)q^m$. Then $a_F(1) = a_G(1)$ holds.

Finally we mention the following application, which is almost characterizing the Eisenstein series E_2 of weight 2 (see (22)). The proof is straightforward.

Lemma 2. *Let* $G(\tau) = \sum_{m=0}^\infty a(m)q^m \in V(\mathfrak{H})$ *be normalized with* $a(1) = 1$. *Then*

$$n\left(T_\Sigma^{(0)}(n)(G)\right) = \sigma(n)(G) \text{ for all } n \in \mathbb{N}$$ (35)

if and only if

$$G = a_0 + \sum_{m=1}^\infty \sigma_{-1}(m)q^m.$$ (36)

This is implied by Corollary 3. If $a_0 := -\frac{1}{24}$, then $G = -\frac{1}{24}E_2$.

5. Final Proofs of the Results

Before we give the proofs of the main results of this paper, we note the following useful observation. Let $f(\tau)$ be given as in Theorem 2. Let $f_1(\tau) := q^{\frac{1}{t}+m_0} a_{m_0}$. Since f_1 is not period $f_1(\tau + 1) \ne f_1(\tau)$, we cannot apply the Hecke operators introduced in the last section, since they are not well-defined. Nevertheless we can proceed in two ways. By abuse of notation, up to the value of ε_p the action is well defined. Or we fix representatives of the involved matrices. Actually in this way the functional equation $(*_n)$ is stated. Then $f_1(\tau)$ satisfies the functional equation

$$T_\Pi^!(p)(f_1) := f(p\tau) \prod_{b=0}^{p-1} f\left(\frac{\tau + b}{p}\right) = \varepsilon_p f_1^{p+1}.$$

Since $f(\tau) = f_1(\tau) f_2(\tau)$ with

$$f_2(\tau) := 1 + \sum_{m=1}^{\infty} \frac{a_{m_0+m}}{a_{m_0}} q^m, \tag{37}$$

we have that f satisfies a functional equation of our particular type if and only if f_2 satisfies a functional equation. Here the nontrivial factor ε_p is induced by f_1.

5.1. *Proof of Theorem 2*

In view of the observation above, it is sufficient to prove the Theorem in the following variation. Let $f(\tau) = 1 + \sum_{m=1}^{\infty} a_m q^m$ satisfy

$$T_\Pi(n)(f) = f^{\sigma(n)} \text{ for all } n \in \mathbb{N}, \tag{38}$$

then f is uniquely determined by a_1. There are several ways to do this. We first give a direct way, which reveals a recursion property of the Fourier coefficients a_m.

$$\prod_{\substack{a \cdot d = n \\ b \pmod n}} \left(1 + \sum_{m=1}^{\infty} a_m e^{2\pi i m \frac{a\tau+b}{d}} \right) = \left(1 + \sum_{m=1}^{\infty} a_m q^m \right)^{\sigma(n)} \tag{39}$$

$$\prod_{\substack{a \cdot d = n \\ b \pmod n}} \left(1 + \sum_{m=1}^{\infty} a_m e^{2\pi i \left(ma^2 \tau + \frac{mab}{n} \right)} \right) = \left(1 + \sum_{m=1}^{\infty} a_m q^{nm} \right)^{\sigma(n)}. \tag{40}$$

Comparing the coefficients of q^n on both sides leads to the recursion formula

$$n a_n + Q_n(a_1, \dots, a_{n-1}) = \sigma(n) a_1. \tag{41}$$

Here Q_n is a polynomial with integer coefficients in $n-1$ variables.

A second proof (sketch) can be given in the following way. Let f be as before and $f = \exp(g)$, with $g = q + \mathcal{O}(q^2)$. Then the functional equations lead to the situation studied in Lemma 2, where a_1 is more general. Finally we obtain $g = \frac{a_1}{24} (E_2 - 1)$.

5.2. *Proofs of Theorem 3 and Corollary 1*

We start with the proof of the Theorem. We have to show that the function

$$f(\tau) = \prod_{n=1}^{\infty} (1 - q^n)^a \quad (a \in \mathbb{Z}) \tag{42}$$

is satisfying the functional equations for all $n \in \mathbb{N}$. Applying (28) and the fact that the functional equations are compatible with taking powers, we are finally left with the claim:

$$T_\Pi(p)(h) = h^{p+1} \text{ for all prime numbers } p, \tag{43}$$

where $h(\tau) := \prod_{m=1}^{\infty}(1 - q^m)$. We have

$$T_{\Pi}(p)(h)(\tau) = h(p\tau) \prod_{b=0}^{p-1} h\left(\frac{\tau + b}{p}\right) \tag{44}$$

$$= \prod_{m=1}^{\infty}(1 - q^{pm}) \prod_{b=0}^{p-1} \prod_{m=1}^{\infty} \left(1 - q^{\frac{m}{p}} \xi_p^{mb}\right). \tag{45}$$

Here $\xi_p := e^{2\pi i/p}$. We interchange the order of the products $\prod_{b=0}^{p-1} \prod_{m=1}^{\infty}$. Then we consider the two cases $p \mid m$ and $p \nmid m$. The first case leads to the expression h^p. The second case is related to the p-th cyclotomic polynomial as defined in the introduction and leads to the expression

$$\prod_{\substack{m=1 \\ (m,p)=1}}^{\infty}(1 - q^m). \tag{46}$$

This term matches perfectly with the first term. Putting everything together proves the Theorem since $f(\tau) = 1 - aq + \mathcal{O}(q^2)$ is the unique solution of the functional equations for all $a \in \mathbb{Z}$. As a byproduct we have also proven the Corollary.

5.3. *Proofs of Theorem 4 and Corollary 2*

The Dedekind eta function satisfies the functional equation $(*_n)$ for all $n \in \mathbb{N}$. Let f be an automorpic form on $\mathfrak{H}$ with respect to $\mathrm{SL}_2(\mathbb{Z})$ satisfying

$$T_{\Pi}(p)(f) = \varepsilon_p \, f^{p+1} \tag{47}$$

for at least one prime p. We show that this already implies that f has no zeros and hence has to be proportional to $\eta(\tau)^a$ $(a \in \mathbb{N})$.

Let $f(\tau_0) = 0$. Let $\tau_0 \in \mathcal{F}$. Note that $|p\tau_0| > 1$ for all prime numbers p. Let $\tau_1 := p\tau_0$. Then it follows from

$$T_{\Pi}(p)(f)(\tau_1) = f(p\tau_1) \prod_{b=0}^{p-1} f\left(\frac{\tau_1 + b}{p}\right) = 0, \tag{48}$$

that also $f(\tau_1) = 0$. This procedure can be repeated and shows that $p^n \tau_0$ are infinitely many nonequivalent zeros of f. Note that for every $n \in \mathbb{N}$ an $s \in \mathbb{Z}$ exists that $p^n \tau_0 + s \in \mathcal{F}$. But since every automorphic form has only finite many zeros in $\mathcal{F}$ we have a contraction, which implies that that f has no zero. Hence f has to be of the prescribed type. If $f \in M_k$ of integer weight k and trivial multiplier, and normalized. Then k is divisible by 12 and $f = \Delta^{\frac{k}{12}}$.

Acknowledgements

The author would like to thank Aloys Krieg and Atsushi Murase for useful conversations related to the topic.

References

[1] R. E. Borcherds, *Automorphic forms on $O_{s+2,2}(\mathbb{R})$ and infinite products*, Invent. Math. **120** (1995), 161–213.

[2] J. H. Bruinier, *Borcherds Products on $O(2,l)$ and Chern Classes of Heegner Divisors*, Lecture Notes in Math. **1780** (2002), Springer Verlag.

[3] D. A. Cox, *Primes of the form $x^2 + ny^2$. Fermat, class field theory and complex multiplication*, A Wiley-Interscience Publication, New York, 1989.

[4] A. Dabholkar, S. Murthy, and D. Zagier, *Quantum Black Holes, Wall Crossing, and Mock Modular Forms*, arXiv:1208.4074v1 [hep-th] 20 Aug 2012.

[5] B. Heim and A. Murase, *A characterization of Holomorphic Borcherds Lifts by Symmetries*, International Mathematics Research Notices (2014). doi: 10.1093/imm/rnv021.

[6] B. Heim, C. Kaiser, and A. Murase, *Borcherds lifts, Symmetries and the Andre-Oort conjecture*, manuscript 23.02.2015.

[7] M. Koecher and A. Krieg, *Elliptische Funktionen und Modulformen*, 2nd edition, Springer, Berlin-Heidelberg-New York (2007).

[8] G. Köhler, *Eta Products and Theta Series Identities*, Springer Monographs in Mathematics, Springer, Berlin-Heidelberg-New York (2011).

[9] M. Kontsevich, *Product formulas for modular forms on $O(2,n)$*, Seminaire Bourbaki 821 (1996).

[10] A. Leutbecher, *Zahlentheorie, Eine Einfürung in die Algebra*, Springer Verlag, Springer, Berlin-Heidelberg-New York (1996).

[11] E. Neher, *Jacobis Tripleprodukt-Identität und η-Identitäten in der Theorie affiner Lie-Algebren*, Jahresbericht d. Dt. Math.-Verein **87** (1985), 164–181.

[12] K. Ono, *The web of modulariy: arithmetic of the coefficients of modular forms and q-series*, CBMS Regional Conference Series in Mathematics **102** Amer. Math. Soc., Providence, RI, (2004).

[13] R. Remmert, G. Schumacher, *Funktionentheorie 2*, Dritte Auflage. Springer Berlin-Heidelberg-New York (2007).

[14] G. Shimura, *Introduction to the arithmetic theory of automorphic functions*, Reprint of the 1971 original. Publications of the Math. Soc. of Japan, 11. Kan Memorial Lectures, 1. Princeton Univ. Press, Princeton, NJ, 1994.

Chapter 7

The Hexagonal Lattice and the Epstein Zeta Function

Andreas Henn

Lehrstuhl II – Geometrie, TU Dortmund
E-mail: andreas.henn@math.tu-dortmund.de

I have been familiar with the name of Armin Leutbecher ever since I discovered his book *Zahlentheorie* during my student years. "In algebra there are many definitions; some of them are needed." This remark of a fellow student of Professor Leutbecher, quoted in the preface to that book, perfectly echoes my own sentiments when confronted with abstract algebra for the first time. His book proved valuable to me by providing concrete mathematical material to which the abstract theory could be applied and from which it had grown, especially from the field of quadratic number fields. For this reason I feel honored to have been asked to contribute this article, which may be said to have as its subject the Eisenstein integers, to this volume. I dedicate it to Professor Leutbecher with my best wishes. To state the contents more precisely, this is an expository article on the hexagonal lattice, the Epstein zeta function, and an application to physics. The properties of the lattice and the zeta function are reviewed, then the proof of Montgomery of a minimality property of the Epstein zeta function of the hexagonal lattice is outlined. Finally, the connection to a problem in mathematical physics is sketched.

Contents

1. Introduction

This chapter is motivated by an application of results from number theory to a problem in mathematical physics. In the 2012 paper [23] by E. Sandier and S. Serfaty, configurations of charged particles in the plane under the presence of an external field are studied. In the first part of the paper, they introduce a renormalized energy of such configurations and prove that the hexagonal lattice is a minimizer of this energy for lattice configurations (Theorem 2 of [23]); the corresponding property for all configurations of given average density and satisfying a certain bounded-ness condition is formulated as a conjecture. A connection to superconductivity, which is the main motivation for their work, is established in the second part: They show that minimizers of the Ginzburg–Landau energy of a superconductor with an applied magnetic field can asymptotically be described by minimizers of this renor-malized energy. This is a first step towards rigorously explaining the appearance of hexagonal lattices as *Abrikosov lattices*, densely packed vortex lattices in super-conductors, predicted by the theoretical physicist A. A. Abrikosov in the 1950s and later experimentally observed.[1] An account of Abrikosov's discoveries can be found in his 2003 Nobel lecture.

Sandier's and Serfaty's proof of this minimality property of the hexagonal lattice makes use of results from analytic number theory: For lattice configurations, the renormalized energy defined in their paper can be expressed by the Epstein zeta function. So minimizing this renormalized energy is reduced to minimizing the Epstein zeta function. This problem has been studied by many mathematicians beginning with Rankin in the 1950s; the solution is given by the hexagonal lattice.

In the first two sections of this article, the properties of the hexagonal lattice and the Epstein zeta function are reviewed. Then we sketch a proof by Montgomery of the minimality of the Epstein zeta function of the hexagonal lattice. In the subsequent section, the application of this result in the proof by Sandier and Serfaty is described; a minor error in their computations is corrected.

2. The hexagonal lattice

For the material of this section, the reference is [6, Chap. 4.6]. For $n \geq 1$, the lattice A_n is defined as

$$A_n = \{(x_0, x_1, \ldots, x_n) \in \mathbb{Z}^{n+1} \mid x_0 + x_1 + \ldots + x_n = 0\}.$$

The lattice A_2 is similar to the *hexagonal lattice* (sometimes also called the *trian-gular lattice*), given by the generator matrix

$$N = \begin{pmatrix} 1 & 0 \\ \frac{1}{2} & \frac{\sqrt{3}}{2} \end{pmatrix}.$$

From this representation one can see that the hexagonal lattice can be viewed as the ring of the Eisenstein integers

$$\mathcal{O} = \mathbb{Z} + \mathbb{Z}\omega,$$

where

$$\omega = \frac{1 + \sqrt{-3}}{2},$$

---

[1]For images see http://www.fys.uio.no/super/vortex/

the ring of integers of the field $\mathbb{Q}(\sqrt{-3})$. The Gram matrix with regard to this generator matrix is

$$T = \begin{pmatrix} 1 & \frac{1}{2} \\ \frac{1}{2} & 1 \end{pmatrix}$$

of determinant $3/4$. The minimal norm is 1 and the set R of minimal vectors is

$$R = \left\{ (\pm 1, 0), (\pm 1/2, \pm \sqrt{3}/2) \right\}$$

(corresponding to $\mathcal{O}^\times$). The Voronoi cells of this lattice are regular hexagons. The automorphism group is of order 12 and generated by rotation by $\pi/3$ (corresponding to multiplication by ω) and reflection in the x-axis (corresponding to complex conjugation).

The hexagonal lattice is the solution to the sphere packing problem in two dimensions. That means that the densest way of packing non-overlapping equal discs into the plane is realized by centering the discs (of radius $1/2$) in the points of the hexagonal lattice; the density achieved is then

$$\Delta = \frac{\pi}{12} \approx 0.91.$$

Moreover, the hexagonal lattice is also the solution to the covering problem in two dimensions, the problem of covering the plane with equal overlapping discs so that the thickness of the covering, i.e. the average number of spheres that contain a point of the plane, is minimized. The thickness of the covering associated to the hexagonal lattice is

$$\Theta = \frac{2\pi}{\sqrt{3}} \approx 1.21.$$

For proofs the reader is referred to [14, pp. 58-61].

The quadratic form associated to the hexagonal lattice is

$$f(u_1, u_2) = u_1^2 + u_1 u_2 + u_2^2,$$

so the elliptic theta series is given by

$$\theta(z) = \sum_{m,n=-\infty}^{\infty} e^{\pi i (m^2 + mn + n^2) z}$$

for z in the complex upper half-plane $\mathbb{H}$. This series can be expressed by the classical theta series

$$\theta_3(z) = \sum_{n=-\infty}^{\infty} e^{\pi i n^2 z}$$

and

$$\theta_2(z) = \sum_{n=-\infty}^{\infty} e^{\pi i (n+1/2)^2 z}$$

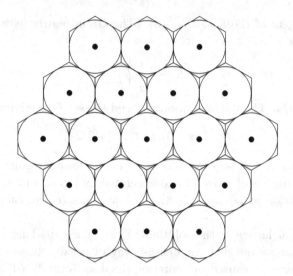

Fig. 1. The hexagonal lattice with its Voronoi cells and the associated sphere packing.

with $z \in \mathbb{H}$, namely, we have

$$\theta(z) = \theta_3(z)\theta_3(3z) + \theta_2(z)\theta_2(3z),$$

cf. [6, 4.6.2]. In [6, Table 4.4], the first 50 nonzero Fourier coefficients are given; for their computation see also the following section.

3. The Epstein zeta function

Throughout this section, we will switch between the languages of lattices, quadratic forms, and positive definite matrices. We denote the transpose of a matrix S by S'. The zeta functions we consider were introduced by P. Epstein in 1903 (cf. [13]).

Definition 1. Let $T \in \operatorname{Sym}(n, \mathbb{R})$ positive definite. The *Epstein zeta function* of T is defined by

$$\zeta(T, s) = \sum_{g \in \mathbb{Z}^n \setminus \{0\}} (g'Tg)^{-s}, \quad s \in \mathbb{C}, \operatorname{Re}(s) > n/2.$$

This series is absolutely convergent for $\operatorname{Re}(s) > n/2$ (for a modern proof we refer the reader to [19, Satz V.5.1]), and we have

$$\zeta(U'TU, s) = \zeta(T, s)$$

for $U \in \operatorname{GL}(n, \mathbb{Z})$. Analogously to the Riemann zeta function, $\zeta(T, s)$ can be continued to a meromorphic function on $\mathbb{C}$ (cf. [19, Satz V.5.2]): Consider the completed zeta function

$$\xi(T, s) = \pi^{-s}\Gamma(s)\zeta(T, s),$$

which satisfies

$$
\begin{aligned}
\xi(T, s) &= \int_0^\infty (\vartheta(T, iy) - 1) y^{s-1} dy \\
&= \int_1^\infty \left[(\vartheta(T^{-1}, iy) - 1)(\det T)^{-1/2} y^{n/2-s} + (\vartheta(T, iy) - 1) y^s \right] \frac{dy}{y} \\
&\quad + \left(\frac{(\det T)^{-1/2}}{s - n/2} - \frac{1}{s} \right)
\end{aligned}
\tag{1}
$$

for $\mathrm{Re}(s) > n/2$, with

$$
\vartheta(T, z) = \sum_{g \in \mathbb{Z}^n} e^{\pi i g' T g z}
$$

for $z \in \mathbb{H}$. The last integral converges for all $s \in \mathbb{C}$, so this yields analytic continuation of $\xi(T, s)$ to $\mathbb{C} \setminus \{0, n/2\}$ and thus of $\zeta(T, s)$ to $\mathbb{C} \setminus \{n/2\}$ with a single pole (of first order) in $s = n/2$ and residue

$$
\mathrm{res}_{n/2} \zeta(T, s) = \frac{\pi^{n/2}}{\Gamma(n/2)} \sqrt{\det T}.
$$

We have the functional equation

$$
(\det T)^{-1/2} \xi(T^{-1}, n/2 - s) = \xi(T, s).
\tag{2}
$$

For a quadratic form f, one can analogously define $\zeta(f, s)$; for a lattice Γ in a euclidean space one defines

$$
\zeta(\Gamma, s) = \sum_{\gamma \in \Gamma \setminus \{0\}} \langle \gamma, \gamma \rangle^{-s}.
$$

Then $\zeta(\Gamma, s) = \zeta(T, s)$ for any Gram matrix T of Γ. As to the functional equation, note that if T is a Gram matrix for Γ, then T^{-1} is a Gram matrix for the dual lattice Γ^*.

The ring of Eisenstein integers $\mathcal{O}$ is a principal ideal domain with $|\mathcal{O}^\times| = 6$. Thus for for the hexagonal lattice Λ, we have

$$
\zeta(\Lambda, s) = 6 \zeta_{\mathbb{Q}(\sqrt{-3})}(s)
$$

with the Dedekind zeta function of the number field $\mathbb{Q}(\sqrt{-3})$, defined by

$$
\zeta_{\mathbb{Q}(\sqrt{-3})}(s) = \sum_{\mathfrak{a}} N(\mathfrak{a})^{-s}
$$

for $\mathrm{Re}(s) > 1$, where summation runs over the nonzero (integral) ideals of $\mathcal{O}$. With the well-known factorization of the zeta function of a quadratic number field (cf. [26, §11]) we get

$$
\zeta(\Lambda, s) = 6 \zeta(s) L(\chi_{-3}, s)
$$

with the Riemann zeta function and the Dirichlet L-series of the primitive Dirichlet character χ_{-3} mod 3 given by

$$\chi_{-3}(2) = -1, \ \chi_{-3}(p) = \left(\frac{-3}{p}\right),$$

so

$$L(\chi_{-3}, s) = 1 - \frac{1}{2^s} + \frac{1}{4^s} - \frac{1}{5^s} + \frac{1}{7^s} - \frac{1}{8^s} + \dots.$$

In particular, we have

$$|\{(x, y) \in \mathbb{Z}^2 \mid x^2 + xy + y^2 = n\}| = 6 \sum_{m|n} \chi_{-3}(m).$$

4. A minimum problem

Two binary quadratic forms f_1, f_2 are called equivalent if there is a matrix

$$\begin{pmatrix} a_{11} & a_{12} \\ a_{21} & a_{22} \end{pmatrix} \in \mathrm{SL}(2, \mathbb{Z})$$

such that

$$f_2(u_1, u_2) = f_1(a_{11}u_1 + a_{12}u_2, a_{21}u_1 + a_{22}u_2).$$

The Epstein zeta function of the (rescaled) hexagonal lattice has the following minimality property:

Theorem 1. *Let f be a positive definite binary quadratic form of discriminant -1 and*

$$h(u_1, u_2) = \frac{1}{\sqrt{3}}(u_1^2 + u_1 u_2 + u_2^2).$$

We have

$$\zeta(f, s) - \zeta(h, s) \geq 0 \quad \text{for all real } s > 0,$$

where equality occurs if and only if f and h are equivalent forms.

Note that $\zeta(f, s) - \zeta(h, s)$ is regular in $s = 1$, so the statement also makes sense in this case.

This theorem has a long history. In the paper [18], D. G. Kendall and R. A. Rankin found a connection between the number of lattice points in a random circle and the Epstein zeta function. More precisely, the variance was shown to be a constant multiple of the value of the respective Epstein zeta function at $s = \frac{3}{2}$. The property of the hexagonal lattice as solution to both the sphere packing problem and the covering problem led them to suppose that the minimum value would be attained for this lattice. This motivated the 1953 paper [22] by Rankin, where he proved the theorem for $s \geq 1.035$. In 1956, J. W. S. Cassels published a proof extending Rankin's result to $s > 0$ in [4]. However, in 1963, V. Ennola pointed out an error in Cassels's argumentation, which he corrected in the paper [11]. In [10], P. H. Diananda simplified Cassels's proof. Finally, a simpler proof was published by Montgomery in 1988 (see [20]). He proved a corresponding minimality result for the elliptic theta series of the hexagonal lattice and then used the fact that the Epstein zeta function is the Mellin transform of the theta series. The remainder of this section is devoted to an outline of Montgomery's proof of the theorem.

Consider, for a positive definite binary quadratic form

$$f(u_1, u_2) = au_1^2 + bu_1u_2 + cu_2^2$$

of discriminant $b^2 - 4ac = -1$, the matrix associated to $2f$, that is

$$T = \begin{pmatrix} 2a & b \\ b & 2c \end{pmatrix}.$$

So we have $\det T = 1$ and

$$T^{-1} = \begin{pmatrix} 2c & -b \\ -b & 2a \end{pmatrix} = U'TU$$

with

$$U = \begin{pmatrix} 0 & -1 \\ 1 & 0 \end{pmatrix} \in \mathrm{GL}(2, \mathbb{Z}).$$

Thus by (1), for the function on $\mathbb{R}_+$ given by

$$\theta_f(\alpha) := \vartheta(T, i\alpha),$$

we have

$$2^{-s}\xi(f, s) = \xi(2f, s) = \frac{1}{s-1} - \frac{1}{s} + \int_1^\infty (\theta_f(\alpha) - 1)(\alpha^s + \alpha^{1-s})\frac{d\alpha}{\alpha} \qquad (3)$$

for all $s \in \mathbb{C} \setminus \{0, 1\}$ (note that Montgomery writes $\xi_f(s) = 2^{-s}\xi(f, s)$). This means that to prove the claimed minimality property of the Epstein zeta function, it suffices to show

Theorem 2. *Let f and h be as above. For any $\alpha > 0$ we have*

$$\theta_f(\alpha) \geq \theta_h(\alpha).$$

If there is an $\alpha > 0$ with $\theta_f(\alpha) = \theta_h(\alpha)$, then f and h are equivalent forms.

Montgomery's paper is devoted to demonstrating this result. For the proof, note that

$$\theta_f(\alpha) = \sum_{m,n=-\infty}^\infty e^{-2\pi\alpha f(m,n)}$$

with

$$\theta_f(1/\alpha) = \alpha\theta_f(\alpha)$$

due to the theta transformation formula. Now factorize f as

$$f(u_1, u_2) = a(u_1 + zu_2)(u_1 + \bar{z}u_2)$$

where $z = x + iy$; we can assume that $z \in \mathbb{H}$. Since $b^2 - 4ac = -1$, we have $a = \frac{1}{2y}$, so

$$f(u_1, u_2) = \frac{1}{2y}(u_1 + xu_2)^2 + \frac{1}{2}yu_2^2.$$

In the case $f = h$, we get $z = \frac{1}{2} + \frac{1}{2}i\sqrt{3}$.

In the following, we write $\theta(\alpha; x, y) = \theta_f(\alpha)$ and $\zeta(s; x, y) = \zeta(f, s)$. We have

$$\theta(\alpha; x, y) = \sum_{n=-\infty}^{\infty} e^{-\pi\alpha y n^2} \sum_{m=-\infty}^{\infty} e^{-\pi\alpha(m+nx)^2/y}.$$

Note that

$$\theta(\alpha; -x, y) = \theta(\alpha; x, y). \tag{4}$$

The previous proofs of Theorem 1 combined the two following lemmata:

Lemma 1. *If $y \geq 3/2$ and $s > 0$, then $\frac{\partial}{\partial y}\zeta(s; x, y) > 0$.*

Lemma 2. *If $0 \leq x \leq 1/2$, $y \geq 3/5$, and $0 \leq s \leq 3$, then $\frac{\partial}{\partial x}\zeta(s; x, y) > 0$.*

Montgomery derives corresponding results on $\theta(\alpha; x, y)$ in two main lemmata. Then the statement of the second theorem follows by elementary calculus.

If f_1 and f_2 are equivalent forms given by $z_1, z_2 \in \mathbb{H}$ and f_1 is transformed into f_2 by $(a_{ij}) \in SL(2, \mathbb{Z})$ as above, we have

$$z_2 = \frac{a_{22}z_1 + a_{12}}{a_{21}z_1 + a_{11}}.$$

A form f is called reduced if $-a < b \leq a < c$ or $0 \leq b \leq a = c$. Thus a form f given by $z \in \mathbb{H}$ is reduced if and only if z lies in

$$\mathcal{D} = \left\{ z \in \mathbb{H} \;\middle|\; -\frac{1}{2} < x \leq \frac{1}{2}, \; |z| \geq 1 \text{ and } |z| > 1 \text{ for } -\frac{1}{2} < x < 0 \right\},$$

the usual fundamental domain of the modular group $SL(2, \mathbb{Z})$. Since every form is equivalent to a reduced form (cf. [26, §8, proof of Satz 1]) (or equivalently every point of the upper half-plane is equivalent to a point in $\mathcal{D}$ under $SL(2, \mathbb{Z})$) and equivalent forms have the same theta series and zeta function, we may restrict to the case of reduced forms. By (4), we may assume that $x \geq 0$.

Now consider Jacobi's theta function given by

$$\theta(t, \beta) = \sum_{k=-\infty}^{\infty} e^{-\pi k^2 t + 2\pi i k \beta}$$

for $t \in \mathbb{R}_+$, $\beta \in \mathbb{R}$. Montgomery proves two lemmata on this function involving several estimations. They are crucial to the proof of his first main lemma:

Lemma 3. *If $\alpha > 0$, $0 < x < 1/2$, and $y \geq 1/2$, then*

$$\frac{\partial}{\partial x}\theta(\alpha; x, y) < 0.$$

The proof makes use of the identity

$$\theta(\alpha; x, y) = (y/\alpha)^{1/2} \sum_{n=-\infty}^{\infty} e^{-\pi \alpha y n^2} \theta(y/\alpha, nx).$$

This equation is differentiated; then some elementary estimations and the estimation from the preceding lemma in [20] (denoted there as Lemma 3) yield the result. To prove his second main lemma (Lemma 6 below), Montgomery proves (by direct calculations and estimations involving $\theta(t, \beta)$ from above) the following differential inequality:

Lemma 4. *If $\alpha > 0$, $0 \leq x \leq 1/2$, and $x^2 + y^2 \geq 1$, then*

$$\frac{\partial^2}{\partial y^2}\theta(\alpha; x, y) + \frac{2}{y}\frac{\partial}{\partial y}\theta(\alpha; x, y) > 0.$$

The following elementary analytic lemma is also needed for the second main lemma:

Lemma 5. *Suppose that*

$$f'(y) + \frac{2}{y}f(y) > 0$$

for all $y \geq y_0 > 0$, and that $f(y_0) \geq 0$. Then $f(y) > 0$ for all $y > y_0$.

Now Montgomery proves his second main lemma:

Lemma 6. *If $\alpha > 0$, $0 \leq x \leq 1/2$, $x^2 + y^2 \geq 1$, then*

$$\frac{\partial}{\partial y}\theta(\alpha; x, y) \geq 0$$

with equality if and only if (x, y) is one of the points $(0, 1)$, $(1/2, \sqrt{3}/2)$.

Note that for our uses showing this inequality for $x = 1/2$ would suffice, but Montgomery needs this more general version for another theorem he proves in the paper. In view of Lemma 4 all that remains to be shown in order to apply Lemma 5 is that

$$\frac{\partial}{\partial y}\theta(\alpha; x, y) \geq 0$$

for $x^2 + y^2 = 1$, $0 \leq x \leq 1/2$, where equality holds only for the points $(0, 1)$, $(1/2, \sqrt{3}/2)$. This is done by reparametrization to polar coordinates, differentiating, and then using Lemma 3.

5. Application to physics

We now describe the part of the paper by Sandier and Serfaty related to the Epstein zeta function in more detail. As described in the introduction, they consider a configuration of charged particles in the plane. Since the Coulombian energy of the electric field generated by a charged particle is infinite, one has to introduce a "renormalized" energy in order to make use of variational methods. Sandier and Serfaty define such a renormalized energy which is described by a vector field. For a configuration given by a lattice Λ, they derive in several steps the expression

$$W(\Lambda) = \frac{1}{2} \lim_{x \to 0} \left(\sum_{p \in \Lambda^* \setminus \{0\}} \frac{e^{2\pi i \langle p, x \rangle}}{4\pi^2 \|p\|^2} + \log \|x\| \right)$$

(cf. [23, Lemma 3.2]) for this renormalized energy W. Now they restrict to lattices Λ whose fundamental parallelogram has volume 2π; the set of these lattices is denoted by $\mathcal{L}$. After rotation, one can assume that a basis of the dual lattice Λ^* is given by vectors of the form

$$\frac{1}{\sqrt{2\pi b}}(1, 0) \text{ and } \frac{1}{\sqrt{2\pi b}}(a, b)$$

with $b > 0$. Then they show

Lemma 7. *(Lemma 3.3 of [23]) Let $\Lambda \in \mathcal{L}$. Then*

$$W(\Lambda) = -\frac{1}{2} \log(\sqrt{2\pi b} |\eta(\tau)|^2)$$

with $\tau = a + ib$ and the Dedekind eta function η.

To demonstrate this, they rewrite for $x = (x_1, x_2)$

$$\frac{1}{2} \sum_{p \in \Lambda^* \setminus \{0\}} \frac{e^{2\pi i \langle p, x \rangle}}{4\pi^2 \|p\|^2} = \frac{1}{2\pi} \sum_{(m,n) \in \mathbb{Z}^2 \setminus \{0\}} e^{\frac{2\pi i}{\sqrt{2\pi b}}(m(ax_1 + bx_2) + nx_1)} \frac{b}{|m\tau + n|^2}.$$

Applying the second Kronecker limit formula (cf. [16, §20.5]) to this series and letting $x \to 0$ yield the result.

This leads to

Lemma 8. *(Lemma 3.4 of [23]) There exists $C \in \mathbb{R}$ such that for any lattice $\Lambda \in \mathcal{L}$, we have*

$$W(\Lambda) = C + \lim_{x \to 0} \left(\frac{1}{8\pi^2} \zeta(\Lambda^*, 1 + x/2) - \int_{\mathbb{R}^2} \frac{\pi}{1 + 4\pi^2 |y|^{2+x}} dy \right).$$

Sandier and Serfaty give an analytic proof for this result, but they note that it can also be proved in the following way: One has

$$(2\pi)^s \zeta(\Lambda^*, s) = \sum_{(m,n) \in \mathbb{Z}^2 \setminus \{0\}} \frac{b^s}{|m\tau + n|^{2s}}$$

for $\operatorname{Re}(s) > 1$. The first Kronecker limit formula (cf. [19, Satz V.5.5]) states

$$(2\pi)^s \zeta(\Lambda^*, s) = \frac{\pi}{s-1} + 2\pi(\gamma_0 - \log 2 - \log(\sqrt{b}|\eta(\tau)|^2)) + O(s-1)$$

as $s \to 1$ with the Euler-Mascheroni constant γ_0. This can be combined with Lemma 7 to prove Lemma 8.

Now Sandier and Serfaty use the results described in our previous section to deduce that the rescaled hexagonal lattice minimizes this expression. In the following, their argumentation is slightly modified to correct a minor error (which is of no importance with regard to the desired result). The discriminant of $\sqrt{\pi}\Lambda^*$ equals $1/4$. Let f denote a quadratic form (of discriminant -1) corresponding to this rescaled lattice. By Λ_0 we denote the rescaled hexagonal lattice with basis vectors

$$\sqrt{\frac{4\pi}{\sqrt{3}}}(1,0), \quad \sqrt{\frac{4\pi}{\sqrt{3}}}(1/2, \sqrt{3}/2),$$

so $\Lambda_0 \in \mathcal{L}$. As above, we denote the quadratic form corresponding to $\sqrt{\pi}\Lambda_0^*$ by h. Then one has

$$\zeta(\Lambda^*, s) = \pi^s \zeta(\sqrt{\pi}\Lambda^*, s)$$

for $\operatorname{Re}(s) > 1$ and thus, using (3),

$$\frac{\Gamma(1+x/2)}{(2\pi)^{1+x/2}} \zeta(\Lambda^*, 1 + x/2)$$

$$= \frac{\Gamma(1+x/2)}{(2\pi)^{1+x/2}} \pi^{1+x/2} \zeta(\sqrt{\pi}\Lambda^*, 1 + x/2)$$

$$= \pi^{1+x/2} \left(\frac{2}{x} - \frac{1}{1+\frac{x}{2}} + \int_1^\infty (\theta_f(\alpha) - 1)(\alpha^{1+x/2} + \alpha^{-x/2}) \frac{d\alpha}{\alpha} \right)$$

for $x > 0$. With dominated convergence, we get

$$4(W(\Lambda) - W(\Lambda_0)) = \int_1^\infty ((\theta_f(\alpha) - \theta_h(\alpha))(\alpha + 1) \frac{d\alpha}{\alpha}.$$

(differing slightly from the corresponding equation in [23]). Then Theorem 2 implies:

Theorem 3. *(Theorem 2 of [23]) Let $\Lambda \in \mathcal{L}$. Then we have $W(\Lambda) \geq W(\Lambda_0)$, where equality holds if and only if Λ is isometric under rotation to Λ_0.*

6. Further developments

In higher dimensions, corresponding results on global minima of the Epstein zeta function have not yet been found. Since the densest sphere packing in three dimensions is given by the face centered cubic lattice, Ennola conjectured in [12] that the Epstein zeta function of this lattice (rescaled to discriminant 1) had the corresponding minimality property. He could show that the zeta function has a local minimum at this lattice for all $s > 0$. P. Sarnak and A. Strömbergsson showed in 2006 in [25] that Ennola's conjecture is false, using (2) and the fact that the dual of the face centered cubic lattice is the body centered cubic lattice. However, numerical evidence suggests that the Epstein zeta function might have a global minimum at the face centered cubic lattice for all $s \geq 3/4$. As to higher dimensions, they show that the Epstein zeta function has local minima for $s > 0$ at the (rescaled) lattices D_4, E_8 and the Leech lattice (giving the densest lattice sphere packings in the respective dimensions) in dimensions 4, 8, and 24, respectively, and they conjecture that these are global minima. Furthermore, they prove that the elliptic theta series has local minima for all iy, $y > 0$, at these lattices and conjecture that these minima are global. They also prove corresponding results on heights of flat tori $\mathbb{R}^n/\Lambda$, given by

$$h(\mathbb{R}^n/\Lambda) = 2 \log 2\pi + \frac{\partial}{\partial s}\zeta(\Lambda^*, s)|_{s=0}.$$

In [7], R. Coulangeon explained Sarnak and Strömbergsson's results in terms of spherical designs: If each layer of an n-dimensional lattice Λ of discriminant 1 is a 4-design, then the Epstein zeta function has a strict local minimum at Λ for $s > n/2$; if moreover $\zeta(\Lambda, s) < 0$ for $0 < s < n/2$, this holds for all $s > 0$; there are analogous results on the elliptic theta series. The connection between spherical designs in lattices and extremality of heights is pursued further in the paper [8].

A connection between local minima of the Epstein zeta function and spherical designs had previously been established by a result of B. N. Delone and S. S. Ryshkov from 1967 (cf. [9]). They showed (in a different terminology) that $\zeta(\Lambda, s)$ is minimal for all sufficiently large $s > 0$ if and only if Λ is perfect and each layer of Λ is eutactic or, more precisely, if each layer is a spherical 2-design.

P. Gruber characterized stationary points of the Epstein zeta function in [15]. In particular, he showed that an n-dimensional lattice is a stationary point for all $s > n$ if and only if each layer is strongly eutactic. In dimension 2, this is the case for the hexagonal and the square lattice, in dimension 3 for the primitive cubic, the face centered cubic, and the body centered cubic lattice; in higher dimensions, the number of such lattices is still known to be finite.

Another approach is due to H. Cohn and A. Kumar: In [5], they formulate a conjecture on the minimization of certain potential energies of periodic point configurations in $\mathbb{R}^2$, $\mathbb{R}^8$, and $\mathbb{R}^{24}$ by the lattices A_2, E_8 and the Leech lattice. This conjecture implies that these lattices are *universally optimal* configurations in the respective dimensions, i.e. they minimize the potential energy among all periodic configurations of points for each completely monotonic potential function. A consequence of this would be that the elliptic theta function (and thus the Epstein zeta function) has a global minimum at the respective lattices, since this is equivalent to universal optimality among lattice configurations. The conjecture also implies that the sphere packings given by E_8 and the Leech lattice are the densest sphere packings in dimension 8 and 24, respectively.

The problem of the minimization of the renormalized energy W from section 5 has been related to the statistical mechanics of Coulomb gases in [24] and [1]. Moreover, L. Bétermin and E. Sandier establish a connection between the minimality of the triangular lattice for the renormalized energy and the value of the n-th order term in the asymptotic expansion of the minimal logarithmic energy of n points on the unit sphere in [1]. They compute the exact value of the renormalized energy of the hexagonal lattice, using the Chowla-Selberg formula; furthermore, they give the following alternative proof of Theorem 3, applying previous results by B. Osgood, R. Phillips, and P. Sarnak: It is shown in [21, Section 4] that for a lattice Λ with Λ^* as in Lemma 7, the height of the corresponding flat torus satisfies

$$h(\mathbb{R}^2/\Lambda) = -\log(b|\eta(\tau)|^4) + \log 2\pi.$$

By [21, Cor. 1(b)], this height is minimized by the (rescaled) hexagonal lattice. Then the theorem follows from the expression for the renormalized energy in Lemma 7. P. Zhang used the method of Sandier and Serfaty in [27] to show optimality of the hexagonal lattice for a different renormalized energy, also related to the Epstein zeta function. Bétermin and Zhang prove a minimality property of the hexagonal lattice for Lennard-Jones and Thomas-Fermi potentials in [2].

For further applications of the Epstein zeta function to physics and more results concerning its (local) minima, we refer the reader to the paper [17] and its extensive bibliography; we also want to mention the related recent survey [3] on the crystallization conjecture.

Acknowledgements

I wish to thank Professors Christof Melcher and Aloys Krieg for their interest in the subject matter, out of which the talk on which this article is based has grown, and Professor Florian Rupp, who attended that talk, for inviting me to write this article. Furthermore, I wish to thank the referees as well as Professors Sylvia Serfaty and Etienne Sandier for bringing recent developments to my attention.

References

[1] Bétermin, L., Sandier, E.: Renormalized energy and asymptotic expansion of optimal logarithmic energy on the sphere. arXiv:1404.4485 [math.AP]

[2] Bétermin, L., Zhang, P.: Minimization of energy per particle among Bravais lattices in $\mathbb{R}^2$: Lennard-Jones and Thomas-Fermi cases. *Communications in Contemporary Mathematics* **17**:6 (2015)

[3] Blanc, X., Lewin, M.: The crystallization conjecture – a review. arXiv:1504.01153 [cond-mat.stat-mech]

[4] Cassels, J. W. S.: On a problem of Rankin about the Epstein zeta function. *Proc. Glasgow Math. Assoc.* **4** (1956), 73-80; **6** (1963), 116

[5] Cohn, H., Kumar, A.: Universally optimal distribution of points on spheres. *J. Am. Math. Soc.* **20** (2007), 99-148,

[6] Conway, J. H., Sloane, N. J. A: *Sphere Packings, Lattices and Groups.* Springer-Verlag, New York, third edition 1999

[7] Coulangeon, R.: Spherical designs and zeta functions of lattices. *Int. Math. Res. Not.* 2006, Art. ID 49620

[8] Coulangeon, R., Lazzarini, G.: Spherical designs and heights of euclidean lattices. *Journal of Number Theory* **141** (2014), 288-315

[9] Delone, B. N., Ryshkov, S. S.: A contribution to the theory of the extrema of a multidimensional ζ-function. *Dokl. Akad. Nauk. SSSR* **173** (1967), 991-994. English version in *Sov. Math. Dokl.* **8** (1967), 499-503

[10] Diananda, P. H.: Notes on two lemmas concerning the Epstein zeta function. *Proc. Glasgow Math. Assoc.* **6** (1964), 202-204

[11] Ennola, V.: A lemma about the Epstein zeta function. *Proc. Glasgow Math. Assoc.* **6** (1964), 198-201

[12] Ennola, V.: On a problem about the Epstein zeta function. *Proc. Cambridge Philos. Soc.* **60** (1964), 855-875

[13] Epstein, P.: Zur Theorie allgemeiner Zetafunctionen I, II. *Math. Ann.* **56** (1903), 615-644; **63** (1907), 205-216

[14] Fejes Tóth, L.: *Lagerungen in der Ebene, auf der Kugel und im Raum.* Springer-Verlag, Berlin, second edition 1972

[15] Gruber, P.: Application of an idea of Voronoi to lattice zeta functions. *Trudy Mat. Sbornik, Proc. Steklov Inst. Math.* **276** (2012), 103-124

[16] Lang, Serge: *Elliptic Functions.* Springer-Verlag, second edition 1987

[17] Lim, S. C., Teo, L. P.: On the minima and convexity of the Epstein zeta function. *J. Math. Phys.* **49**, 073513 (2008)

[18] Kendall, D. G., Rankin, R. A.: On the number of points of a given lattice in a random hypersphere. *Quart. J. Math. Oxford* (2) **4** (1953), 178-189

[19] Koecher, M., Krieg, A.: *Elliptische Funktionen und Modulformen.* Springer-Verlag, Berlin, second edition 2007

[20] Montgomery, H. L.: Minimal theta functions. *Glasgow Math. J.* **30** (1988), 75-85

[21] Osgood, B., Phillips, R., Sarnak, P.: *Extremals of determinants of Laplacians.* J. Functional Analysis **80** (1988), 148-211

[22] Rankin, R. A.: A minimum problem for the Epstein zeta function. *Proc. Glasgow Math. Assoc.* **1** (1953), 148-211

[23] Sandier, E., Serfaty, S.: From the Ginzburg-Landau model to vortex lattice problems. *Comm. Math. Phys.* **313** (2012), 635-743

[24] Sandier, E., Serfaty, S.: 2D Coulomb gases and the renormalized energy. *Ann. Prob.* **43** (2015), 2026-2083

[25] Sarnak, P., Strömbergsson, A.: Minima of Epstein's zeta function and heights of flat tori. *Invent. Math.* **165** (2006), 115-151

[26] Zagier, D. B.: *Zetafunktionen und quadratische Körper.* Springer-Verlag, Berlin, 1981

[27] Zhang, P.: On the minimizer of a renormalized energy related to the Ginzburg-Landau model. *C. R. Acad. Sci. Paris* Ser. I **353** (2015), 255-260

Chapter 8

On Putative q-Analogues of the Fano Plane and Related Combinatorial Structures

Thomas Honold & Michael Kiermaier

*Department of Information Science and Electronics Engineering, Zhejiang
University, 38 Zheda Road, 310027 Hangzhou, China*
E-mail: honold@zju.edu.cn

Mathematisches Institut, Universität Bayreuth, D-95440 Bayreuth, Germany
E-mail: michael.kiermaier@uni-bayreuth.de

A set $\mathcal{F}_q$ of 3-dimensional subspaces of $\mathbb{F}_q^7$, the 7-dimensional vector space over
the finite field $\mathbb{F}_q$, is said to form a q-analogue of the Fano plane if every 2-
dimensional subspace of $\mathbb{F}_q^7$ is contained in precisely one member of $\mathcal{F}_q$. The
existence problem for such q-analogues remains unsolved for every single value of
q. Here we report on an attempt to construct such q-analogues using ideas from
the theory of subspace codes, which were introduced a few years ago by Koetter
and Kschischang in their seminal work on error-correction for network coding.
Our attempt eventually fails, but it produces the largest subspace codes known
so far with the same parameters as a putative q-analogue. In particular we find
a ternary subspace code of new record size 6977, and we are able to construct
a binary subspace code of the largest currently known size 329 in an entirely
computer-free manner.

Herrn Professor Armin Leutbecher zum 80. Geburtstag

Contents

1. Introduction

The Fano plane $\mathcal{F} = \mathrm{PG}(2, \mathbb{F}_2) = \mathrm{PG}(\mathbb{F}_2^3/\mathbb{F}_2)$, the coordinate geometry derived
from a 3-dimensional vector space over the binary field $\mathbb{F}_2$, is the smallest nontrivial

model of an abstract projective geometry. It has 7 points and 7 lines, represented by the one- and two-dimensional subspaces of $\mathbb{F}_2^3/\mathbb{F}_2$, respectively; each line contains 3 points and each point is on 3 lines; any two distinct points are contained in a unique line and any two distinct lines intersect in a unique point. Myriads of other finite models of a projective geometry exist—for each integer $n \geq 2$ and prime power $q > 1$ the n-dimensional coordinate geometry $\mathrm{PG}(n, \mathbb{F}_q) = \mathrm{PG}(\mathbb{F}_q^{n+1}/\mathbb{F}_q)$ over the finite field $\mathbb{F}_q$, and in the planar case many additional examples with the same parameters as some $\mathrm{PG}(2, \mathbb{F}_q)$.

The Fano plane $\mathcal{F} = \mathrm{S}(2, 3, 7)$ is also the smallest nontrivial example of a *Steiner system* $\mathrm{S}(t, k, v)$, which refers to a v-set V (*point set*) and a set of k-subsets of V (*blocks*) having the property that any t-subset of V is contained in exactly one block. The more general concept of a combinatorial t-(v, k, λ) design relaxes the requirement "exactly one block" to a "constant number λ of blocks". Many constructions of t-(v, k, λ) designs are known (including the construction of nontrivial t-designs for all positive integers t by Teirlinck [1]), but comparatively few Steiner systems and still no one at all with $t > 5$).[1]

This article is concerned with vector space analogues of $\mathcal{F}$ in the following sense:

Definition 1. Let $q > 1$ be a prime power. A set $\mathcal{F}_q$ of 3-dimensional subspaces of $\mathbb{F}_q^7/\mathbb{F}_q$ (or any other 7-dimensional vector space V over $\mathbb{F}_q$) is said to be a *q-analogue of the Fano plane* if every 2-dimensional subspace of $\mathbb{F}_q^7$ (respectively, V) is contained in a unique member of $\mathcal{F}_q$.

In projective geometry language, a q-analogue of the Fano plane is a set $\mathcal{F}_q$ of planes in $\mathrm{PG}(6, \mathbb{F}_q)$ such that any pair of distinct points (equivalently, any line) is contained in exactly one plane $E \in \mathcal{F}_q$. In other words, the planes in $\mathcal{F}_q$, when identified with sets of lines, should form an exact cover (i.e., a partition) of the line set of $\mathrm{PG}(6, \mathbb{F}_q)$.

Before going any further, we should remark that at the time of writing this article virtually nothing is known about the existence of such structures—neither existence nor non-existence of a q-analogue of the Fano plane has been proved for a single instance of q. Even in the smallest case $q = 2$, where a putative 2-analogue $\mathcal{F}_2$ would have to contain 381 of the 11811 planes of $\mathrm{PG}(6, \mathbb{F}_2)$, a computer search seems infeasible at present.

P. Cameron [3] introduced the concept of a *design over a finite field* as a vector space analogue ("*q-analogue*", if the underlying field is $\mathbb{F}_q$) of combinatorial designs: A t-(v, k, λ) design over $\mathbb{F}_q$ is a set $\mathcal{C}$ of k-dimensional subspaces of $\mathbb{F}_q^v/\mathbb{F}_q$ (or any other v-dimensional vector space V over $\mathbb{F}_q$) with the property that every t-dimensional subspace of $\mathbb{F}_q^v$ (respectively, V) is contained in exactly λ members of $\mathcal{C}$. The first nontrivial examples of such designs were constructed by S. Thomas [4]. These "Thomas designs" have $q = 2$ and form an infinite family with parameters

---

[1] We should note here that recently Keevash [2] has given a non-constructive proof of the existence of Steiner systems for all values of t.

2-$(v, 3, 7)$, where $v \equiv \pm 1 \pmod 6$ and $v \geq 7$. Taking the ambient space as the finite field $\mathbb{F}_{2^v}$, one may construct the 2-$(v, 3, 7)$ Thomas design $\mathcal{T}_v$ as the set of all 3-dimensional $\mathbb{F}_2$-subspaces $\langle x, y, z \rangle \subset \mathbb{F}_{2^v}$ spanned by the $2^v - 2$ non-rational points in $\mathrm{PG}(2, \mathbb{F}_{2^v})$ of a rational conic (relative to $\mathbb{F}_2$). For example, we can take all points $(x : y : z) \neq (1 : 0 : 0), (0 : 1 : 0), (0 : 0 : 1)$ on the conic $xy + yz + zx = 0$, resulting in $\mathcal{T}_v = \left\{ \langle x, y, \frac{xy}{x+y} \rangle; x, y \in \mathbb{F}_{2^v}^\times \text{ distinct} \right\}$.[2] Although several further constructions of designs over finite fields are now known (including the existence of nontrivial t-designs over $\mathbb{F}_q$ for arbitrarily large t in [5]), the subject has turned out considerably more difficult than ordinary combinatorial design theory. For example, no example of a nontrivial 4-design over a finite field is known at present.

At the end of [4] Thomas briefly discussed q-analogues $\mathrm{S}_q(t, k, v)$ of Steiner systems (i.e. t-$(v, k, 1)$ designs over $\mathbb{F}_q$) and in particular the smallest feasible parameter case $\mathrm{S}_2(2, 3, 7)$. Such a 2-analogue of the Fano plane would consist of $381 = 3 \times 127$ three-dimensional subspaces of $\mathbb{F}_2^7$ (cf. Lemma 2), and it was conceivable to construct it as the union of 3 orbits of a Singer subgroup of $\mathrm{GL}(7, \mathbb{F}_2)$. However, as Thomas reported, this construction is impossible.

A few years ago interest in designs over finite fields was revived through the observation by R. Koetter and F. Kschischang [6] that sets of subspaces of a vector space over a finite field (*subspace codes*) can be used as "distributed channel codes" for error-resilient transmission of information in packet networks. Considering q (*symbol alphabet of the packet network*) and the ambient vector space dimension v (*packet length*) as fixed and restricting attention to *constant-dimension* codes (i.e the dimension k of all codewords is the same), the best performance is achieved by using subspace codes $\mathcal{C}$ that have simultaneously large size $\#\mathcal{C} = |\mathcal{C}|$ and small maximum dimension of an intersection between distinct codewords. Denoting this dimension by $t - 1$, we have that t is the smallest positive integer such that every t-dimensional subspace of $\mathbb{F}_q^v$ is contained in at most one codeword of $\mathcal{C}$. Subspace codes thus satisfy a weaker form of the defining condition for Steiner systems over finite fields.[3] A standard double-counting argument gives $\#\mathcal{C} \times \begin{bmatrix} k \\ t \end{bmatrix}_q \leq \begin{bmatrix} v \\ t \end{bmatrix}_q$ with equality if and only if each t-dimensional subspace of $\mathbb{F}_q^v$ is contained in precisely one codeword of $\mathcal{C}$. Hence Steiner systems over finite fields are optimal as subspace codes.[4]

---

[2] Checking the design property is somewhat tedious, but at least we can see immediately from the definition that $\mathcal{T}_v$ has the required $(2^v - 2)/6 \times (2^v - 1) = (2^v - 1)(2^{v-1} - 1)/3$ blocks.

[3] The difference is quite similar to that between *linear spaces* (two distinct points are connected by exactly one line) and *partial linear spaces* (two distinct points are connected by at most one line), as defined in Incidence Geometry. Subspace codes could thus be called "partial Steiner systems over finite fields".

[4] From this we also see that the parameters q, t, k, v of an $\mathrm{S}_q(t, k, v)$, like those of an ordinary Steiner system, must obey certain *integrality conditions*. In fact the existence of an $\mathrm{S}_q(t, k, v)$ implies the existence of an $\mathrm{S}_q(t - 1, k - 1, v - 1)$. The so-called *derived designs* [32], which are formed by the blocks through a fixed 1-dimensional subspace of $\mathbb{F}_q^v$, have these parameters. Hence a necessary condition for the existence of an $\mathrm{S}_q(t, k, v)$ is that $\begin{bmatrix} v-s \\ t-s \end{bmatrix}_q / \begin{bmatrix} k-s \\ t-s \end{bmatrix}_q$ must be an integer for $1 \leq s \leq t$.

In the sequel we will exclusively be concerned with subspace codes of constant dimension $k = 3$, so-called *plane subspace codes*, and packet length $v = 7$. Plane subspace codes with $t = 3$ are trivial—the whole plane set of $PG(6, \mathbb{F}_q)$ forms such a code. Plane subspace codes with $t = 1$ consist of pairwise skew planes and are known as *partial plane spreads* in Finite Geometry. The maximum size of a partial plane spread in $PG(6, \mathbb{F}_q)$ is known to be $q^4 + 1$ from the work of Beutelspacher [7, Th. 4.1].[5] This leaves the case $t = 2$ considered so far as the only unresolved case. Restricting attention to this case, we will from now on tacitly assume that "subspace code" includes the assumption $t = 2$.

More than 25 years have passed since Thomas' fundamental work [4] and the existence problem for q-analogues of the Fano plane is still undecided. On the other hand, serious attempts, often relying on quite sophisticated computational methods, have been made to construct large subspace codes—including the parameter set of a putative 2-analogue. These will now be briefly reviewed. Accordingly, let $\mathcal{C}$ be a binary plane subspace code with $v = 7$ or, in geometric terms, a set of planes in $PG(6, \mathbb{F}_2)$ mutually intersecting in at most a point. As discussed above, we have $\#\mathcal{C} \leq 381$ with equality if and only if $\mathcal{C}$ is a 2-analogue of the Fano plane. The first nontrivial lower bound on the maximum size of $\mathcal{C}$ was established by Koetter and Kschischang [6], who showed that $\#\mathcal{C} = 256$ is realized by a so-called lifted maximum-rank distance code (LMRD code). Kohnert and Kurz [9] improved this to $\#\mathcal{C} = 304$, employing a computer search for plane subspace codes in $PG(6, \mathbb{F}_2)$ with an automorphism of order 21 acting irreducibly on a hyperplane. The current record is $\#\mathcal{C} = 329$ and was established by Braun and Reichelt in [10] using a refinement of this method. In [11], as part of the classification of all optimal plane subspace codes in the smaller geometry $PG(5, \mathbb{F}_2)$, an optimal $\#\mathcal{C} = 77$ subspace code was constructed by first expurgating an LMRD code (size 64) to a particular subspace code of size 56 and then augmenting this code by 21 further planes. As shown in [12], the underlying idea can be used to provide an alternative construction of a plane subspace code of size 329 in $PG(6, \mathbb{F}_2)$.

In this paper we will develop a general framework for constructing large plane subspace codes in $PG(6, \mathbb{F}_q)$ along the lines of [11, 12], but also introduce several new ideas (in Sections 4 and 5). Our main results are the construction of a general q-ary subspace code $\mathcal{C}$ of size $q^8 + q^5 + q^4 - q - 1$, whose planes meet a fixed solid (3-flat) of $PG(6, \mathbb{F}_q)$ in at most a point (Theorem 3 in Section 5), and a detailed analysis of the extension problem for $\mathcal{C}$ (or rather, a distinguished subcode $\mathcal{C}_0 \subset \mathcal{C}$) by planes meeting S in a line, which enables us to give the first computer-free construction of a plane subspace code of size 329 in $PG(6, \mathbb{F}_2)$ (see above) and a computer-aided construction of a plane subspace code of size 6977 in $PG(6, \mathbb{F}_3)$ (Section 6, in particular Theorem 4). Theorem 3 improves the best previously known construction for general q [13], and the ternary subspace code of size 6977

---

[5]In general, the maximum size of a partial plane spread in $PG(v-1, \mathbb{F}_q)$ is known for $v \equiv 0, 1 \bmod 3$ (all q) and for $q = 2$ (all v); for the latter see [8].

is by way the largest known code with its parameters. In order to make the paper self-contained, we provide a general introduction to the combinatorics of subspace codes in Section 2 and an account of related previous subspace code constructions in Section 3.

In the sequel $\mathcal{F}_q$ always denotes a putative q-analogue of the Fano plane. The term "dimension" refers to vector space dimension, but otherwise geometric language will be extensively used. When referring to the geometric dimension of a t-dimensional subspace of $\mathbb{F}_q^v/\mathbb{F}_q$, we use the term "$(t-1)$-flat of $\mathrm{PG}(v-1, \mathbb{F}_q)$".

Let us close this introduction with a remark on vector space analogues of the Fano plane over infinite fields. Using transfinite recursion, it is fairly easy to show that for any field K with $|K| = \infty$ a K-analogue $\mathcal{F}_K$, defined as in Def. 1, does exist. For example, in the case $K = \mathbb{Q}$ we can enumerate the lines of $\mathrm{PG}(6, \mathbb{Q})$ as $L_0, L_1, L_2, \ldots$ and recursively define sets $\mathcal{E}_0 = \emptyset$, $\mathcal{E}_1, \mathcal{E}_2, \ldots$ of planes as follows: If $\mathcal{E}_n$ already contains a plane $E \supset L_n$, we set $\mathcal{E}_{n+1} = \mathcal{E}_n$; otherwise, among the planes containing L_n there exists a plane E that has no line in common with any of the planes in $\mathcal{E}_n$, and we set $\mathcal{E}_{n+1} = \mathcal{E}_n \cup \{E\}$.[6] It is then readily verified that $\mathcal{F}_{\mathbb{Q}} = \bigcup_{n=0}^{\infty} \mathcal{E}_n$ is the required $\mathbb{Q}$-analogue of $\mathcal{F}$.

In fact it is even true that the plane set of any geometry $\mathrm{PG}(v-1, K)$, $|K| = \infty$, $v \geq 5$, can be partitioned into Steiner systems $\mathrm{S}_K(2, 3, v)$; see [14] for details.

2. Counting Preliminaries

Let us first recall that the number of k-dimensional subspaces of an n-dimensional vector space over $\mathbb{F}_q$ is given by the Gaussian binomial coefficient

$$\begin{bmatrix} n \\ k \end{bmatrix}_q = \frac{(q^n - 1)(q^{n-1} - 1) \cdots (q^{n-k+1} - 1)}{(q^k - 1)(q^{k-1} - 1) \cdots (q - 1)},$$

which is polynomial in q of degree $k(n-k)$ and satisfies $\begin{bmatrix} n \\ k \end{bmatrix}_q = \begin{bmatrix} n \\ n-k \end{bmatrix}_q = q^{k(n-k)} \cdot \begin{bmatrix} n \\ k \end{bmatrix}_{q^{-1}}$. In particular the number of points (and hyperplanes) of $\mathrm{PG}(n-1, \mathbb{F}_q)$ is equal to

$$\begin{bmatrix} n \\ 1 \end{bmatrix}_q = \begin{bmatrix} n \\ n-1 \end{bmatrix}_q = \frac{q^n - 1}{q - 1} = 1 + q + \cdots + q^{n-1}.$$

Subspaces U of $\mathbb{F}_q^n/\mathbb{F}_q$ of dimension k are in one-to-one correspondence with matrices $\mathbf{U} = \mathrm{cm}(U) \in \mathbb{F}_q^{k \times n}$ in reduced row-echelon form via $U = \langle \mathrm{cm}(U) \rangle$, the row space of the matrix $\mathrm{cm}(U)$, and $\mathbf{U} = \mathrm{cm}(\langle \mathbf{U} \rangle)$.[7] If $\mathrm{cm}(U)$ has pivot columns in positions $1 \leq j_1 < j_2 < \cdots < j_k \leq n$ then the number of unspecified entries ("wildcards") in $\mathrm{cm}(U)$ is $i = 1(j_2 - j_1 - 1) + 2(j_3 - j_2 - 1) + \cdots + (k-1)(j_k - j_{k-1} - 1) + k(n - j_k)$

---

[6]More precisely, the first plane with this property, according to some predefined order $E_0, E_1, E_2, \ldots$ on the set of planes of $\mathrm{PG}(6, \mathbb{Q})$, is chosen. The existence of such a plane follows from the fact that L_n and the finitely many solids (3-flats) $E' + L_n$, $E' \in \mathcal{E}_n$ a plane intersecting L_n (necessarily in a point), cannot cover all points of $\mathrm{PG}(6, \mathbb{Q})$.

[7]The name 'cm' resembles "canonical matrix".

and determines a partition of the integer i into at most $n - k$ parts of size at most k.[8] The coefficient a_i of q^i in $\begin{bmatrix} n \\ k \end{bmatrix}_q$ counts the number of such partitions, and consequently the monomial $a_i q^i$ counts the k-dimensional subspaces of $\mathbb{F}_q^n$ having exactly i unspecified entries in their canonical matrix.

These and a few additional observations allow for "almost everything" in $\mathrm{PG}(n - 1, \mathbb{F}_q) = \mathrm{PG}(\mathbb{F}_q^n / \mathbb{F}_q)$ to be counted. Consider, for example, any solid (3-flat) S in $\mathrm{PG}(6, \mathbb{F}_q)$ and count the planes of $\mathrm{PG}(6, \mathbb{F}_q)$ according to their intersection size with S. There are q^{12} planes disjoint from S, corresponding to the q^{12} canonical matrices

$$\begin{pmatrix} 1 & 0 & 0 & * & * & * & * \\ 0 & 1 & 0 & * & * & * & * \\ 0 & 0 & 1 & * & * & * & * \end{pmatrix}$$

(for this arrange coordinates such that $S = (0, 0, 0, *, *, *, *)$); there are $q^6 \cdot \begin{bmatrix} 3 \\ 2 \end{bmatrix}_q \begin{bmatrix} 4 \\ 1 \end{bmatrix}_q = q^6(q^2 + q + 1)(q^3 + q^2 + q + 1)$ planes E meeting S in a point (considering the hyperplane $H = E + S$ and the intersection point $P = E \cap S$ as fixed, these correspond to lines disjoint from the plane S/P in $H/P \cong \mathrm{PG}(4, \mathbb{F}_q)$, of which there are q^6 corresponding to the canonical matrix shape $\begin{pmatrix} 1 & 0 & * & * & * \\ 0 & 1 & * & * & * \end{pmatrix}$); there are $q^2 \cdot \begin{bmatrix} 3 \\ 1 \end{bmatrix}_q \begin{bmatrix} 4 \\ 2 \end{bmatrix}_q = q^2(q^2 + q + 1)(q^4 + q^3 + 2q^2 + q + 1)$ planes E meeting S in a line (considering the 4-flat $T = E + S$ and the line $L = E \cap S$ as fixed, these correspond to points outside the line S/L in $T/L \cong \mathrm{PG}(2, \mathbb{F}_q)$);[9] and finally, there are $\begin{bmatrix} 4 \\ 3 \end{bmatrix}_q = \begin{bmatrix} 4 \\ 1 \end{bmatrix}_q = q^3 + q^2 + q + 1$ planes contained in S.

Now let $\mathcal{C}$ be a set of planes in $\mathrm{PG}(6, \mathbb{F}_q)$ mutually intersecting in at most a point (a plane subspace code in the terminology of Section 1). Fixing any solid S in $\mathrm{PG}(6, \mathbb{F}_q)$, we can count how many planes in $\mathcal{C}$ intersect S in a subspace of dimension $i \in \{0, 1, 2, 3\}$. This leads to the concept of "spectra" (or "intersection vectors") with respect to solids, which already capture a great deal of structural information about $\mathcal{C}$.

Definition 2. The *spectrum* (or *intersection vector*) of $\mathcal{C}$ with respect to S is defined as the 4-tuple $\alpha(S) = (\alpha_0(S), \alpha_1(S), \alpha_2(S), \alpha_3(S))$, $\alpha_i(S) = \#\{E \in \mathcal{C}; \dim(E \cap S) = i\}$, of non-negative integers.

The example counting problem discussed above amounts to determining the spectrum of the whole plane set of $\mathrm{PG}(6, \mathbb{F}_q)$ with respect to any solid, which turned out to be a constant independent of S.[10]

Lemma 1. *Let $\mathcal{C}$ be a plane subspace code of size M in $\mathrm{PG}(6, \mathbb{F}_q)$ and S any solid in $\mathrm{PG}(6, \mathbb{F}_q)$. The spectrum $\alpha = \alpha(S)$ of $\mathcal{C}$ with respect to S satisfies $\alpha_0 + \alpha_1 + \alpha_2 + \alpha_3 =$*

---

[8]The number of (positive) parts is $\sum_{\nu^i=1}^{k-1}(j_{\nu^i+1} - j_{\nu^i} - 1) + n - j_k = n - k - (j_1 - 1)$.

[9]Here we have used $\begin{bmatrix} 4 \\ 2 \end{bmatrix}_q = q^4 + q^3 + 2q^2 + q + 1$, which follows from counting the partitions into at most 2 parts of size ≤ 2 according to their sum: $0 = 0$, $1 = 1$, $2 = 2 = 1 + 1$, $3 = 2 + 1$, $4 = 2 + 2$.

[10]The latter also follows from the observation that $\mathrm{GL}(7, \mathbb{F}_q)$ acts transitively on the set of all plane-solid pairs (E, S) with fixed intersection dimension i.

M and the following system of linear inequalities:

$$\begin{aligned}
\begin{bmatrix}3\\1\end{bmatrix}_q \cdot \alpha_0 + \quad q^2\alpha_1 \quad &\leq q^8 \cdot \begin{bmatrix}3\\1\end{bmatrix}_q \\
(q+1)\alpha_1 + (q^2+q)\alpha_2 \quad &\leq q^3 \cdot \begin{bmatrix}3\\1\end{bmatrix}_q \cdot \begin{bmatrix}4\\1\end{bmatrix}_q \\
\alpha_2 + \begin{bmatrix}3\\1\end{bmatrix}_q \cdot \alpha_3 &\leq \begin{bmatrix}4\\2\end{bmatrix}_q \\
\alpha_3 &\leq 1
\end{aligned}$$

The explicit form of all four inequalities is obtained by inserting $\begin{bmatrix}3\\1\end{bmatrix}_q = q^2 + q + 1$, $\begin{bmatrix}4\\1\end{bmatrix}_q = q^3 + q^2 + q + 1$ and $\begin{bmatrix}4\\2\end{bmatrix}_q = q^4 + q^3 + 2q^2 + q + 1 = (q^2+1)(q^2+q+1)$.

Proof. The equation $\alpha_0 + \alpha_1 + \alpha_2 + \alpha_3 = M$ is clear from the definition of the spectrum. The first three inequalities are proved by counting the line-plane pairs (L, E) with $E \in \mathcal{C}$, $L \subset E$ and $\dim(L \cap S) = i$ for $i = 0, 1, 2$, respectively, in two ways and using the fact that every line is contained in at most one plane of $\mathcal{C}$ (and hence counted at most once on the left-hand side). The right-hand side of the corresponding inequality gives the total number of lines L with $\dim(L \cap S) = i$. Finally, since two distinct planes of $\mathcal{C}$ generate an at least 5-dimensional space, S can contain at most one plane of $\mathcal{C}$ and thus $\alpha_3 \in \{0, 1\}$. $\qquad\square$

Lemma 1 can be used to derive quite restrictive conditions on the parameters of a putative q-analogue of the Fano plane. This is the subject of Lemma 2. For the statement of the lemma recall that the cyclotomic polynomials $\Phi_n(X) \in \mathbb{Z}[X]$, defined recursively by $X^n - 1 = \prod_{d|n} \Phi_d(X)$ for $n \in \mathbb{N}$, satisfy $\Phi_p(X) = X^{p-1} + X^{p-2} + \cdots + X + 1$ for prime numbers p, as well as $\Phi_6(X) = X^2 - X + 1$. In terms of cyclotomic polynomials the number of points of $\mathrm{PG}(n-1, \mathbb{F}_q)$ is $\begin{bmatrix}n\\1\end{bmatrix}_q = \frac{q^n - 1}{q - 1} = \prod_{d|n, d\neq 1} \Phi_d(q)$.

Lemma 2. *If a q-analogue $\mathcal{F}_q$ of the Fano plane exists, it must have the following properties:*

(1) The number of planes in $\mathcal{F}_q$ is

$$\begin{aligned}
\#\mathcal{F}_q = \Phi_7(q)\Phi_6(q) &= (q^6 + q^5 + q^4 + q^3 + q^2 + q + 1)(q^2 - q + 1) \\
&= q^8 + q^6 + q^5 + q^4 + q^3 + q^2 + 1,
\end{aligned}$$

with $\Phi_6(q)\Phi_3(q) = q^4 + q^2 + 1$ planes passing through each point of $\mathrm{PG}(6, \mathbb{F}_q)$.

(2) The spectrum of $\mathcal{F}_q$ with respect to solids takes the two values

$$\begin{aligned}
\alpha_0 &= (q^8 - q^7 + q^3, q^7 + q^6 + q^5 - q^3 - q^2 - q, q^4 + q^3 + 2q^2 + q + 1, 0), \\
\alpha_1 &= (q^8 - q^7, q^7 + q^6 + q^5, q^4 + q^3 + q^2, 1)
\end{aligned}$$

with corresponding frequencies

$$\begin{aligned}
f_0 &= q^{12} + q^{10} + q^9 + q^8 + q^7 + q^6 + q^4, \\
f_1 &= q^{11} + q^{10} + 2q^9 + 3q^8 + 3q^7 + 4q^6 + 4q^5 + 3q^4 + 3q^3 + 2q^2 + q + 1.
\end{aligned}$$

Proof. For a q-analogue of the Fano plane the first three inequalities in Lemma 1 are in fact equalities (for any solid S) and, conversely, this property (even if it holds only for one particular solid S) implies that C must be a q-analogue of the Fano plane.

Further, the triangular shape of the system implies that each of the two possible choices $\alpha_3 \in \{0, 1\}$ leads to a unique solution for $\alpha_1, \alpha_2, \alpha_3$.

In the first case ($\alpha_3 = 0$) we obtain

$$\alpha_2 = q^4 + q^3 + 2q^2 + q + 1 = \Phi_4(q)\Phi_3(q),$$

$$\alpha_1 = \frac{1}{q+1}\left(q^3 \begin{bmatrix} 3 \\ 1 \end{bmatrix}_q \begin{bmatrix} 4 \\ 1 \end{bmatrix}_q - q(q+1)\alpha_2\right)$$

$$= \frac{1}{q+1}\left(q^3 \cdot \Phi_3(q) \cdot \Phi_4(q)\Phi_2(q) - q \cdot \Phi_2(q) \cdot \Phi_4(q)\Phi_3(q)\right)$$

$$= (q^3 - q)\Phi_4(q)\Phi_3(q) = q \cdot \Phi_4(q)\Phi_3(q)\Phi_2(q)\Phi_1(q)$$

$$= q(q^4 - 1)(q^2 + q + 1) = q^7 + q^6 + q^5 - q^3 - q^2 - q,$$

$$\alpha_0 = q^8 - \frac{q^2}{q^2 + q + 1} \cdot \alpha_1 = q^8 - q^7 + q^3,$$

as asserted. The second case ($\alpha_3 = 1$) is done similarly.

Finally, a solid S of $\mathrm{PG}(6, \mathbb{F}_q)$ has $\alpha_3(S) = 1$ iff it contains a plane of $\mathcal{F}_q$. The number of such solids is

$$f_1 = \#\mathcal{F}_q \cdot \begin{bmatrix} 4 \\ 1 \end{bmatrix}_q = \Phi_7(q)\Phi_6(q)\Phi_4(q)\Phi_2(q)$$

$$= q^{11} + q^{10} + 2q^9 + 3q^8 + 3q^7 + 4q^6 + 4q^5 + 3q^4 + 3q^3 + 2q^2 + q + 1,$$

and the number of solids with $\alpha_3(S) = 0$ is

$$f_0 = \begin{bmatrix} 7 \\ 4 \end{bmatrix}_q - f_1 = \begin{bmatrix} 7 \\ 3 \end{bmatrix}_q - f_1$$

$$= \Phi_7(q)\Phi_6(q)\Phi_5(q) - \Phi_7(q)\Phi_6(q)\Phi_4(q)\Phi_2(q)$$

$$= \#\mathcal{F}_q \cdot q^4 = q^{12} + q^{10} + q^9 + q^8 + q^7 + q^6 + q^4,$$

completing the proof.   □

Remark 1. More general results on the intersection structure of a putative q-analogue of the Fano plane can be found in [15, Sect. 4].

Performing the same computations, mutatis mutandis, for putative Steiner systems $S_q(2, 3, v)$ with arbitrary ambient space dimension v yields non-integral solutions and hence excludes the existence of an $S_q(2, 3, v)$ for $v \equiv 0, 2, 4, 5 \pmod 6$. Thus an $S_q(2, 3, v)$ can exist only for $v \in \{7, 9, 13, 15, 19, 21, 25, 27, \dots\}$. For the particular case $q = 2$, $v = 13$ existence has been proved in [16], providing the only known nontrivial example of a Steiner system over a finite field. This remarkable result was the outcome of a computer search for Steiner systems $S_2(2, 3, 13)$ invariant under the normalizer of a Singer subgroup of $\mathrm{GL}(13, \mathbb{F}_2)$, a group of order

$(2^{13} - 1) \cdot 13 = 106483$, and of course facilitated by the fact that Steiner systems $S_2(2, 3, 13)$ with this additional structure exist.[11]

3. Augmented LMRD Codes

The initial subspace code constructions by Koetter, Kschischang and Silva [6, 17] were based on the observation that the dimension of the intersection of two k-dimensional subspaces U, V of $\mathbb{F}_q^v / \mathbb{F}_q$ with canonical matrices of the special form $(\mathbf{I}_k | \mathbf{A})$, $(\mathbf{I}_k | \mathbf{B})$ can be expressed through the rank of the matrix $\mathbf{A} - \mathbf{B} \in \mathbb{F}_q^{k \times (v-k)}$. In fact it is easily seen that $U \cap V = \{(\mathbf{x} | \mathbf{x} \mathbf{A}); \mathbf{x} \in \text{Ker}(\mathbf{A} - \mathbf{B})\} \cong \text{Ker}(\mathbf{A} - \mathbf{B})$ (the left kernel of $\mathbf{A} - \mathbf{B}$) and thus $\dim(U \cap V) = k - \text{rank}(\mathbf{A} - \mathbf{B})$.

From earlier work of Delsarte [18] (and independently Gabidulin and Roth [19, 20]) the maximum number of matrices in $\mathbb{F}_q^{m \times n}$ having pairwise rank distance at least d is known to be $q^{(m-d+1)n}$, provided that $m \leq n$.[12] Subsets $\mathcal{A} \subseteq \mathbb{F}_q^{m \times n}$ of size $q^{(m-d+1)n}$ with $\text{rank}(\mathbf{A} - \mathbf{B}) \geq d$ for all pairs of distinct $\mathbf{A}, \mathbf{B} \in \mathcal{A}$ are known as $(m, n, m - d + 1)$ *maximum rank distance (MRD)* codes. Via the *lifting construction* $\mathcal{A} \to \mathcal{L} \subseteq \mathbb{F}_q^{m \times (m+n)}$, $\mathbf{A} \mapsto \langle (\mathbf{I}_m | \mathbf{A}) \rangle$ they give rise to subspace codes $\mathcal{L}$ in $\text{PG}(m + n - 1, \mathbb{F}_q)$ of size $\#\mathcal{L} = \#\mathcal{A} = q^{(m-d+1)n}$, constant dimension m and maximum intersection dimension $m - d$, as we have indicated above. These subspace codes are called *lifted maximum rank distance (LMRD)* codes.

In the case of interest to us we can find q^8 matrices in $\mathbb{F}_q^{3 \times 4}$ at pairwise rank distance ≥ 2 and lift these to a plane LMRD code in $\text{PG}(6, \mathbb{F}_q)$ of size q^8 with maximum intersection dimension 1. This gives the lower bound $\#\mathcal{C} \geq q^8$ for the maximum size of a plane subspace code in $\text{PG}(6, \mathbb{F}_q)$, which is already of the same asymptotic order as a putative 2-analogue of the Fano plane ($\#\mathcal{F}_q = q^8 + q^6 + q^5 + q^4 + q^3 + q^2 + 1$).

Following the work in [6, 17], several constructions have been proposed for augmenting LMRD codes without increasing t. (Note that increasing t sacrifices the error-correction capabilities of the original subspace code.) All these constructions are variants of the so-called *echelon-Ferrers* construction introduced in [21], which combines subspace codes in different Schubert cells of the corresponding Grassmannian in a certain way.[13] We will not delve into this further, but instead only mention that the maximum size of an augmented LMRD code obtained in this way is $\#\mathcal{C} = q^8 + \begin{bmatrix} 4 \\ 2 \end{bmatrix}_q = q^8 + q^4 + q^3 + 2q^2 + q + 1$ and provide a different construction of such a code below.

---

[11] An $S_2(2, 3, 13)$ contains as many as $\frac{(2^{13}-1)(2^{12}-1)}{21} = 1597245$ planes out of a total of $\begin{bmatrix} 13 \\ 3 \end{bmatrix}_2 = 3269560515$ planes in $\text{PG}(12, \mathbb{F}_2)$, rendering any unrestricted search for such a structure completely infeasible.

[12] The assumption $m \leq n$ imposes no essential restriction, since matrices can be transposed without changing the rank.

[13] "Schubert cell" refers to the set of all subspaces whose canonical matrices have their pivot columns fixed.

In fact the bound $\#\mathcal{C} \leq q^8 + \begin{bmatrix} 4 \\ 2 \end{bmatrix}_q$ holds for any plane subspace code in $\mathrm{PG}(6, \mathbb{F}_q)$ containing an LMRD code. This is a consequence of the following lemma, which could be easily generalized to arbitrary packet length v.

Lemma 3. *Let $\mathcal{L}$ be a plane LMRD code in $\mathrm{PG}(6, \mathbb{F}_q)$ and $S = (0, 0, 0, *, *, *, *)$ the special solid defined by $x_1 = x_2 = x_3 = 0$. Then the planes in $\mathcal{L}$ cover all lines that are disjoint from S (and no other lines).*

Proof. A line L disjoint from S has a canonical matrix of the form $(\mathbf{Z}|\mathbf{B})$ with $\mathbf{Z} \in \mathbb{F}_q^{2 \times 3}$ in canonical form and $\mathbf{B} \in \mathbb{F}_q^{2 \times 4}$ arbitrary. Now let $\mathcal{A}$ be the matrix code corresponding to $\mathcal{L}$ and consider the map $\mathcal{A} \to \mathbb{F}_q^{2 \times 4}$, $\mathbf{A} \mapsto \mathbf{ZA}$. Since $\mathrm{rank}(\mathbf{Z}) = 2$ and the minimum nonzero rank in $\mathcal{A}$ is 2, this map must be injective, hence also surjective. Thus there exists $\mathbf{A} \in \mathcal{A}$ such that $\mathbf{B} = \mathbf{ZA}$, implying $\mathrm{cm}(L) = \mathbf{Z}(\mathbf{I}_3|\mathbf{A})$. The latter just says that L is contained in the plane $\langle (\mathbf{I}_3|\mathbf{A}) \rangle \in \mathcal{L}$. $\qquad\square$

With the aid of this lemma the bound $\#\mathcal{C} \leq q^8 + \begin{bmatrix} 4 \\ 2 \end{bmatrix}_q$ is established as follows: A fortiori $\mathcal{C}$ covers every line disjoint from S and hence cannot contain a plane meeting S in a point (as such a plane would contain lines disjoint from S). Thus, apart from the planes in $\mathcal{L}$, it contains only planes meeting S in a line or planes entirely contained in S. The number of such planes is bounded by the total number of lines in S, yielding the bound. (Moreover, the bound can be achieved only if no plane of $\mathcal{C}$ is contained in S.)

We close this section with an alternative construction of an augmented plane LMRD code in $\mathrm{PG}(6, \mathbb{F}_q)$ of size $q^8 + \begin{bmatrix} 4 \\ 2 \end{bmatrix}_q$. Such a code was first constructed in [13]. Our construction uses the existence of a *line packing* of $\mathrm{PG}(3, \mathbb{F}_q)$, which refers to a partition of the line set into line spreads, where a *line spread* is itself defined as a partition of the point set into lines (the same as a partial line spread that covers all points).[14] Line packings of $\mathrm{PG}(3, \mathbb{F}_q)$ exist for all prime powers $q > 1$; cf. [22, 23]. Since line spreads of $\mathrm{PG}(3, \mathbb{F}_q)$ have size $q^2 + 1$ and $\begin{bmatrix} 4 \\ 2 \end{bmatrix}_q = (q^2 + 1)(q^2 + q + 1)$, the number of line spreads in a line packing is $q^2 + q + 1$.

Theorem 1. *Any plane LMRD code $\mathcal{L}$ in $\mathrm{PG}(6, \mathbb{F}_q)$ can be augmented by $\begin{bmatrix} 4 \\ 2 \end{bmatrix}_q = q^4 + q^3 + 2q^2 + q + 1$ further planes to yield a plane subspace code $\mathcal{C}$ of size $\#\mathcal{C} = q^8 + \begin{bmatrix} 4 \\ 2 \end{bmatrix}_q = q^8 + q^4 + q^3 + 2q^2 + q + 1.$[15]*

Proof. Choose a packing $\mathscr{P} = \{\mathcal{P}_1, \ldots, \mathcal{P}_{q^2+q+1}\}$ of $\mathrm{PG}(S/\mathbb{F}_q) \cong \mathrm{PG}(3, \mathbb{F}_q)$, and let $\{P_1, \ldots, P_{q^2+q+1}\}$ be a set of points in $\mathrm{PG}(6, \mathbb{F}_q)$ forming a set of representatives for the $q^2 + q + 1$ 4-flats containing S.[16] For $1 \leq i \leq q^2 + q + 1$ connect the point P_i

---

[14]Line packings form a projective analogue of the standard resolution of the line set of an affine plane into parallel classes.

[15]We remind the reader one last time that all subspace codes considered (including LMRD codes) have $t = 2$ (maximum intersection dimension 1).

[16]By this we mean $P_i \notin S$ and the 4-flats $F_i = P_i + S$, $1 \leq i \leq q^2 + q + 1$, account for all 4-flats above S.

to all $q^2 + 1$ lines L_{ij} in $\mathcal{P}_i$ to form a set of $(q^2 + 1)(q^2 + q + 1)$ planes $E_{ij} = P_i + L_{ij}$. We claim that $\mathcal{C} = \mathcal{L} \cup \{E_{ij}\}$ has the required property.

Clearly the "new" planes E_{ij} cover no line disjoint from S and each line in S exactly once. Now suppose, for contradiction, that L is a line meeting S in a point P and contained in two different new planes $E = P_i + L_{ij}$, $E' = P_{i'} + L_{i'j'}$. Then L must meet both L_{ij} and $L_{i'j'}$ in P, whence L_{ij} and $L_{i'j'}$ intersect and $i \neq i'$. But the 4-flats $F_i = L + S = F_{i'}$ coincide, contradiction! $\qquad\Box$

The subspace code $\mathcal{C}$ of Theorem 1 is quite small in comparison with the codes constructed later in our main theorems. But we feel that the construction method is of independent interest and have included it for this reason.

4. First Expurgating and Then Augmenting

In this section we describe the basic idea used in [11] to overcome the size restriction imposed on subspace codes containing LMRD codes, tailored (and generalized) to the case of plane subspace codes in $\mathrm{PG}(6, \mathbb{F}_q)$ with arbitrary q.

Given a plane LMRD code $\mathcal{L}$ in $\mathrm{PG}(6, \mathbb{F}_q)$, we must obviously remove some of the q^8 planes in $\mathcal{L}$ first and then augment the resulting subcode $\mathcal{L}_0 \subset \mathcal{L}$ as far as possible. What is the best way to do this? The "removed" set of planes $\mathcal{L}_1$, of size $\#\mathcal{L}_1 = M_1$ say, covers $(q^2 + q + 1)M_1$ lines disjoint from the special solid $S = (0, 0, 0, 0, *, *, *)$, which become *free lines* of $\mathcal{L}_0$ in the sense that any *new plane* added to $\mathcal{L}_0$, which contains only lines disjoint from S that are free, will not increase t (i.e., introduce a multiple line cover). Of course we are only interested in adding new planes which meet S in a point at this stage, since this is the only way to go beyond the construction in Section 3. In this case, provided an exact rearrangement of the free lines into new planes is possible, the subspace code size will increase to

$$q^8 - M_1 + \frac{(q^2 + q + 1)M_1}{q^2} = q^8 + \frac{(q + 1)M_1}{q^2}, \qquad (1)$$

since new planes contain only q^2 lines disjoint from S. It is clear that M_1 must be a multiple of q^2, and it has been shown in [11] that $M_1 = q^2$ is not feasible but $M_1 = q^3$ can be realized for a particular choice of $\mathcal{L}$ and as far as only the rearrangement of lines disjoint from S matters. (As an additional requirement, the chosen new planes must not introduce a multiple cover of a line meeting S in a point.) We will now develop the technical machinery needed to derive this result, adapted to the case $v = 7$.

Since the ambient space of $\mathrm{PG}(6, \mathbb{F}_q)$ does not matter (as long as it is 7-dimensional over $\mathbb{F}_q$), we take it as $V = W \times \mathbb{F}_{q^4}$, where W denotes the trace-zero subspace of $\mathbb{F}_{q^4}/\mathbb{F}_q$ (consisting of all $x \in \mathbb{F}_{q^4}$ satisfying $\mathrm{trace}(x) = \mathrm{trace}_{\mathbb{F}_{q^4}/\mathbb{F}_q}(x) = x + x^q + x^{q^2} + x^{q^3} = 0$). This allows us to use the additional structure of $\mathrm{PG}(6, \mathbb{F}_q)$ imposed by the extension field $\mathbb{F}_{q^4}$. In this model our special solid is $S = \{0\} \times \mathbb{F}_{q^4} \cong \mathbb{F}_{q^4}$ (naturally); likewise, we make the identification $W \times \{0\} \cong W$.

Subspaces of $V/\mathbb{F}_q$ can be parametrized in the form

$$U = \{(x, f(x) + y); x \in Z, y \in T, f \in \mathrm{Hom}(Z, \mathbb{F}_{q^4}/T)\}, \tag{2}$$

where

$$Z = \{x \in W; \exists y \in \mathbb{F}_{q^4} \text{ such that } (x, y) \in U\},$$
$$T = \{y \in \mathbb{F}_{q^4}; (0, y) \in U\}$$

and $f: Z \to \mathbb{F}_{q^4}$ is any $\mathbb{F}_q$-linear map whose graph (in the sense of Real Analysis) $\Gamma_f = \{(x, f(x)); x \in Z\}$ is contained in U. The $\mathbb{F}_q$-subspaces $Z \subseteq W$ (projection of U onto W) and $T \subseteq \mathbb{F}_{q^4}$ (naturally isomorphic to the kernel $U \cap S$ of this projection) are uniquely determined by U, while f is only determined up to addition of an $\mathbb{F}_q$-linear map with values in T and may therefore be replaced by any element in the coset $f + \mathrm{Hom}(Z, T) \in \mathrm{Hom}(Z, \mathbb{F}_{q^4})/\mathrm{Hom}(Z, T) \cong \mathrm{Hom}(Z, \mathbb{F}_{q^4}/T)$.[17] We denote this parametrization by $U = U(Z, T, f)$, using sometimes the subspaces $Z \times \{0\}$, $\{0\} \times T$ of $\mathrm{PG}(V/\mathbb{F}_q)$ in place of Z, T, as indicated above.

Observe that the subspaces disjoint from S are precisely the graphs $\Gamma_f = U(Z, \{0\}, f)$ of $\mathbb{F}_q$-linear maps $f: Z \to \mathbb{F}_{q^4}$. At the other extreme, the subspaces containing S are of the form $U(Z, S, 0) = Z \times S$.

The incidence relation between subspaces of $V/\mathbb{F}_q$ can also be described within this setting: $U(Z', T', f') \subseteq U(Z, T, f)$ if and only if $Z' \subseteq Z$, $T' \subseteq T$ and $f|_{Z'} - f' \in \mathrm{Hom}(Z', T)$.

Now recall from Galois Theory that the powers $\mathrm{id}, \varphi, \varphi^2, \varphi^3$ of the Frobenius automorphism $\varphi: \mathbb{F}_{q^4} \to \mathbb{F}_{q^4}$, $x \mapsto x^q$ of $\mathbb{F}_{q^4}/\mathbb{F}_q$ form a basis of $\mathrm{End}(\mathbb{F}_{q^4}/\mathbb{F}_q)$ over $\mathbb{F}_{q^4}$. This says that every $\mathbb{F}_q$-linear map $f: \mathbb{F}_{q^4} \to \mathbb{F}_{q^4}$ is evaluation of a unique linearized polynomial $a(X) = a_0 X + a_1 X^q + a_2 X^{q^2} + a_3 X^{q^3} \in \mathbb{F}_{q^4}[X]$ of symbolic degree ≤ 3. For simplicity we write $x \mapsto f(x)$ as $a_0 x + a_1 x^q + a_2 x^{q^2} + a_3 x^{q^3}$. The restriction map $f \mapsto f|_W$ then gives that every element of $\mathrm{Hom}(W, \mathbb{F}_{q^4})$ is represented uniquely as $a_0 x + a_1 x^q + a_2 x^{q^2}$ for some $a_0, a_1, a_2 \in \mathbb{F}_{q^4}$ (since the linear maps vanishing on W are of the form $a(x + x^q + x^{q^2} + x^{q^3})$ with $a \in \mathbb{F}_{q^4}$).

Next we name various subspaces of $\mathrm{Hom}(W, \mathbb{F}_{q^4})$, which will subsequently play an important role:

$$\mathcal{G} = \{a_0 x + a_1 x^q; a_0, a_1 \in \mathbb{F}_{q^4}\},$$
$$\mathcal{R} = \{ax^q - a^q x; a \in \mathbb{F}_{q^4}\},$$
$$\mathcal{T} = \{ax^q - a^q x; a \in W\},$$
$$\mathcal{D}(Z, P) = r(ab^q - a^q b)^{-1}\langle ax^q - a^q x, bx^q - b^q x\rangle$$

for a 2-dimensional subspace $Z = \langle a, b \rangle$ of W and a point $P = \mathbb{F}_q(0, r)$ of the special solid S (i.e. $r \in \mathbb{F}_{q^4}^\times$). The space $\mathcal{G}$ has minimum rank distance 2 (since

---

[17]It goes without saying that "Hom" denotes the set of $\mathbb{F}_q$-linear maps between the indicated $\mathbb{F}_q$-spaces, which forms an $\mathbb{F}_q$-space of its own with respect to the point-wise operations.

$a_0 x + a_1 x^q \neq 0$ has at most q zeros in W) and size $\#\mathcal{G} = q^8$. It is therefore an MRD code. We call it the *Gabidulin code*, since it is a basis-free version of a member of the family of MRD codes constructed in [19], which are nowadays commonly called Gabidulin codes. Further we have $\mathcal{D}(Z,P) \subset \mathcal{T} \subset \mathcal{R} \subset \mathcal{G}$, $\mathcal{T}$ has constant rank 2 (since $ax^q - a^q x$ has 1-dimensional kernel $\mathbb{F}_q a$ if $a \in W \setminus \{0\}$), $\mathcal{R} \setminus \mathcal{T}$ has constant rank 3, and $\mathcal{D}(Z,P)$ consists of all linear maps $f \in \mathcal{G}$ satisfying $f(Z) \subseteq \mathbb{F}_q r$.[18]

Finally we fix $\mathcal{L} = \{\Gamma_f; f \in \mathcal{G}\}$ for the remainder of this article and call $\mathcal{L}$ the *lifted Gabidulin code*. The reader may check that $f \mapsto \Gamma_f$ provides a basis-free description of the lifting construction (passing from matrix codes to subspace codes) and hence $\mathcal{L}$ is a plane LMRD code as needed for the subsequent discussion.

Lemma 4. *For a set of planes $\mathcal{L}_1 \subseteq \mathcal{L}$ let $\mathcal{G}_1 \subseteq \mathcal{G}$ be the corresponding set of linear maps in the Gabidulin code. In order that the free lines determined by $\mathcal{L}_1$ can be rearranged into new planes meeting S in a point, it is necessary and sufficient that $\#\mathcal{G}_1 = mq^2$ is a multiple of q^2 and for each 2-dimensional subspace $Z \subset W$ there exist (not necessarily distinct) points $P_1, \ldots, P_m$ on S and linear maps $f_1, \ldots, f_m \in \mathcal{G}$ such that*

$$\mathcal{G}_1 = \biguplus_{i=1}^{m} \left(f_i + \mathcal{D}(Z, P_i) \right).$$

Note that the condition requires $\mathcal{G}_1$ to be a union of cosets of spaces $\mathcal{D}(Z,P)$ simultaneously in $q^2 + q + 1$ different ways, one for each 2-dimensional subspace $Z \subset W$. The number of new planes in the rearrangement must be $m(q^2 + q + 1)$, but the rearrangement itself is perhaps not uniquely determined by $\mathcal{G}_1$. Moreover, the lemma does not say anything about whether the rearrangement introduces a multiple cover of some line meeting S in a point.

Proof of the lemma. Lines L disjoint from S as well as new planes N meeting S in a point are contained in a unique hyperplane H above S ($H = L + S$ resp. $H = N + S$). "Old" planes $E \in \mathcal{L}$ are transversal to these hyperplanes and the H-section $E \mapsto E \cap H$ identifies $\mathcal{L}$ with the set of q^8 lines in $\mathrm{PG}(H)$ disjoint from S (since $\mathcal{L}$ is an LMRD code). In terms of the parametrization $H = H(Z, \mathbb{F}_{q^4}, 0)$, $E = \Gamma_f$, $L = \Gamma_g$ the corresponding H-section is just restriction $g = f|_Z$. Thus we can look at each hyperplane above S separately.

Let H be such a hyperplane and Z the corresponding 2-dimensional subspace of W. Planes in H meeting S in the point $P = \mathbb{F}_q(0, r)$ have the form $N = N(Z, \mathbb{F}_q r, g)$ with $g \in \mathrm{Hom}(Z, \mathbb{F}_{q^4})$ and contain the q^2 lines $L = \Gamma_h$, $h \in g + \mathrm{Hom}(Z, \mathbb{F}_q r)$, disjoint from S. Denoting by $f \in \mathcal{G}$ the unique linear map such that $f|_Z = g$, we have that $f + \mathcal{D}(Z, P)$ restricts to $g + \mathrm{Hom}(Z, \mathbb{F}_q r)$ on Z. Hence the mq^2 free lines in H determined by the planes in $\mathcal{L}_1$ can be rearranged into new planes $N(Z, \mathbb{F}_q r, g)$ iff $\mathcal{G}_1$ is a disjoint union of cosets of the form $f + \mathcal{D}(Z, P)$ with $P \in S$, $f \in \mathcal{G}_1$. □

---

[18]Of course $0 \in \mathcal{T}$ has rank $0 \neq 2$, but it is custom to refer to a matrix space as a constant-rank space if all nonzero matrices in the matrix space have the same rank.

Now observe that our distinguished space $\mathcal{T}$ contains one space $\mathcal{D}(Z, P)$ for each $Z = \langle a, b \rangle \subset W$, viz. $\mathcal{D}(Z, P)$ with $P = \mathbb{F}_q(0, ab^q - a^q b)$. Hence $\mathcal{G}_1 = \mathcal{T}$ satisfies the conditions of Lemma 4 with $m = q$, $P_1 = \cdots = P_q = P = \mathbb{F}_q(0, ab^q - a^q b)$ and $f_1, \ldots, f_q$ a system of coset representatives for $\mathcal{T}/\mathcal{D}(Z, P)$. A fortiori the same is true for any coset of $\mathcal{T}$ in $\mathcal{G}$, and even for any disjoint union of "rotated" cosets $\biguplus_{j=1}^r (f_j + r_j \mathcal{T})$ with $r_j \in \mathbb{F}_{q^4}^\times$ and $f_j \in \mathcal{G}$.[19]

The next theorem, which closes this section, shows that if we take $\mathcal{G}_1 = \mathcal{R}$, the distinguished subspace of order q^4 defined along with $\mathcal{T}$, then the corresponding rearrangement into new planes does not introduce a multiple line cover and hence results in a plane subspace code with $t = 2$.

Theorem 2. *Let $\mathcal{C}$ be the set of planes in $\mathrm{PG}(W \times \mathbb{F}_{q^4}) \cong \mathrm{PG}(6, \mathbb{F}_q)$ obtained by removing all planes $E = \Gamma_f$, $f \in \mathcal{R}$, from $\mathcal{L}$ and adding all planes of the form $N = N(Z, P, g)$ with $Z = \langle a, b \rangle \subset W$ 2-dimensional, $P = \mathbb{F}_q(0, ab^q - a^q b)$ (so P depends on Z) and $g = f|_Z$ for some $f \in \mathcal{R}$. Then $\mathcal{C}$ forms a subspace code (i.e., $t = 2$) of size $\#\mathcal{C} = q^8 + q^3 + q^2$. Moreover, $\mathcal{C}$ can be augmented by $\begin{bmatrix} 4 \\ 2 \end{bmatrix}_q$ further planes meeting S in a line to a subspace code $\widehat{\mathcal{C}}$ of size $\#\widehat{\mathcal{C}} = q^8 + q^4 + 2q^3 + 3q^2 + q + 1$.*

Proof. Since $M_1 = \#\mathcal{R} = q^4$, the rearrangement increases the size of the subspace code by $(q + 1)M_1/q^2 = q^3 + q^2$. Thus $\#\mathcal{C} = q^8 + q^3 + q^2$, and it remains to show that $\mathcal{C}$ still has $t = 2$.

By Lemma 4 and the definition of $\mathcal{C}$, the new planes $N = N(Z, P, g)$ added to $\mathcal{L}_0 = \mathcal{L} \setminus \mathcal{L}_1$ cover each free line exactly once. Hence it suffices to check that no line meeting S in a point is covered more than once.

To this end we first we show that the map $\langle a, b \rangle \mapsto \mathbb{F}_q(ab^q - a^q b)$ (i.e. $Z \mapsto P$) is one-to-one. This implies that new planes in different hyperplanes above S do not meet on S and hence cannot intersect in a line. Suppose, by contradiction, that different subspaces Z_1, Z_2 of W correspond to the same point P. Since $\dim(Z_1 \cap Z_2) = 1$, we can write $Z_1 = \langle a, b_1 \rangle$, $Z_2 = \langle a, b_2 \rangle$. The $\mathbb{F}_q$-linear map $ax^q - a^q x \in \mathrm{Hom}(W, \mathbb{F}_{q^4})$ has kernel $\mathbb{F}_q a$ and hence maps Z_1, Z_2 to different 1-dimensional subspaces $\mathbb{F}_q(ab_1^q - a^q b_1) \neq \mathbb{F}_q(ab_2^q - a^q b_2)$; contradiction!

Next let $N_1 = N(Z, P, g_1)$, $N_2 = N(Z, P, g_2)$, $g_i = f_i|_Z$, be different new planes meeting S in the same point P (and hence with the same Z). Write $Z = \langle a, b \rangle$ and $f_1(x) - f_2(x) = u_0 x + u_1 x^q$. The planes N_1, N_2 have a point outside S (and hence a line through P) in common iff there exists $x \in Z \setminus \{0\}$ such that $f_1(x) - f_2(x) \in \mathbb{F}_q(ab^q - a^q b)$. Setting $x = \lambda a + \mu b$, this is equivalent to a nontrivial solution $(\lambda, \mu, \nu^i) \in \mathbb{F}_q^3$ of the equation

$$\lambda(u_0 a + u_1 a^q) + \mu(u_0 b + u_1 b^q) + \nu^i(ab^q - a^q b) = 0.$$

Thus $f_1, f_2 \in \mathcal{G}$ determine new planes N_1, N_2 satisfying $N_1 \cap N_2 = \{P\}$ for those choices of $Z = \langle a, b \rangle \subset W$ (equivalently, for those choices of the hyperplane $H =$

---

[19]For the latter the points P_i vary not only with Z but also with j.

$Z + S$) for which $u_0 a + u_1 a^q$, $u_0 b + u_1 b^q$, $ab^q - a^q b$ are linearly independent over $\mathbb{F}_q$.[20]

With these preparations we can now prove that $\mathcal{C}$ still has $t = 2$. For $f_1, f_2 \in \mathcal{R}$ we have $f_1 - f_2 \in \mathcal{R}$ and hence of the form $ux^q - u^q x$. If f_1, f_2 are in different cosets of $\mathcal{D}(Z, P)$ then $u \notin Z$. The equation $\lambda(ua^q - u^q a) + \mu(ub^q - u^q b) + \nu^i(ab^q - a^q b) = 0$ can be rewritten as

$$\begin{vmatrix} a & b & u \\ a^q & b^q & u^q \\ -\mu & \lambda & \nu^i \end{vmatrix} = 0.$$

If (λ, μ, ν^i) is nonzero then using the linear dependence of the rows of this matrix we can express the conjugates $(a^{q^i}, b^{q^i}, u^{q^i})$ as linear combinations (with coefficients in $\mathbb{F}_{q^4}$) of (a, b, u) and $(-\mu, \lambda, \nu^i) \in \mathbb{F}_q^3$. This shows that the 4×3 matrix formed from the conjugates of (a, b, u) has rank 2 and implies that a, b, u are linearly dependent over $\mathbb{F}_q$; contradiction. Thus $\mathcal{C}$ has the required property.

The augmented subspace code $\widehat{\mathcal{C}}$ is constructed in the same way as in the proof of Theorem 1. The only thing that needs to be checked is that each 4-flat F above S contains a point $Q \notin S$ that is not covered by any new plane $N \in \mathcal{C}$. Equivalently, for any $x \in W \setminus \{0\}$ the new planes $N = N(Z, \mathbb{F}_q r, g)$ with $x \in Z$ do not cover all q^4 points $\mathbb{F}_q(x, y)$, $y \in \mathbb{F}_{q^4}$. This property will now be verified through explicit computation.

A 2-dimensional subspace $Z \subset W$ containing x has the form $Z = \langle a, x \rangle$ with $a \in W$ and $ax^q - a^q x \neq 0$. The points $\mathbb{F}_q(x, y)$ covered by the q^2 new planes corresponding to Z have $y = ux^q - u^q x + \mu(ax^q - a^q x) = (u + \mu a)x^q - (u + \mu a)^q x$ for $u \in \mathbb{F}_{q^4}/Z$, $\mu \in \mathbb{F}_q$. It follows that y takes precisely the q^3 values in the image I of the linear map $c \mapsto cx^q - c^q x$, which has kernel $\mathbb{F}_q x$. In other words, the points in the 4-flat $F = (\mathbb{F}_q x) \times S$ covered by the new planes in $\mathcal{C}$ form the affine part of a solid, viz. $(\mathbb{F}_q x) \times I$, with plane at infinity $\{0\} \times I$. In particular, there are $q^4 - q^3$ valid choices for the point Q. This completes the proof of the Theorem 2. $\qquad \square$

In the binary case $q = 2$ the size of the augmented subspace code in Theorem 2 is $\#\widehat{\mathcal{C}} = 303$, falling short by 1 of the corresponding code in [9]. On the other hand, $\#\widehat{\mathcal{C}}$ strictly exceeds the bound imposed on codes containing an LMRD code for every q, showing already the effectiveness of our approach. However, this is not the end of the story; Theorem 2 will be improved upon later.

5. An Attempt to Construct a q-Analogue and its Failure

In this section we apply the method developed in the previous section to the construction problem for q-analogues of the Fano plane. The attempt eventually fails for every q but produces the largest known plane subspace codes in $\mathrm{PG}(6, \mathbb{F}_q)$.

---

[20]Viewed projectively, this requires that $f(x) = u_0 x + u_1 x^q$ maps the line $Z = \langle a, b \rangle$ to another line $Z' = f(Z)$ of $\mathrm{PG}(\mathbb{F}_{q^4}/\mathbb{F}_q)$ and the point $\mathbb{F}_q(ab^q - a^q b)$ corresponding to Z is not on Z'.

We start with a few words on automorphisms of subspace codes in $\mathrm{PG}(V/\mathbb{F}_q)$. The group $G = \mathrm{GL}(V/\mathbb{F}_q)$ obviously acts on plane subspace codes in $\mathrm{PG}(V/\mathbb{F}_q)$, but is by way too large for our purpose. The stabilizer G_S of our special solid S in $\mathrm{GL}(V/\mathbb{F}_q)$ consists of all maps L of the form $(x,y)L = (xL_{11}, xL_{12} + yL_{22})$ with $L_{11} \in \mathrm{GL}(W/\mathbb{F}_q)$, $L_{22} \in \mathrm{GL}(\mathbb{F}_{q^4}/\mathbb{F}_q)$ and $L_{12} \in \mathrm{Hom}(W, \mathbb{F}_{q^4})$. The map L sends a plane $E = \Gamma_f$ disjoint from S to Γ_g with $g = L_{11}^{-1} f L_{22} + L_{11}^{-1} L_{12}$ (composition of maps is from left to right for the moment), so that on the corresponding maps $f \in \mathrm{Hom}(W, \mathbb{F}_{q^4})$ it affords the group of all "affine" transformations $f \mapsto A \circ f \circ B + C$ with $A \in \mathrm{GL}(\mathbb{F}_{q^4}/\mathbb{F}_q)$, $B \in \mathrm{GL}(W/\mathbb{F}_q)$ and $C \in \mathrm{Hom}(W, \mathbb{F}_{q^4})$.

The group G_S is still too large for our purpose, but we have that the Gabidulin code $\mathcal{G}$ is invariant under the subgroup consisting of all maps $f \mapsto rf$ with $r \in \mathbb{F}_{q^4}^\times$, which acts as a Singer group on the projective space $\mathrm{PG}(S/\mathbb{F}_q) \cong \mathrm{PG}(3, \mathbb{F}_q)$. This group, or rather the corresponding subgroup $\Sigma \leq \mathrm{GL}(V/\mathbb{F}_q)$ consisting of all maps $(x,y) \mapsto (x, ry)$ with $r \in \mathbb{F}_{q^4}^\times$, is suitable for our purpose.[21] It is our next goal to make the expurgation-augmentation process of Section 4 invariant under Σ.

How large should the set $\mathcal{L}_1$ of removed planes be for a putative q-analogue $\mathcal{F}_q$? We can arrange coordinates in such a way that S does not contain a block of $\mathcal{F}_q$ and hence $q^8 - q^7 + q^3$ blocks are disjoint from S; cf. Lemma 2. This requires

$$\#\mathcal{L}_1 = q^7 - q^3 = q^3(q^4 - 1) = (q^4 - q^3)(q^3 + q^2 + q + 1)$$

and the number of new planes through each point $P \in S$ to be $(q^4 - q^3)(q^2 + q + 1)/q^2 = q^4 - q$.[22] Hence a Σ-invariant construction of $\mathcal{F}_q$ is at least conceivable and, even better, there is a canonical candidate for a Σ-invariant subset $\mathcal{G}_1 \subset \mathcal{G}$ of the appropriate size, viz. the union of all "rotated" cosets $r(f + \mathcal{T})$ with $f \in \mathcal{R} \setminus \mathcal{T}$ and $r \in \mathbb{F}_{q^4}^\times$.[23] A moment's reflection shows that this set $\mathcal{G}_1$ consists precisely of all binomials $a_0 x + a_1 x^q$ with 1-dimensional kernel in $\mathbb{F}_{q^4}/\mathbb{F}_q$ complementary to W (thus the rank in $\mathrm{Hom}(W, \mathbb{F}_{q^4})$ is 3). The complementary subset $\mathcal{G}_0 = \mathcal{G} \setminus \mathcal{G}_1$ consists of 0, the $2(q^4 - 1)$ monomials rx, rx^q with $r \in \mathbb{F}_{q^4}^\times$, the $(q^4 - 1)(q^2 + q + 1)$ binomials $r(ux^q - u^q x)$ with $r \in \mathbb{F}_{q^4}^\times$ and $u \in W \setminus \{0\}$ (these have rank 2 in $\mathrm{Hom}(W, \mathbb{F}_{q^4})$) and $(q^4 - 1)(q^3 + q^2 + q + 1)(q - 2)$ binomials $a_0 x + a_1 x^q$ with no nontrivial zero in $\mathbb{F}_{q^4}$. The set $\mathcal{G}_1$ decomposes as

$$\mathcal{G}_1 = \biguplus_{r \in \mathbb{F}_{q^4}^\times/\mathbb{F}_q^\times} r(\mathcal{R} \setminus \mathcal{T}),$$

showing that the $(q-1) \times \frac{q^4 - 1}{q - 1} = q^4 - 1$ cosets $r(f + \mathcal{T})$ used are pairwise disjoint, as needed for the construction.

New planes are defined by connecting the free lines L in the planes corresponding to $f + \mathcal{T}$ to the points $P = \mathbb{F}_q(0, ab^q - a^q b)$, where $Z = \langle a, b \rangle \subset W$ is the 2-

---

[21] Viewed as collineation group, Σ has order $q^4 - 1$ (not the same as the Singer group).

[22] As a consistency check, use that this number can also be obtained by subtracting from the total number $q^4 + q^2 + 1$ of blocks through P (cf. Lemma 2) the number $q^2 + q + 1$ of blocks that meet S in a line through P. Indeed, $q^4 + q^2 + 1 - (q^2 + q + 1) = q^4 - q$.

[23] The spaces $r\mathcal{T}$ itself cannot be used, since these are not disjoint.

dimensional subspace determined by the hyperplane $H = L + S = Z \times S$ (the same definition as in Section 4), and rotating: Free lines in the planes corresponding to $r(f + T)$, $r \in \mathbb{F}_{q^4}^{\times}$, are connected to $rP = \mathbb{F}_q(0, r(ab^q - a^q b))$ in the same way. The collection $\mathcal{N}$ of $(q^4 - q)(q^3 + q^2 + q + 1)$ new planes determined in this way is certainly Σ-invariant and contains $q^4 - q$ planes meeting S in any particular point P. By construction, $\mathcal{N}$ forms an exact cover of the free lines determined by $\mathcal{L}_1$ (and $\mathcal{L}_0 \cup \mathcal{N}$ forms an exact cover of all lines disjoint from S), but $\mathcal{N}$ may cover some lines meeting S in a point more than once.

If for some value of q the set $\mathcal{L}_0 \cup \mathcal{N}$ still had $t = 2$, then the present construction would have been a big step towards the desired q-analogue $\mathcal{F}_q$, leaving only the task to augment it by $\begin{bmatrix} 4 \\ 2 \end{bmatrix}_q$ further planes meeting S in a line. Unfortunately, however, it turns out that $\mathcal{L}_0 \cup \mathcal{N}$ never has $t = 2$, rendering a construction of a q-analogue $\mathcal{F}_q$ in this way impossible. This negative result will follow from our subsequent analysis, which on the other hand will tell us precisely how many planes should be removed from $\mathcal{L}_0 \cup \mathcal{N}$ in order to restore $t = 2$. Fortunately, this number turns out to be rather small.

Let $\mathcal{N}_1 \subset \mathcal{N}$ be the set of $q^4 - q = (q-1)(q^3 + q^2 + q)$ new planes passing through the special point $P_1 = \mathbb{F}_q(0, 1)$. We are interested in finding the largest subset(s) $\mathcal{N}_1' \subseteq \mathcal{N}_1$ consisting of planes mutually intersecting in P_1. Denoting by M_1' the maximum size of such a subset $\mathcal{N}_1'$, it is clear from the preceding development and Σ-invariance of $\mathcal{N}$ that $\mathcal{L}_0$ can then be augmented by a subset $\mathcal{N}'$ of size $M' = M_1'(q^3 + q^2 + q + 1)$ without increasing t. If $\mathcal{N}_1'$ is invariant under the subgroup of Σ corresponding to $\mathbb{F}_q^{\times}$ then $\mathcal{N}'$ may be taken in the form $\mathcal{N}' = \biguplus_{L \in \Sigma} L(\mathcal{N}_1') = \biguplus_{r \in \mathbb{F}_{q^4}^{\times}/\mathbb{F}_q^{\times}} r\mathcal{N}_1'$, making the augmented subspace code $\mathcal{C} = \mathcal{L}_0 \cup \mathcal{N}'$ again Σ-invariant.[24] If $\mathcal{N}_1'$ is not uniquely determined then there are many further choices for $\mathcal{N}'$, which could lead to better overall subspace codes during the final augmentation step.[25]

Before writing down $\mathcal{N}_1$ in explicit form we will introduce some further terminology. Relative to a 2-dimensional subspace $Z \subset W$, the letters a, b, c, d will henceforth denote a basis of $\mathbb{F}_{q^4}/\mathbb{F}_q$ such that $Z = \langle a, b \rangle$, $W = \langle a, b, c \rangle$ and $\mathrm{Tr}(d) = 1$.[26] Further we set $\delta(x, y) = xy^q - x^q y = \left| \begin{smallmatrix} x & y \\ x^q & y^q \end{smallmatrix} \right|$ for $x, y \in \mathbb{F}_{q^4}$, which constitutes an $\mathbb{F}_q$-bilinear, antisymmetric map with right annihilators $\{y \in \mathbb{F}_{q^4}; \delta(x, y) = 0\} = \mathbb{F}_q x$ and corresponding right images $\delta(x, \mathbb{F}_{q^4}) = \{z \in \mathbb{F}_{q^4}; \mathrm{Tr}(x^{-q-1} z) = 0\} = x^{q+1} W$ (provided that $x \neq 0$). The latter follows from Hilbert's Satz 90, using $z = xy^q - x^q y \iff x^{-q-1} z = (y/x)^q - y/x$. Since $\delta(x, y) = x \prod_{\lambda \in \mathbb{F}_q} (y - \lambda x)$, we also have that $\mathbb{F}_q \delta(x, y)$ depends only on the line $L = \langle x, y \rangle$ of $\mathrm{PG}(\mathbb{F}_{q^4}/\mathbb{F}_q)$ (provided that $\mathbb{F}_q x \neq \mathbb{F}_q y$) and is computed as the product of all points of L in $\mathbb{F}_{q^4}^{\times}/\mathbb{F}_q^{\times}$. Accordingly, we can write $\delta(L)$ for $\mathbb{F}_q \delta(x, y)$ and thus have a well-defined map $L \mapsto \delta(L)$ from lines to points of $\mathrm{PG}(\mathbb{F}_{q^4}/\mathbb{F}_q)$. As shown above, $L \mapsto \delta(L)$ maps

---

[24] "$r\mathcal{N}_1'$ refers to the image of $\mathcal{N}_1'$ under $(x, y) \mapsto (x, ry)$.

[25] Later we will see that the number of choices for $\mathcal{N}_1'$ is at least $(q^2)^{q^3 + q^2 + q + 1}$; cf. Section 6.

[26] The element d can be fixed once and for all, but c depends on Z, of course.

the line pencil through $\mathbb{F}_q x$ bijectively onto the plane $x^{q+1}W$,[27] but we also have the following

Lemma 5. *$L \mapsto \delta(L)$ maps the lines contained in any plane E of $\mathrm{PG}(\mathbb{F}_{q^4}/\mathbb{F}_q)$ bijectively onto the points of another plane E'. If $\epsilon \in \mathbb{F}_{q^4}^\times$ satisfies $\epsilon^q = -\epsilon$ then $(aW)' = a^{q+1}\epsilon W$ for $a \in \mathbb{F}_{q^4}^\times$.*

Note that $\epsilon^q = -\epsilon$, or $\epsilon^{q-1} = -1$, is equivalent to $\epsilon \in \mathbb{F}_q^\times$ for even q and to $\epsilon \notin \mathbb{F}_q^\times \wedge \epsilon^2 \in \mathbb{F}_q^\times$ for odd q. In the latter case $\mathbb{F}_q^\times \epsilon$ is the unique element of order 2 in $\mathbb{F}_{q^4}^\times/\mathbb{F}_q^\times$. Further note that every plane of $\mathrm{PG}(\mathbb{F}_{q^4}/\mathbb{F}_q)$ has the form aW for some $a \in \mathbb{F}_{q^4}^\times$ (by Singer's Theorem).

Proof. Since any two lines in E intersect and $L \mapsto \delta(L)$ is injective on line pencils, it is clear that the $q^2 + q + 1$ points $\delta(L)$ for $L \subset E$ are distinct.

Now consider the special plane $E = W = \{x \in \mathbb{F}_{q^4}; x + x^q + x^{q^2} + x^{q^3} = 0\}$. For $x, y \in W$ we have

$$\mathrm{Tr}\big(\epsilon\delta(x,y)\big) = \mathrm{Tr}(\epsilon xy^q - \epsilon x^q y)$$
$$= \epsilon xy^q - \epsilon x^q y^{q^2} + \epsilon x^{q^2} y^{q^3} - \epsilon x^{q^3} y - (\epsilon x^q y - \epsilon x^{q^2} y^q + \epsilon x^{q^3} y^{q^2} - \epsilon xy^{q^3})$$
$$= \epsilon(x + x^{q^2})(y^q + y^{q^3}) - \epsilon(x^q + x^{q^3})(y + y^{q^2})$$
$$= \epsilon(x + x^{q^2})(y^q + y^{q^3}) + \epsilon(x + x^{q^2})(y + y^{q^2})$$
$$= \epsilon(x + x^{q^2})(y + y^q + y^{q^2} + y^{q^3}) = 0$$

and hence $\delta(x,y) \in \epsilon^{-1}W = \epsilon W$. Thus $W' = \epsilon W$, and then $\delta(ax, ay) = a^{q+1}\delta(x,y)$ yields $(aW)' = a^{q+1}\epsilon W$. $\qquad\square$

Finally, for a plane E in $\mathrm{PG}(\mathbb{F}_{q^4}/\mathbb{F}_q)$ we define $\delta(E)$ as the product of all points on E in $\mathbb{F}_{q^4}^\times/\mathbb{F}_q^\times$ (this yields a map $E \mapsto \delta(E)$ from planes to points of $\mathrm{PG}(\mathbb{F}_{q^4}/\mathbb{F}_q)$) and is completely analogous to the case of lines, and in the case $E \neq W$ another projective invariant $\sigma(E)$ as

$$\sigma(E) = \frac{\delta(E)}{\delta(Z)^{q+1}}, \quad \text{where } Z = E \cap W. \tag{3}$$

The reason for this extra definition will become clear in a moment (cf. the subsequent Lemma 6).

Now we turn to the description of the new planes in $\mathcal{N}_1$. By the reasoning in Section 4 and since $\mathcal{D}(Z, P_1) = \mathcal{D}(\langle a, b\rangle, P_1) = \delta(a,b)^{-1}\langle ax^q - a^q x, bx^q - b^q x\rangle$, the planes in $\mathcal{N}_1$ are parametrized as $N = N(Z, P_1, g)$, where $Z \subset W$ is 2-dimensional and $g\colon Z \to \mathbb{F}_{q^4}$ is of the form

$$g(x) = \delta(a,b)^{-1}\left(\lambda(dx^q - d^q x) + \mu(cx^q - c^q x)\right) = \frac{\delta(\lambda d + \mu c, x)}{\delta(a,b)}$$

---
[27]This fact was already used implicitly in some proofs.

with $\lambda \in \mathbb{F}_q^\times$, $\mu \in \mathbb{F}_q$ ($q^4 - q$ choices for N), and cover the $(q+1)q$ points

$$\mathbb{F}_q \left(x, \frac{\delta(\lambda d + \mu c, x)}{\delta(a,b)} + \nu^i \right), \quad \mathbb{F}_q x \in Z, \ \nu^i \in \mathbb{F}_q \tag{4}$$

outside S.

We call a pair of new planes $N, N' \in \mathcal{N}_1$ a *collision* if N, N' have a point outside S (and hence a line through P_1) in common. Such collisions are precisely the obstructions to adding N, N' simultaneously to the expurgated LMRD code $\mathcal{L}_0 = \mathcal{L} \setminus \mathcal{L}_1$ of size $q^8 - q^7 + q^3$. From Theorem 2 (and its "rotated" analogues, so-to-speak) we know that collisions between $N = N(Z, P_1, g)$ and $N' = N(Z', P_1, g')$ can occur only if $Z \neq Z'$. In this case $Z \cap Z' = \mathbb{F}_q z$ is a single point, so that every collision takes the form

$$\frac{\delta(\lambda d + \mu c, z)}{\delta(a, z)} + \nu^i = \frac{\delta(\lambda' d + \mu' c, z)}{\delta(a', z)} + \nu^{i'} \tag{5}$$

with z, a, a' spanning W. Rewriting the denominator as $\delta(a, z)$ makes the actual correspondence $(Z, \lambda, \mu) \mapsto N$ depend on z. However, since $\delta(a, b)$ and $\delta(a, z)$ differ only by a factor in $\mathbb{F}_q^\times$, this dependence disappears in the projective view, where Z and the point $\mathbb{F}_q(\lambda d + \mu c)$ correspond collectively to a set of $q - 1$ new planes, viz. $N(Z, P_1, \mathbb{F}_q^\times g)$ with $g(x) = \delta(\lambda d + \mu c, x)/\delta(a, b)$.[28]

Further note that setting $E = Z + \mathbb{F}_q(\lambda d + \mu c)$ gives a parametrization of the $q^4 - q = (q-1)(q^3 + q^2 + q)$ new planes in $\mathcal{N}_1$, $q - 1$ planes at a time, by the $q^3 + q^2 + q$ planes $E \neq W$ of $\mathrm{PG}(\mathbb{F}_{q^4}/\mathbb{F}_q)$.[29]

Lemma 6. *Let $N = N(Z, P_1, g)$, $N' = N(Z', P_1, g')$ be planes in $\mathcal{N}_1$ parametrized by distinct planes E, E' of $\mathrm{PG}(\mathbb{F}_{q^4}/\mathbb{F}_q)$ in the fashion just described. Collisions between any of the $2(q-1)$ planes in $N(Z, P_1, \mathbb{F}_q^\times g) \uplus N(Z', P_1, \mathbb{F}_q^\times g')$ fall into the following two cases:*

(1) $\sigma(E) \neq \sigma(E')$. In this case there are no collisions among the planes in $N(Z, P_1, \mathbb{F}_q^\times g) \uplus N(Z', P_1, \mathbb{F}_q^\times g')$.

(2) $\sigma(E) = \sigma(E')$. In this case any new plane in $N(Z, P_1, \mathbb{F}_q^\times g)$ collides with a unique new plane in $N(Z', P_1, \mathbb{F}_q^\times g')$ and vice versa, and we can select a maximum of $q - 1$ mutually non-colliding planes from $N(Z, P_1, \mathbb{F}_q^\times g) \uplus N(Z', P_1, \mathbb{F}_q^\times g')$.

Proof. First suppose $Z = Z'$. In this case there are no collisions, and we must show $\sigma(E) \neq \sigma(E')$ or, equivalently, $\delta(E) \neq \delta(E')$. The planes of $\mathrm{PG}(\mathbb{F}_{q^4}/\mathbb{F}_q)$ have the form rW with r running through a system of coset representatives for $\mathbb{F}_q^\times$ in $\mathbb{F}_{q^4}^\times$, and clearly $\delta(rW) = r^{q^2+q+1}\delta(W)$. Since $\gcd(q^3 + q^2 + q + 1, q^2 + q + 1) = 1$, $E \mapsto \delta(E)$ is a bijection and the result follows.

---

[28]Of course this remark also applies when changing the generators a, b of Z.

[29]Since the line $\langle c, d \rangle$ is skew to Z, the q points $\mathbb{F}_q(d + \mu c)$, $\mu \in \mathbb{F}_q$, determine the q planes $E \neq W$ above Z. Replacing $\lambda d + \mu c$ by $\lambda d + \mu c + \alpha a + \beta b$ has no effect on the plane $N(Z, P_1, g)$, since $\delta(a, x), \delta(b, x) \in \mathbb{F}_q \delta(Z)$ for $x \in Z$ and hence g is only changed inside the coset $g + \mathrm{Hom}(Z, \mathbb{F}_q)$.

Now suppose $Z \neq Z'$ and set $Z \cap Z' = \mathbb{F}_q z$. Assuming w.l.o.g. $g(x) = \delta(d + \mu c, x)/\delta(a, x)$, we have from (4) that the points on $N(Z, P_1, \lambda g)$ of the form $\mathbb{F}_q(z, y)$ are those with $y \in \lambda g(z) + \mathbb{F}_q$, i.e. the q points $\neq P_1$ on the line through $\mathbb{F}_q(z, \lambda g(z))$ and P_1. Hence the points $\mathbb{F}_q(z, y)$ on the planes in $N(Z, P_1, \mathbb{F}_q^\times g)$ are those with $y \in \mathbb{F}_q^\times g(z) + \mathbb{F}_q$, an orbit of the affine group $\mathrm{AGL}(1, \mathbb{F}_q) = \{u \mapsto \lambda u + \nu^i; \lambda \in \mathbb{F}_q^\times, \nu^i \in \mathbb{F}_q\}$ acting on $\mathbb{F}_{q^4}$. The orbits corresponding to N, N' are either disjoint and there are no collisions, or the orbits coincide and the planes in $N(Z, P_1, \mathbb{F}_q^\times g)$ and $N(Z', P_1, \mathbb{F}_q^\times g')$ are matched up in pairs covering the same line L through P_1 and a point of the form $\mathbb{F}_q(z, \lambda g(z))$. In this case we can select at most one plane from each matching pair without introducing collisions. If we do so, the selected planes will cover the same lines L as the corresponding planes in $N(Z, P_1, \mathbb{F}_q^\times g)$, say, and hence there is no obstruction to selecting exactly one plane from each pair.

It remains to show that the two cases just described are characterized by $\sigma(E) \neq \sigma(E')$ and $\sigma(E) = \sigma(E')$, respectively. For this we use the fact that $\mathbb{F}_q^\times u_1 + \mathbb{F}_q = \mathbb{F}_q^\times u_2 + \mathbb{F}_q$, or $\mathbb{F}_q u_1 + \mathbb{F}_q = \mathbb{F}_q u_2 + \mathbb{F}_q$, is equivalent to $\mathbb{F}_q(u_1^q - u_1) = \mathbb{F}_q(u_2^q - u_2)$. This is an instance of the equivalence $\delta(L_1) = \delta(L_2) \iff L_1 = L_2$ for lines L_1, L_2 through the same point (in this case the point $\mathbb{F}_q = \mathbb{F}_q 1$).[30] Using this fact and $u^q - u = \prod_{\lambda \in \mathbb{F}_q}(u + \lambda)$ we can rewrite the collision criterion (5) as

$$g(z)^q - g(z) = \prod_{\nu^i \in \mathbb{F}_q} \frac{\delta(d + \mu c, z) + \nu^i \delta(a, z)}{\delta(a, z)}$$

$$= \delta(a, z)^{-q} \prod_{\nu^i \in \mathbb{F}_q} \delta(d + \mu c + \nu^i a, z)$$

$$\in \delta(Z)^{-q} \prod_{\substack{L \subset E \\ \mathbb{F}_q z \in L \wedge L \neq Z}} \delta(L)$$

$$= \delta(Z)^{-q-1} \prod_{\substack{L \subset E \\ \mathbb{F}_q z \in L}} \delta(L)$$

$$= \frac{\delta(E)}{\delta(Z)^{q+1}} \cdot (\mathbb{F}_q z)^q$$

$$= \sigma(E) \cdot (\mathbb{F}_q z)^q = \sigma(E') \cdot (\mathbb{F}_q z)^q,$$

where we have used that the product of all points in E on the $q + 1$ lines through $\mathbb{F}_q z$ involves $\mathbb{F}_q z$ exactly $q + 1$ times and all other points exactly once. Cancelling the factor $(\mathbb{F}_q z)^q$ completes the proof of the lemma. $\qquad \square$

As a consequence of Lemma 6 we obtain that there exist subsets $\mathcal{N}_1' \subseteq \mathcal{N}_1$ of size $\#\mathcal{N}_1' = (q - 1) \cdot \#\mathrm{Im}(\sigma)$ which can be added to the expurgated LMRD code $\mathcal{L}_0$ while still maintaining $t = 2$. For this we choose for each point Q in the image of σ a plane $E \neq W$ with $\sigma(E) = Q$ and take $\mathcal{N}_1'$ as the union of all sets

---

[30]It is also straightforward to show directly that $u \mapsto (u^q - u)^{q-1}$ is a separating invariant for the orbits of $\mathrm{AGL}(1, \mathbb{F}_q)$ on $\mathbb{F}_{q^4}$, i.e. $\mathbb{F}_q^\times u_1 + \mathbb{F}_q = \mathbb{F}_q^\times u_2 + \mathbb{F}_q$ iff $(u_1^q - u_1)^{q-1} = (u_2^q - u_2)^{q-1}$.

$N(Z, P_1, \mathbb{F}_q^\times g)$ parametrized by these planes. In the smallest case $q = 2$, where $\#N(Z, P_1, \mathbb{F}_q^\times g) = 1$, such a set $\mathcal{N}_1'$ is clearly maximal.[31]

Hence our next goal is to obtain more detailed information on the map $E \mapsto \sigma(E)$ with domain the set of $q^3 + q^2 + q$ planes $E \neq W$ in $\mathrm{PG}(\mathbb{F}_{q^4}/\mathbb{F}_q)$, and in particular determine its image size. As a first step towards this we establish an explicit formula for $\sigma(E)$. The formula is stated in terms of the absolute invariant $\sigma(E)^{q-1} \in \mathbb{F}_{q^4}^\times$, which is obtained by composing $E \mapsto \sigma(E)$ with the group isomorphism $\mathbb{F}_{q^4}^\times/\mathbb{F}_q^\times \to (\mathbb{F}_{q^4}^\times)^{q-1}$, $r\mathbb{F}_q^\times \mapsto r^{q-1}$.

Lemma 7. *For a plane* $E = aW \neq W$ *of* $\mathrm{PG}(\mathbb{F}_{q^4}/\mathbb{F}_q)$ *we have*

$$\sigma(E)^{q-1} = 1 - \frac{a^{(q-1)(q^2+1)} - 1}{a^{q-1} - 1}.$$

Proof. First we show $\delta(W) = \mathbb{F}_q\epsilon$ or, equivalently, $\delta(W)^{q-1} = -1$, with ϵ as in Lemma 5. From $X^{q^3} + X^{q^2} + X^q + X = \prod_{w \in W}(X - w)$ the product of all elements in $W \setminus \{0\}$ is 1. For a point $P = \mathbb{F}_q x$ the quantity $\delta(P)^{q-1} = x^{q-1}$ differs from $\prod_{x \in P \setminus \{0\}} x = \prod_{\lambda \in \mathbb{F}_q^\times}(\lambda x) = -x^{q-1}$ just by its sign. Hence we have $\delta(W)^{q-1} = (-1)^{q^2+q+1} \prod_{w \in W \setminus \{0\}} w = (-1)^{q^2+q+1} = -1$ as claimed.[32]

This gives $\delta(aW) = \mathbb{F}_q a^{q^2+q+1}\epsilon$ and $\delta(aW)^{q-1} = -a^{q^3-1}$ for any $a \in \mathbb{F}_{q^4}^\times$.

Since $\sigma(aW)^{q-1} = \delta(aW)^{q-1}/\delta(Z)^{q^2-1}$, where $Z = W \cap aW$, we also need to compute $\delta(W \cap aW)$. This can be done as follows:

The $\mathbb{F}_q$-space $W \cap aW$ is the set of zeros of the polynomial

$$X^{q^3} + X^{q^2} + X^q + X - a^{q^3}\left((a^{-1}X)^{q^3} - (a^{-1}X)^{q^2} - (a^{-1}X)^q - a^{-1}X\right)$$

$$= (1 - a^{q^3-q^2})X^{q^2} + (1 - a^{q^3-q})X^q + (1 - a^{q^3-1})X$$

$$= (1 - a^{q^3-q^2})\left(X^{q^2} + \frac{1 - a^{q^3-q}}{1 - a^{q^3-q^2}}X^q + \frac{1 - a^{q^3-1}}{1 - a^{q^3-q^2}}X\right),$$

and hence

$$\delta(W \cap aW)^{q-1} = \frac{1 - a^{q^3-1}}{1 - a^{q^3-q^2}},$$

$$\sigma(aW)^{q-1} = \frac{-a^{q^3-1}(1 - a^{q^3-q^2})^{q+1}}{(1 - a^{q^3-1})^{q+1}}$$

$$= -\frac{a^{q^3-1}(1 - a^{1-q^3})(1 - a^{q^3-q^2})}{(1 - a^{1-q})(1 - a^{q^3-1})}$$

$$= \frac{1 - a^{q^3-q^2}}{1 - a^{1-q}}$$

$$= \frac{a^{q-1} - a^{q^3-q^2+q-1}}{a^{q-1} - 1}$$

---

[31] Whether such sets $\mathcal{N}_1'$ are maximal in general remains an open problem.
[32] Note that the last equality is trivially true for even q.

$$= 1 - \frac{a^{q^3 - q^2 + q - 1} - 1}{a^{q-1} - 1}$$

$$= 1 - \frac{a^{(q-1)(q^2+1)} - 1}{a^{q-1} - 1},$$

as asserted. □

From Lemma 7 it is clear that $\sigma(E) = \mathbb{F}_q$ for the planes of the form $E = a^{q+1}W \neq W$ and no other planes. Since there are q^2 such planes, we have that $\#\operatorname{Im}(\sigma) \leq q^3 + q^2 + q - (q^2 - 1) = q^3 + q + 1$. It turns out that equality holds in this bound, and hence a maximum of $\#\mathcal{N}_1' = (q-1)(q^3 + q + 1)$ planes passing through any given point $P \in S$ can be added to $\mathcal{L}_0$ without increasing t. Before proving this theorem, we note that the existence of collisions already implies that a q-analogue of the Fano plane cannot be constructed by our present method.

Theorem 3. *Let $\mathcal{L}_0$ be the plane subspace code of size $q^8 - q^7 + q^3$ obtained from the lifted Gabidulin code $\mathcal{L}$ by removing all planes Γ_f corresponding to binomials $f(x) = r(ux^q - u^q x)$ with $r \in \mathbb{F}_{q^4}^\times$, $u \in \mathbb{F}_{q^4} \setminus W$. Then $\mathcal{L}_0$ can be augmented by $(q^4 - 1)(q^3 + q + 1)$ new planes meeting S in a point, $(q-1)(q^3 + q + 1)$ of them passing through any point $P \in S$, to a subspace code $\mathcal{C}$ with size $\#\mathcal{C} = q^8 + q^5 + q^4 - q - 1$. Moreover, $\mathcal{C}$ may be chosen as a Σ-invariant code.*

Proof. As discussed above, we need only show that the values of σ on the $q^3 + q$ planes not of the form $a^{q+1}W$ are distinct. This is equivalent to

$$\frac{x - 1}{y - 1} \neq \frac{x^{q^2+1} - 1}{y^{q^2+1} - 1} \tag{6}$$

for any pair of distinct elements $x, y \in \mathbb{F}_{q^4}^\times$ that are $(q-1)$-th powers but not $(q^2 + 1)$-th roots of unity.

Assume by contradiction that equality holds in (6) for some pair x, y. Then, since the right-hand side is in the subfield $\mathbb{F}_{q^2}$, we can conclude that also

$$\frac{x - 1}{y - 1} = \left(\frac{x - 1}{y - 1}\right)^{q^2} = \frac{x^{q^2} - 1}{y^{q^2} - 1}.$$

The two equations can be rewritten as

$$\sum_{i=0}^{q^2} x^i = \frac{x^{q^2+1} - 1}{x - 1} = \frac{y^{q^2+1} - 1}{y - 1} = \sum_{i=0}^{q^2} y^i,$$

$$\sum_{i=0}^{q^2-1} x^i = \frac{x^{q^2} - 1}{x - 1} = \frac{y^{q^2} - 1}{y - 1} = \sum_{i=0}^{q^2-1} y^i,$$

and together imply $x^{q^2} = y^{q^2}$ and hence $x = y$; contradiction. □

Remark 2. The map $\mathbb{F}_q a \mapsto \sigma(aW)$ leaves each coset of the subgroup consisting of the $(q+1)$-th powers (or (q^2+1)-th roots of unity) in $\mathbb{F}_{q^4}^\times/\mathbb{F}_q^\times$ invariant and induces bijections on all nontrivial cosets; in particular, the set of values excluded from $\mathrm{Im}(\sigma)$ consists of the q^2 points $\neq \mathbb{F}_q$ in $\mathrm{PG}(\mathbb{F}_{q^4}/\mathbb{F}_q)$ that are of the form $\mathbb{F}_q a^{q+1}$.

This refinement of Theorem 3 follows from

$$\sigma(aW)^{(q-1)(q^2+1)} = \left(\frac{a^{q-1} - a^{(q-1)(q^2+1)}}{a^{q-1} - 1} \right)^{q^2+1} = \left(\frac{a^{q-1}(1 - a^{q^3-q^2})}{a^{q-1} - 1} \right)^{q^2+1}$$

$$= \left(\frac{a^{q^3-q^2}(1 - a^{q-1})}{a^{q^3-q^2} - 1} \right) \left(\frac{a^{q-1}(1 - a^{q^3-q^2})}{a^{q-1} - 1} \right)$$

$$= a^{q^3-q^2+q-1} = a^{(q-1)(q^2+1)},$$

which shows the claimed coset invariance, and the known behaviour of $\mathbb{F}_q a \mapsto \sigma(aW)$ on the subgroup of $(q+1)$-th powers and its complement. In the next section we will discuss the geometric significance of this subgroup.

6. Extensions

The subspace code $\mathcal{C}$ of Theorem 3 is far from being unique—we can select the $q-1$ new planes in one of the q^2 "collision classes" independently at each point of S and even mix planes from different collision classes for $q > 2$, resulting in at least $(q^2)^{q^3+q^2+q+1}$ different choices for $\mathcal{C}$ (exactly 4^{15} different choices for $q = 2$).

On the other hand, if we omit the selection of a collision class at every point of S then no ambiguity is introduced. The resulting subspace code, we call it $\mathcal{C}_0$, has size $\#\mathcal{C}_0 = \#\mathcal{C} - (q^4 - 1) = q^8 + q^5 - q$ and is clearly Σ-invariant. Moreover, the size of a maximal[33] extension $\overline{\mathcal{C}_0}$ of $\mathcal{C}_0$ is no less than the size of a maximal extension $\overline{\mathcal{C}}$ of $\mathcal{C}$.

The planes we should consider for augmenting $\mathcal{C}_0$ are essentially of two types—at most $q^4 - 1$ planes meeting S in a point and at most $\begin{bmatrix} 4 \\ 2 \end{bmatrix}_q = q^4 + q^3 + 2q^2 + q + 1$ planes meeting S in a line.[34] Hence the size of $\overline{\mathcal{C}_0}$ is a priori bounded by $q^8 + q^5 + q^4 - q - 1 \leq \#\overline{\mathcal{C}_0} \leq q^8 + q^5 + 2q^4 + q^3 + 2q^2$. For large q one may consider this as a satisfactory answer to the extension problem for $\mathcal{C}_0$, but for small values of q this is certainly not true.

For more precise results we need to describe the free lines of $\mathcal{C}_0$ meeting S in a point. Prior to this description, we collect a few geometric facts about the coset partition of $\mathbb{F}_{q^4}^\times$ relative to the subgroup O of $(q+1)$-th powers, and we prove two further auxiliary results, which seem to be of independent interest.

The point set $\mathcal{O} = \{\mathbb{F}_q a^{q+1}; a \in \mathbb{F}_{q^4}^\times\} = \{\mathbb{F}_q x; x \in \mathbb{F}_{q^4}^\times, x^{(q-1)(q^2+1)} = 1\}$ corresponding to O defines an elliptic quadric and hence an ovoid in $\mathrm{PG}(\mathbb{F}_{q^4}/\mathbb{F}_q) \cong \mathrm{PG}(3, \mathbb{F}_q)$. This can be seen by rewriting $x^{(q-1)(q^2+1)} = x^{q^3-q^2+q-1} = 1$ as

---

[33] "Maximal" refers to "maximal size", not the weaker "maximal with respect to set inclusion".

[34] Adding planes contained in S to $\mathcal{C}_0$ is not an option.

$x^{q^3+q} - x^{q^2+1} = 0$ and further as $\epsilon x^{q^3+q} - \epsilon x^{q^2+1} = 0$, where $\epsilon^{q-1} = -1$. The map $x \mapsto \epsilon x^{q^3+q} - \epsilon x^{q^2+1}$ takes values in $\mathbb{F}_q$ and hence constitutes a quadratic form on $\mathbb{F}_{q^4}/\mathbb{F}_q$. Since $\#\mathcal{O} = q^2 + 1$, the corresponding quadric must be elliptic.

Hence the coset partition with respect to O determines a partition $\mathscr{O}$ of the point set of $\mathrm{PG}(\mathbb{F}_{q^4}/\mathbb{F}_q)$ into $q + 1$ ovoids, which are transitively permuted by $\mathbb{F}_{q^4}^\times$ (acting as a Singer group).[35]

It is well-known (see e.g. [25], [26] or [27]) that $\mathcal{O}$ has a unique tangent plane in each of its points and meets the remaining $q^3 + q$ planes of $\mathrm{PG}(\mathbb{F}_{q^4}/\mathbb{F}_q)$ in $q + 1$ points (the points of a non-generate conic). The tangent plane to $\mathcal{O}$ in $\mathbb{F}_q = \mathbb{F}_q 1$ is $W' = \epsilon W$ (the plane with equation $\mathrm{Tr}(\epsilon x) = 0$), where as before $\epsilon^{q-1} = -1$. This follows from $\mathrm{Tr}(\epsilon \cdot 1) = \mathrm{Tr}(\epsilon) = \epsilon - \epsilon + \epsilon - \epsilon = 0$ and

$$\mathrm{Tr}(\epsilon a^{q+1}) = \epsilon a^{q+1} - \epsilon a^{q^2+q} + \epsilon a^{q^3+q^2} - \epsilon a^{1+q^3}$$
$$= \epsilon(a - a^{q^2})(a^q - a^{q^3}) = \epsilon(a - a^{q^2})^{q+1},$$

which shows that $\mathbb{F}_q a^{q+1} \notin \epsilon W$ unless $\mathbb{F}_q a^{q+1} = \mathbb{F}_q$.

It follows that each plane E is tangent to a unique ovoid in $\mathscr{O}$ and meets the remaining q ovoids in $q + 1$ points. More precisely, $E = aW$ is tangent to $a\mathcal{O}$ in $a\epsilon$, as follows from $\epsilon\mathcal{O} = \mathcal{O}$.[36]

In particular, W itself is tangent to $\mathcal{O}$ in $\mathbb{F}_q \epsilon$, and the points of W are partitioned into the singleton $\{\mathbb{F}_q \epsilon\}$ and q ovoid sections $W \cap \alpha^i \mathcal{O}$, $1 \leq i \leq q$, of size $q + 1$.

Now recall from Section 5 that $L \mapsto \delta(L)$ maps the pencil of all lines through $\mathbb{F}_q a$ bijectively onto the plane $a^{q+1} W$. The planes of this form are exactly the tangent planes to $\mathcal{O}$ and represent a dual ovoid $\mathcal{O}^*$ in $\mathrm{PG}(\mathbb{F}_{q^4}/\mathbb{F}_q)$. Hence we can dualize each of the above properties. In particular this gives that the q^2 planes in $\mathcal{O}^* \setminus \{W\}$ (i.e. those with $\sigma(E) = \mathbb{F}_q$, the "colliding planes") intersect W in the q^2 lines not passing through the distinguished point $\mathbb{F}_q \epsilon$.[37]

Our final and most important geometric observation relates the line orbits of the Singer group $\mathbb{F}_{q^4}^\times$ to the ovoidal fibration $\mathscr{O}$. Since $\delta(rL) = r^{q+1}\delta(L)$ for $r \in \mathbb{F}_{q^4}^\times$, every line orbit $[L]$ corresponds to a unique ovoid in $\mathscr{O}$ (the ovoid containing the point $\delta(L)$). The map $[L] \to \delta(L)\mathcal{O}$ must be a bijection, since this is true for $L \mapsto \delta(L)$ at any fixed point $\mathbb{F}_q a$ and every line orbit (resp., ovoid) contains a line through $\mathbb{F}_q a$ (resp., has a nonempty ovoid section in $a^{q+1} W$).

In fact the foregoing shows that there are q regular line orbits $[L]$ (i.e., of length $q^3 + q^2 + q + 1$) and one "short" line orbit of length $q^2 + 1$ represented by the subfield $\mathbb{F}_{q^2}$ (since $\delta(\mathbb{F}_{q^2}) = \mathbb{F}_q \epsilon$). The short orbit contains exactly one line through each point (i.e., it forms a line spread); any regular orbit contains $q + 1$ lines through

---

[35] A partition of $\mathrm{PG}(3, \mathbb{F}_q)$ into $q + 1$ ovoids is often called an *ovoidal fibration*. The ovoidal fibration $\mathscr{O}$ has been further investigated in [24].

[36] Note that $\mathbb{F}_q \epsilon \in \mathcal{O}$. For even q this is trivial. If q is odd and α is a primitive element of $\mathbb{F}_{q^4}$ then $\epsilon = \alpha^{(q^3+q^2+q+1)/2} = (\alpha^{(q^2+1)/2})^{q+1}$ satisfies $\epsilon^{q-1} = -1$ and is a $(q + 1)$-th power in $\mathbb{F}_{q^4}^\times$.

[37] The point $\mathbb{F}_q \epsilon$ represents the dual tangent plane to $\mathcal{O}^*$ in W, and the q^2 lines represent the dual lines connecting $W \in \mathcal{O}^*$ to the remaining points of $\mathcal{O}^*$.

each point $\mathbb{F}_q a$, which form a quadric cone with vertex $\mathbb{F}_q a$; in particular no three of these $q+1$ lines are coplanar.[38]

We have seen in Lemma 5 that $a, b \in W$ implies $\epsilon\delta(a, b) \in W$ (i.e. $\delta(Z) \in W' = \epsilon W$ for any line $Z = \langle a, b \rangle \subset W$). The following similar but less obvious result will be used in the sequel.

Lemma 8. *For $a, b \in W$ we also have $\epsilon a^{q^3}\delta(a, b)^{q+1} \in W$.*

This is easily seen to be equivalent to $z^{q^3}\delta(Z)^{q+1} \in W'$ for all lines Z in W and all points $\mathbb{F}_q z$ on Z.

Proof. First note that W contains a unique line $L_0 = \mathbb{F}_{q^2}\varepsilon$ of the short line orbit, which is determined by $\varepsilon^{q^2-1} = -1$.[39] Since the map $W \to \mathbb{F}_{q^4}$, $b \mapsto \epsilon a^{q^3}\delta(a, b)^{q+1}$ is constant on lines through $\mathbb{F}_q a$, it suffices to consider the cases (1) $\mathbb{F}_q a \notin L_0$, $\mathbb{F}_q b \in L_0$ and (2) $\mathbb{F}_q a \in L_0$, $b \in W$ arbitrary.

(1) Since all nonzero elements $b \in L_0$ satisfy $b^{q^2-1} = -1$, we write $b = \varepsilon$ in this case. Our task is to show that the alternating sum of the conjugates (over $\mathbb{F}_q$) of

$$a^{q^3}\delta(a, \varepsilon)^{q+1} = a^{q^3}(a\varepsilon^q - a^q\varepsilon)(a^q\varepsilon^{q^2} - a^{q^2}\varepsilon^q) = -a^{q^3}(a\varepsilon^q - a^q\varepsilon)(a^q\varepsilon + a^{q^2}\varepsilon^q)$$
$$= -a^{q^3+q+1}\varepsilon^{q+1} + a^{q^3+2q}\varepsilon^2 - a^{q^3+q^2+1}\varepsilon^{2q} + a^{q^3+q^2+q}\varepsilon^{q+1}$$

is equal to zero. Since a^{q^3+q+1} and $a^{q^3+q^2+q}$ are conjugate and $\varepsilon^{q+1} = \epsilon$, the alternating sums of the conjugates of the first and last summand cancel. For the two summands in the middle we obtain likewise

$$a^{q^3+2q}\varepsilon^2 - a^{2q^2+1}\varepsilon^{2q} + a^{2q^3+q}\varepsilon^2 - a^{q^2+2}\varepsilon^{2q}$$
$$- (a^{q^3+q^2+1}\varepsilon^{2q} - a^{q^3+q+1}\varepsilon^2 + a^{q^2+q+1}\varepsilon^{2q} - a^{q^3+q^2+q}\varepsilon^2)$$
$$= a^{q^3+q}(a^q + a^{q^3} + a + a^{q^2})\varepsilon^2 - a^{q^2+1}(a^{q^2} + a + a^{q^3} + a^q)\varepsilon^{2q} = 0,$$

since $a \in W$.

(2) Writing $a = \varepsilon$, we have

$$\varepsilon^{q^3}\delta(\varepsilon, b)^{q+1} = -\varepsilon^q(\varepsilon b^q - \varepsilon^q b)(\varepsilon^q b^{q^2} + \varepsilon b^q)$$
$$= b^{q+1}\varepsilon^{2q+1} - b^{2q}\varepsilon^{q+2} + b^{q^2+1}\varepsilon^{3q} - b^{q^2+q}\varepsilon^{2q+1}.$$

The alternating sum of the conjugates of the third summand is $b^{q^2+1}\varepsilon^{3q} + b^{q^3+q}\varepsilon^3 - b^{q^2+1}\varepsilon^{3q} - b^{q^3+q}\varepsilon^3 = 0$. For the alternating sum of the conjugates of the rest we obtain, using $(\varepsilon^{2q+1})^q = \varepsilon^{2q^2+q} = \varepsilon^{q+2}$, $(\varepsilon^{q+2})^q = \varepsilon^{q^2+2q} = -\varepsilon^{2q+1}$ and $b^{q+1} + b^2 + b^{q^3+1} = (b^q + b + b^{q^3})b = -b^{q^2+1}$, etc.,

$$(b^{q+1} - b^{q^3+q^2} - b^{2q^2} + b^2 - b^{q^2+q} + b^{q^3+1})\varepsilon^{2q+1}$$
$$+ (-b^{q^2+q} + b^{q^3+1} - b^{2q} + b^{2q^3} + b^{q^3+q^2} - b^{q+1})\varepsilon^{q+2}$$
$$= (-b^{q^2+1} + b^{q^2+1})\varepsilon^{2q+1} + (b^{q^3+q} - b^{q^3+q})\varepsilon^{q+2} = 0.$$

This completes the proof of the lemma. □

---

[38] See [28] for more information on this.

[39] Thus L_0 is the $\mathbb{F}_{q^2}$-analogue of the point $\mathbb{F}_q \varepsilon$ and can also be seen as the kernel of the relative trace map $\mathrm{Tr}_{\mathbb{F}_{q^4}/\mathbb{F}_{q^2}}$.

The second auxiliary result is the projective version of Lemma 7.

Lemma 9. *For $a \in \mathbb{F}_{q^4} \setminus \mathbb{F}_q$ we have $\sigma(aW) = \mathbb{F}_q \epsilon a^{-q}(a^q - a)^{q+1}$.*[40]

Proof. By Lemma 7,

$$\sigma(aW)^{q-1} = \frac{a^{q-1} - a^{q^3-q^2+q-1}}{a^{q-1} - 1} = \frac{a^q - a^{q^3-q^2+q}}{a^q - a} = -\frac{a^{q^3} - a^{q^2}}{a^{q^2-q}(a^q - a)}$$

$$= -\frac{(a^q - a)^{q^2-1}}{a^{q^2-q}} = \left(\epsilon \cdot \frac{(a^q - a)^{q+1}}{a^q}\right)^{q-1}.$$

The result follows.   □

Now we are ready to resume the analysis of augmenting $\mathcal{C}_0$. Recall from Section 5 that the $(q-1)(q^3+q)$ planes in $\mathcal{C}_0$ meeting S in $P_1 = \mathbb{F}_q(0,1)$ have the form $N = N(Z, P_1, g) = \{(x, g(x) + v^i); x \in Z, v^i \in \mathbb{F}_q\}$, where $Z = \langle a, b \rangle \subset W$ is 2-dimensional, $g(x) = \delta(\lambda d + \mu c, x)/\delta(a,b)$ and the plane $E = \langle a, b, \lambda d + \mu c \rangle$ is not of the form $u^{q+1}W$.

In what follows, by a *free line* we mean a line not covered by a plane in $\mathcal{C}_0$, and by a *free plane* a plane which contains only free lines and hence can be individually added to $\mathcal{C}_0$ without increasing t. From Section 5 we know that the $(q-1)q^2$ planes $N(Z, P_1, g)$ with E of the form $u^{q+1}W$ and their images under Σ are free. We will denote this set of $(q^4 - 1)q^2$ free planes by $\mathcal{N}''$, so that $\mathcal{N} = \mathcal{N}' \uplus \mathcal{N}''$ in the terminology of Section 5.

For the statement of the next lemma recall that the 4-flats in $\mathrm{PG}(W \times \mathbb{F}_{q^4})$ above S are of the form $F = \mathbb{F}_q x \times \mathbb{F}_{q^4} = \mathbb{F}_q(x, 0) + S$ with $\mathbb{F}_q x$ a point in W (i.e. $x \in W$ is uniquely determined up to scalar multiples in $\mathbb{F}_q^\times$).

Lemma 10. *Let $F = \mathbb{F}_q x \times \mathbb{F}_{q^4}$ be a 4-flat containing S and $P_0 = \mathbb{F}_q(x, 0) = F \cap W$.*

(1) A line $L \subset F$ meeting S in a point is free if and only if either $P_0 \in L$ or the plane generated by P_0 and L meets S in a line L' such that $\delta(L') \in x^q \mathcal{O}$.

(2) A plane $E \subset F$ meeting S in a line L' is free if and only if $P_0 \in E$ and $\delta(L') \in x^q \mathcal{O}$.

Note that, in view of the preceding discussion, the condition $\delta(L') \in x^q \mathcal{O}$ holds precisely for the lines L' in a certain line orbit of $\mathbb{F}_{q^4}^\times$ on $\mathrm{PG}(S/\mathbb{F}_q) \cong \mathrm{PG}(\mathbb{F}_{q^4}/\mathbb{F}_q)$. Points $\mathbb{F}_q x$, $\mathbb{F}_q x'$ in the same ovoid section $W \cap x\mathcal{O} = W \cap x'\mathcal{O}$ are associated with the same line orbit, and the induced map from ovoid sections to line orbits is a bijection.[41] Moreover, the degenerate ovoid section $\{\mathbb{F}_q \epsilon\}$ is associated with the short line orbit $[\mathbb{F}_{q^2}]$ (since $\delta(\mathbb{F}_{q^2}) = \mathbb{F}_q \epsilon \in \mathcal{O} = \epsilon^q \mathcal{O}$).

---

[40]Note that $aW = W$ is equivalent to $a \in \mathbb{F}_q^\times$ (e.g., by Singer's Theorem).

[41]The ovoid $x^q \mathcal{O} = (x\mathcal{O})^q$ differs from $x\mathcal{O}$ only by conjugation with the Frobenius automorphism of $\mathbb{F}_{q^4}/\mathbb{F}_q$. If we choose orbit representatives with $1 \in L'$ then the condition of the lemma becomes $\delta(L') \in W \cap x^q \mathcal{O}$, the conjugate ovoid section in W.

Proof of the lemma. Since the sets of free lines and free planes, as well as the stated conditions, are Σ-invariant, it suffices to consider the cases $P_1 \in L$ and $P_1 \in E$.

(1) The line $L_0 = \langle P_0, P_1 \rangle = \mathbb{F}_q x \times \mathbb{F}_q$ is free, since $N = N(Z, P_1, g) \in \mathcal{C}_0$ has $g(x) = \delta(\lambda d + \mu c, x)/\delta(a, x) \notin \mathbb{F}_q$. The remaining $q^3 - 1$ lines in F meeting S in P_1 have the form $L = \mathbb{F}_q(x, y) + P_1$ with $y \in \mathbb{F}_{q^4} \setminus \mathbb{F}_q$ and correspond to nontrivial additive cosets of $\mathbb{F}_q$ in $\mathbb{F}_{q^4}$. Inspecting the proof of Lemma 6 shows that such a line L is free iff $\mathbb{F}_q(y^q - y) = \mathbb{F}_q \delta(1, y) \neq x^q \sigma(E)$ for all planes E through $\mathbb{F}_q x$ with $E \notin \mathcal{O}^*$. Since this condition depends only on $\mathbb{F}_q^\times y$, the free lines form a union of planes through L_0 whose intersecting lines $L' = \langle 1, y \rangle$ with S are determined by the conditions $\delta(L') \neq x^q \sigma(E)$.[42]

The planes $E = uW$ containing $\mathbb{F}_q x$ are characterized by $x/u \in W$. One such plane is W, which will be excluded from now on. Using Lemma 9, homogeneity of δ and Lagrange's Theorem for the group $\mathbb{F}_{q^4}^\times / \mathbb{F}_q$, we obtain

$$x^q \sigma(uW) = \mathbb{F}_q \epsilon (x/u)^q \delta(1, u)^{q+1} = \mathbb{F}_q \epsilon (u/x)^{q^2+q+1} \delta(x/u, x)^{q+1}$$
$$= \mathbb{F}_q \epsilon (x/u)^{q^3} \delta(x/u, x)^{q+1}.$$

By Lemma 8, $x^q \sigma(uW) \in W$ for all planes $uW \neq W$ containing $\mathbb{F}_q x$. Now we distinguish two cases.

Case 1: $\mathbb{F}_q x = \mathbb{F}_q \epsilon$. In this case, since no plane in $\mathcal{O}^*$ except W passes through $\mathbb{F}_q \epsilon$, all $q^2 + q$ planes $uW \neq W$ containing $\mathbb{F}_q \epsilon$ provide a condition $\delta(L') \neq \epsilon^q \sigma(uW)$. But $\delta(L') \in W$ and the invariants $\sigma(uW)$ are distinct and $\neq 1$. Hence $\delta(L') = \mathbb{F}_q \epsilon^q = \mathbb{F}_q \epsilon$ remains as the only possibility. This implies $L' = \mathbb{F}_{q^2}$ and $\delta(L') \in \epsilon \mathcal{O} = \mathcal{O}$ as asserted.

Case 2: $\mathbb{F}_q x \neq \mathbb{F}_q \epsilon$. In this case exactly q of the planes in $\mathcal{O}^*$ pass through $\mathbb{F}_q x$ and the condition $\delta(L') \neq x^q \sigma(uW)$ applies to q^2 planes. Since $(x/u)^{q^3} \in x^{q^3} \mathcal{O}$ iff $u^{q^3} \in \mathcal{O}$ iff $u \in \mathcal{O}$, we must have $x^q \sigma(uW) \notin x^{q^3} \mathcal{O}$ for these q^2 planes. Hence the q^2 values taken by $x^q \sigma(uW)$ form the complementary set $W \setminus x^{q^3} \mathcal{O}$ and the condition reduces to $\delta(L') \in x^{q^3} \mathcal{O}$. Since $x^{q^3-q} = (x^{q^2-q})^{q+1} \in \mathcal{O}$, this is in turn equivalent to $\delta(L') \in x^q \mathcal{O}$ as asserted.

(2) Clearly any plane satisfying these conditions is free. Conversely, if E is free and $P_0 \in E$ then Part (1) can be applied to any line $L \subset E$ satisfying $P_0 \notin L \neq L'$ and gives $\delta(L') \in x^q \mathcal{O}$. Thus it remains to show that in the case $P_0 \notin E$ the plane E cannot be free.

Consider the solid $T = \langle E, P_0 \rangle$, which meets S in a plane $E' \supset L'$. Connecting P_0 to the $q^2 + q$ lines $\neq L'$ in E and applying Part (1) gives that all lines $\neq L'$ in E' must be in the same line orbit of $\mathbb{F}_{q^4}^\times$ in $\mathrm{PG}(S/\mathbb{F}_q)$. Since E' contains no more than $q + 1$ lines of any line orbit,[43] we have a contradiction, and the proof of the lemma is complete. $\qquad\qquad\square$

---

[42]The points in $\langle L_0, L' \rangle$ on the $q - 1$ lines through P_1 different from L_0, L' are those of the form $\mathbb{F}_q(x, y')$ with y' in the $\mathrm{AGL}(1, \mathbb{F}_q)$-orbit $\mathbb{F}_q^\times y + \mathbb{F}_q$.
[43]More precisely, the lines in E' fall into $q + 1$ orbits—a single line in the short orbit and $q + 1$ lines forming a dual conic in each of the q regular orbits.

In the sequel we write $\mathcal{E}$ for the set of free planes meeting S in a line. Part (2) of Lemma 10 says that the planes in $\mathcal{E}$ have the form $\mathbb{F}_q x \times L'$ ("decomposable" planes) with L' in the line orbit associated to $\mathbb{F}_q x$.

Clearly the largest extension (still having $t = 2$) of $\mathcal{C}_0$ by planes in $\mathcal{E}$ is obtained in the following way: (1) Add all $q^2 + 1$ planes generated by $\mathbb{F}_q(\epsilon, 0)$ and a line in the short line orbit of $\mathbb{F}_{q^4}^\times$ on $\mathrm{PG}(S/\mathbb{F}_q)$. These planes have the form $\mathbb{F}_q \epsilon \times (\mathbb{F}_{q^2})r$ with $r \in \mathbb{F}_{q^4}^\times / \mathbb{F}_{q^2}^\times$. (2) For each ovoid section $W \cap x\mathcal{O}$ of size $q+1$ decompose the associated regular line orbit $[L']$ of $\mathbb{F}_{q^4}^\times$ on $\mathrm{PG}(S/\mathbb{F}_q)$ into $q+1$ mutually disjoint partial spreads and a remainder of minimum size (i.e., the union of the partial spreads, a subset of $[L']$, should have maximum size). Choose a bijection from $W \cap x\mathcal{O}$ to the set of these partial spreads and add all planes $\mathbb{F}_q x' \times L$ with $\mathbb{F}_q x' \in W \cap x\mathcal{O}$ and L a line in the partial spread corresponding to $\mathbb{F}_q x'$.

For small values of q it turns out that the regular Singer line orbits of $\mathrm{PG}(3, \mathbb{F}_q)$ admit decompositions into fairly large partial spreads. As a consequence, maximal extensions of $\mathcal{C}_0$ by planes in $\mathcal{E}$ improve on the code $\mathcal{C}$ of Theorem 3. Below we will discuss in more detail the cases $q = 2, 3$, where the number of additional planes is 29 and 114 respectively.[44]

Of course we are ultimately interested in finding the largest extension of $\mathcal{C}_0$ by planes of any of the two types. For $q = 2$ it turns out that all but one of the theoretical maximum of $15 + 29 = 44$ additional planes can be added to $\mathcal{C}_0$, resulting in the largest presently known subspace code $\overline{\mathcal{C}_0}$ of size $286 + 43 = 329$; cf. [10, 12]. This case will be considered further below, culminating in a computer-free construction of one such code.

It seems difficult, however, to generalize the analysis in the binary case to larger values of q. In the ternary case $q = 3$ the largest extension of $\mathcal{C}_0$ we have found by a computer search has size $\#\overline{\mathcal{C}_0} = 6977$,[45] but we do not yet know whether this is the true maximum.

We summarize our present knowledge about the extension problem for $\mathcal{C}_0$ in the following theorem. Part (1) and (2) are the result of a computer search. For the computation of canonical forms and automorphism groups of subspace codes, the algorithm in [29] (based on [30], see also [31]) is used. Part (3) represents a slight improvement of Theorem 3 for general q.

Theorem 4. *Let $\mathcal{C}_0$ be the plane subspace code of size $q^8 + q^5 - q$ obtained by the expurgation-augmentation process described in Section 5 and with no planes in $\mathcal{N}''$ selected.*

(1) *For $q = 2$ maximal extensions $\overline{\mathcal{C}_0}$ of $\mathcal{C}_0$ have size $\#\overline{\mathcal{C}_0} = 329$. There exist 26 496 different isomorphism types of such extensions, all with trivial automorphism group. Moreover, both possible intersection patterns with S, viz.*

---

[44]Compare this with the number $q^4 - 1 = 15$ resp. 80 of planes in $\mathcal{N}''$ that can be added to $\mathcal{C}_0$, and also with the theoretical maximum of $\begin{bmatrix} 4 \\ 2 \end{bmatrix}_q = 35$ resp. 130 additional planes in $\mathcal{E}$.

[45]Compare this with the upper bound $\#\overline{\mathcal{C}_0} \le 6801 + 80 + 114 = 6995$.

$(a_0, a_1, a_2, a_3) = (136, 164, 29, 0)$ *and* $(136, 165, 28, 0)$, *occur with numbers of isomorphism types* 10 368 *and* 16 128, *respectively*.

(2) *For* $q = 3$ *there exists an extension* $\overline{C_0}$ *of size* $\#\overline{C_0} = 6977$.

(3) *For general* q *there exists an extension* $\overline{C_0}$ *of size* $\#\overline{C_0} = q^8 + q^5 + q^4 + q^2 - q$, *obtained by adding to the subspace code* C *of Theorem 3 the* $q^2 + 1$ *planes in* $\mathcal{E}$ *of the form* $\mathbb{F}_q \epsilon \times (\mathbb{F}_{q^2}) r$, $r \in \mathbb{F}_{q^4}^\times / \mathbb{F}_{q^2}^\times$.

Proof of Part (3). Since W is the tangent plane to $\mathcal{O}$ in $\mathbb{F}_q \epsilon$, W meets the remaining q^2 tangent planes in $\mathcal{O}^*$ in the q^2 lines not passing through $\mathbb{F}_q \epsilon$. This means that the planes in $\mathcal{N}''$ have the form $N(Z, P, g)$ with Z not passing through $\mathbb{F}_q \epsilon$ and hence do not interfere with the $q^2 + 1$ new planes, which have the form $E = E(\mathbb{F}_q \epsilon, (\mathbb{F}_{q^2}) r, 0)$.[46] $\qquad\square$

In the remainder of this section we will present a computer-free construction of a maximal extension $\overline{C_0}$ in the case $q = 2$ and briefly comment on the case $q = 3$, which is remarkable in several respects.[47]

Representing $\mathbb{F}_{16}$ as $\mathbb{F}_2[\alpha]$ with $\alpha^4 + \alpha + 1 = 0$, we have $\mathbb{F}_{16}^\times = \langle \alpha \rangle$ and $W = \{1, \alpha, \alpha^2, \alpha^4, \alpha^5, \alpha^8, \alpha^{10}\}$.[48] The subfield $\mathbb{F}_4 \subset \mathbb{F}_{16}$ represents a line of PG$(3, \mathbb{F}_2)$ and generates the short line orbit $[\mathbb{F}_4] = \{\mathbb{F}_4 \alpha^{3i}; 0 \le i \le 4\}$. In addition there are two regular line orbits represented by $L_1 = \{1, \alpha, \alpha^4\}$ and $L_2 = \{1, \alpha^2, \alpha^8\} = \varphi(L_1)$. The remaining lines through 1 are $\{1, \alpha^3, \alpha^{14}\}$, $\{1, \alpha^{11}, \alpha^{12}\}$ in $[L_1]$ and $\{1, \alpha^6, \alpha^{13}\}$, $\{1, \alpha^7, \alpha^9\}$ in $[L_2]$. The ovoidal fibration is $\mathcal{O} = \{\mathcal{O}, \alpha\mathcal{O}, \alpha^2\mathcal{O}\}$ with $\mathcal{O} = \{\alpha^{3i}; 0 \le i \le 4\}$, and the corresponding W-sections are $\mathcal{O} \cap W = \{1\}$, $\alpha\mathcal{O} \cap W = \{\alpha, \alpha^4, \alpha^{10}\}$, $\alpha^2 \mathcal{O} \cap W = \{\alpha^2, \alpha^5, \alpha^8\}$.

Decomposing $[L_1]$, $[L_2]$ into partial spreads is best done in a graph-theoretic setting. We view the lines in each orbit as vertices of a circulant graph via $\alpha^i L \mapsto i \in \mathbb{Z}_{15}$. Then $\alpha^i L \cap \alpha^j L \ne \emptyset$ iff $j - i \in \{\pm 1, \pm 3, \pm 4\}$ for $L \in [L_1]$, and similarly for $[L_2]$.[49] In this way partial spreads in the line orbits correspond to cocliques (independent sets) of the associated circulant graph, and an optimal decomposition into $q + 1 = 3$ partial spreads corresponds to a 3-colorable (vertex) subgraph of maximum size. In the case under consideration the two graphs are isomorphic (since the orbits are interchanged by φ) and have chromatic number 4. It is readily seen that the maximum cocliques in Γ_1 (the graph corresponding to $[L_1]$) are $S = \{0, 2, 7, 9\}$ and its cyclic shifts modulo 15, and that $\{S, S+1, S+4\}$ forms an optimal decomposition of $[L_1]$ into 3 partial spreads of size 4 (and some remainder of size 3).

---

[46]Viewed geometrically, the planes in $\mathcal{N}''$ contain no points in $(\mathbb{F}_q \epsilon \times \mathbb{F}_{16}) \setminus S$ and hence cannot have a line with a plane $\mathbb{F}_q \epsilon \times (\mathbb{F}_{q^2}) r$ in common.

[47]Verehrter Jubilar, Sie haben sicher schon bemerkt, dass die Ordnung der multiplikativen Gruppe $\mathbb{F}_{3^4}^\times$ gerade 80 ist – eine stattliche Zahl, welche sich bei der hier vorgeschlagenen projektiven Betrachtung jedoch auf 40 verjüngt.

[48]We can represent the points of PG$(3, \mathbb{F}_2)$ by the nonzero elements of $\mathbb{F}_{16}^\times$.

[49]In general the circulant graph associated with a regular line orbit has as its connection set the pairwise differences of the logs in $\mathbb{F}_{q^4}^\times / \mathbb{F}_q^\times$ of the points on a representative line.

At this point we see that $\mathcal{C}_0$ can be extended by $29 = 5 + 4 + 4 + 4 + 4 + 4 + 4$ planes meeting S in a line. In what follows, we choose the corresponding partial line spreads as the short line orbit (a total spread) for $F = \mathbb{F}_{2^6} \times \mathbb{F}_{16} = \mathbb{F}_2 \times \mathbb{F}_{16}$ and the six partial spreads corresponding to S, $S + 1$, $S + 4$ and their images under φ.[50]

We have yet at our disposal the actual "wiring" between the six points $x \in W \setminus \{1\}$ and the six partial line spreads. Since $\delta(L_1) = 1 \cdot \alpha \cdot \alpha^4 = \alpha^5 \in \alpha^2 \mathcal{O}$, Lemma 10 only stipulates that the points in $W \cap \alpha\mathcal{O} = \{\alpha^1, \alpha^4, \alpha^{10}\}$ are connected to the three line spreads in $[L_1]$ and the points in $\{\alpha^2, \alpha^5, \alpha^8\}$ to the three line spreads in $[L_2]$. The actual choice of the bijections (out of six feasible choices for each of the two ovoid sections) should maximize the number of planes in $\mathcal{N}''$ that can be added to further extend the resulting subspace code of size $286 + 29 = 315$.

In order to solve this problem, we must take a closer look at the lines covered by the planes in $\mathcal{N}''$ and how these relate to the lines covered by the extended code of size 315. The "local" situation at $P_1 = \mathbb{F}_2(0, 1)$ is depicted in the following table:

$x \backslash L'$	5, 10	1, 4	2, 8	3, 14	6, 13	11, 12	7, 9
0		×	×	×	×	×	×
5	×	×	c	×		×	
10	×	c	×		×		×
1	×		×	c	×		×
2	×	×		×	c	×	
4	×		×		×	c	×
8	×	×		×		×	c

The rows of the table are indexed with the logs of the elements $x \in W$ (corresponding to the 4-flats F above W), the columns with pairs (i, j) such that $\alpha^i + \alpha^j = 1$ (corresponding to the lines L' in $\mathrm{PG}(\mathbb{F}_{16}/\mathbb{F}_2)$ through 1), and the table entries '×', 'c' indicate that the plane $\mathbb{F}_2 x \times L'$ conflicts with a plane in $\mathcal{C}_0$ (i.e. $\mathbb{F}_2 x \times L' \notin \mathcal{E}$), respectively, with the two planes $N = N(Z, P_1, g) \in \mathcal{N}''$ that have $x \in Z$. For this recall that in general the q planes in $\mathcal{N}''$ of the form $N(Z, P_1, g)$ with $x \in Z$ cover the same set of $q - 1$ lines meeting S in a point, and that these lines are in the plane $\mathbb{F}_q x \times L'$ with L' determined by $\delta(L') = \mathbb{F}_q x^q$.[51]

Now suppose we connect $x \in \{\alpha^1, \alpha^4, \alpha^{10}\}$ to one of the three partial line spreads in $[L_1]$, say S. Then, writing $P_r = \mathbb{F}_2(0, r)$ and using the action of Σ on $\mathcal{N}''$, we see that the planes $N = N(Z, P_r, g) \in \mathcal{N}''$ with $x \in Z$ conflict with $\mathbb{F}_2 x \times (rL')$, where L' is the line through 1 matched to x by the 'c' entries in the table. Thus there are

---

[50]This choice is closely related to the essentially unique packing of the 35 lines of $\mathrm{PG}(3, \mathbb{F}_2)$ into 7 spreads, which represents a solution to Kirkman's Schoolgirl Problem. The packing is obtained by applying a certain cyclic shift modulo 15 to the second orbit decomposition and then adding the 3 lines omitted from each orbit decomposition to the partial spreads in the other set, one at a time.

[51]The planes $E \in \mathcal{O}^* \setminus \{W\}$ parametrizing the planes in $\mathcal{N}''$ have $\sigma(E) = \mathbb{F}_q$, whence $\delta(L') = \sigma(E)x^q = \mathbb{F}_q x^q$; cf. the proof of Lemma 6.

precisely 4 values of r for which the later choice of a plane $N = N(Z, P_r, g) \in \mathcal{N}''$ with $x \in Z$ is forbidden, viz. those r for which $rL' \in \mathcal{S}$.[52] Applying the same reasoning to all $x \in W \setminus \{1\}$ and all valid choices for $\mathcal{S}$, we obtain the following 3×3 arrays of forbidden values for r. As before, elements of $\mathbb{F}_{16}^{\times}$ are represented by their logs with respect to α, and in place of the partial spreads we have listed the corresponding cocliques of the circulant graph.[53] Further, the ordering of $W \setminus \{1\}$ is chosen in such a way that the arrays are symmetric with respect to the main diagonal.[54]

$x \backslash \mathcal{S}$	$0, 2, 7, 9$	$1, 3, 8, 10$	$4, 6, 11, 13$
10	$0, 2, 7, 9$	$1, 3, 8, 10$	$4, 6, 11, 13$
1	$1, 3, 8, 10$	$2, 4, 9, 11$	$5, 7, 12, 14$
4	$4, 6, 11, 13$	$5, 7, 12, 14$	$8, 10, 0, 2$

$x \backslash \mathcal{S}$	$0, 4, 14, 3$	$2, 6, 1, 5$	$8, 12, 7, 11$
5	$0, 4, 14, 3$	$2, 6, 1, 5$	$8, 12, 7, 11$
2	$2, 6, 1, 5$	$4, 8, 3, 7$	$10, 14, 9, 13$
8	$8, 12, 7, 11$	$10, 14, 9, 13$	$1, 5, 0, 4$

The task is now to match, for each of the two tables, the row labels to the column labels in such a way that the number of points P_r that admit a non-conflicting choice $N = N(Z, P_r, g)$, i.e. a choice of Z such that r is is forbidden for no $x \in Z$, is maximized.

A moments reflection shows that the best we can do is to use the main diagonals of the tables (or one of the other two row-and-column transversals without repeated 4-tuples) for the matching, i.e. $10 \mapsto \{0, 2, 7, 9\}$, $1 \mapsto \{1, 3, 8, 10\}$, $4 \mapsto \{4, 6, 11, 13\}$, and similarly for the second table. This ensures that for each P_r at most two points $x_1, x_2 \in W \setminus \{1\}$ are forbidden and leads to a valid choice for Z unless the line through x_1, x_2 contains 1.[55] Since the only such line is $\{1, \alpha^5, \alpha^{10}\}$ and the three 4-tuples in the first row of the first table are transversal to the corresponding 4-tuples of the second table, we can make a non-conflicting choice of Z for all but one P_r. When using the two main diagonals the "bad" point is P_{11}.

In all, we can extend $\mathcal{C}_0$ by $29 + 14 = 43$ planes to a subspace code $\overline{\mathcal{C}_0}$ of size 329 as claimed.

Finally, we consider briefly the case $q = 3$. Here the number of points and lines in S are 40 and 130, respectively, with line orbit sizes 10, 40, 40, 40. Representing $\mathbb{F}_{81}$ as $\mathbb{F}_3[\alpha]$ with $\alpha^4 - \alpha^3 - 1 = 0$ (a generator of $\mathbb{F}_{81}^{\times}$) and the points of $PG(\mathbb{F}_{81}/\mathbb{F}_3)$ as α^i, $0 \leq i < 40$, we obtain

$$W = \{\alpha^5, \alpha^{13}, \alpha^{15}, \alpha^{20}, \alpha^{22}, \alpha^{25}, \alpha^{26}, \alpha^{31}, \alpha^{34}, \alpha^{35}, \alpha^{37}, \alpha^{38}, \alpha^{39}\}$$

with ovoid sections $W \cap \mathcal{O} = \{\alpha^{20}\}$, $W \cap \alpha\mathcal{O} = \{\alpha^5, \alpha^{13}, \alpha^{25}, \alpha^{37}\}$, $W \cap \alpha^2\mathcal{O} = \{\alpha^{22}, \alpha^{26}, \alpha^{34}, \alpha^{38}\}$, $W \cap \alpha^3\mathcal{O} = \{\alpha^{15}, \alpha^{31}, \alpha^{35}, \alpha^{39}\}$ and corresponding line orbit

---

[52] Since $L' \in [L_1]$ and $[L_1]$ is regular, the correspondence $r \mapsto rL'$ is a bijection.

[53] Thus, for example, $0, 2, 7, 9$ refers to the partial spread $\mathcal{S} = \{\alpha^0 L_1, \alpha^2 L_1, \alpha^7 L_1, \alpha^9 L_1\}$ with lines $\alpha^0 L_1 = \{1, \alpha, \alpha^4\}$, $\alpha^2 L_1 = \{\alpha^2, \alpha^3, \alpha^6\}$, $\alpha^7 L_1 = \{\alpha^7, \alpha^8, \alpha^{11}\}$, $\alpha^9 L_1 = \{\alpha^9, \alpha^{10}, \alpha^{13}\}$, and $0, 4, 14, 3$ to the partial spread $\mathcal{S}' = \{\alpha^0 L_2, \alpha^4 L_2, \alpha^{14} L_2, \alpha^3 L_2\} = \varphi(\mathcal{S})$.

[54] This can be done, since the offsets of the cocliques are the same as that of the lines through 1.

[55] This is the only way to block all four lines $Z \subset W$ (the passants to 1) by a 2-set.

representatives

$$L_0 = \mathbb{F}_9 = \{\alpha^0, \alpha^{10}, \alpha^{20}, \alpha^{30}\},$$
$$L_1 = \{\alpha^0, \alpha^2, \alpha^{18}, \alpha^{25}\},$$
$$L_2 = \{\alpha^0, \alpha^1, \alpha^{28}, \alpha^{37}\},$$
$$L_3 = \{\alpha^0, \alpha^5, \alpha^{11}, \alpha^{19}\}.$$

The orbit $[L_2]$ is φ-invariant and admits a decomposition into 4 spreads, corresponding to the cocliques $S = \{0, 2, 8, 10, 16, 18, 24, 26, 32, 34\}$, $S+1$, $S+4$, $S+5$.[56] The other two regular line orbits L_1, L_3 are interchanged by φ and admit an (optimal) decomposition into 5 partial spreads of size 8. For $[L_1]$ the corresponding cocliques are $T = \{1, 2, 11, 12, 21, 22, 31, 32\}$, $T+2$, $T+4$, $T+6$, $T+8$.[57] From this it follows that $\mathcal{C}_0$, of size $\#\mathcal{C}_0 = 6801$, can be extended by $10 + 4 \times 10 + 4 \times 8 + 4 \times 8 = 114$ planes in $\mathcal{E}$ to a subspace code of size 6915.

Proceeding further as in the case $q = 2$, we find that the 4×4 arrays corresponding to $[L_1]$ and $[L_3]$ do not contain row-and-column transversals with all four 8-tuples distinct. Thus the argument used in the case $q = 2$ to extend the intermediate subspace code further by planes in $\mathcal{N}''$ breaks down and the situation becomes considerably more involved. We have conducted a non-exhaustive computer search for maximal extensions of $\mathcal{C}_0$ (a more general approach than only trying to further extend one particular extension of size 6915). As already mentioned, the largest extension found in this way has size $\#\overline{\mathcal{C}_0} = 6977$.

7. Conclusion

We have developed the expurgation-augmentation approach to the construction of good subspace codes, originally presented in [11] and later extended in [12], in greater depth, providing an explicit formula (in terms of the σ-invariant) for the number of new planes meeting the special solid S in a point that can be added to the expurgated lifted Gabidulin code without introducing a multiple cover of some line, and a much refined analysis of the final extension step by planes meeting S in a line.

The existence problem for q-analogues of the Fano plane, which provided a great deal of motivation for the present work, remains grossly open, but this will not discourage us, nor should it discourage anybody else in the audience, from further attempts to resolve it—at least in the case $q = 2$, for which by Moore's Law a computer attack will become feasible in the not too distant future.

Should a q-analogue indeed exist, it may be possible to construct it using a variant of our approach, starting with either a non-Gabidulin MRD code or a smaller

---

[56]This follows from the fact that the differences $0, \pm 2 \pmod 8$ do not occur within the connection set $\{\pm 1, \pm 28, \pm 37, \pm 27, \pm 36, \pm 9\}$ of the circulant graph.

[57]Similarly due to the fact that $0, \pm 1 \pmod{10}$ do not occur within the connection set $\{\pm 2, \pm 18, \pm 25, \pm 16, \pm 23, \pm 7\}$.

set of 3×4 matrices at pairwise rank distance ≥ 2 that cannot be embedded into an MRD code.[58]

The present work may also be continued by investigating, for general q, the sizes of optimal decompositions of Singer line orbits of $\mathrm{PG}(3, \mathbb{F}_q)$ into $q + 1$ partial spreads and how these should be wired to the points of the corresponding ovoid sections in W in order to maximize further extendability by planes in $\mathcal{N}'''$; cf. the end of Section 6. This should narrow down the gap between the lower and upper bound for the size of a maximal extension $\overline{\mathcal{C}_0}$ given at the beginning of Section 6; cf. also Theorem 4 (3).

Finally we believe that large portions of the machinery developed can be generalized to subspace codes of packet lengths $v > 7$. While for larger v there is no analogue of the trace-zero subspace W and hence no canonical choice for the ambient space and its corresponding σ-invariant, it should still be possible to derive by our method some explicit results on the number of planes in $\mathcal{N}'$ that can be added to the expurgated subspace code, and to carry over the extension analysis in Section 6 to some extent.

Acknowledgments

Above all, we are indebted to Prof. Armin Leutbecher for his strong support and encouragement during various stages of our academic careers at TUM. Both of us, who are from rather different student generations, remember vividly his inspiring, seemingly spontaneous (though in fact extremely well-prepared) lectures, which on occasion could also be quite demanding. Picking just another of the many things which come to mind, the first (elder) author also has fond memories of numerous fruitful discussions about current mathematical (and lesser) topics during "Coffee time in Munich at TUM S 4438", an opportunity installed largely on the initiative of Prof. Leutbecher and actively maintained by him until his retirement.

Wir gratulieren Ihnen, lieber Herr Leutbecher, ganz herzlich zu Ihrem runden Geburtstag und wünschen Ihnen gute Gesundheit und viel Freude bei Ihren zukünftigen Unternehmungen – seien sie nun mathematischer oder eher lebensnaher Natur. Und befassen Sie sich bitte bloß nicht mit dem Existenzproblem für q-analoge Fano-Ebenen!

Further, we are grateful to an anonymous referee for the careful reading of our manuscript and valuable remarks, and to our former colleagues at TUM, in particular Jürgen Scheurle and Florian Rupp, for all their efforts to make this volume a true success.

---

[58]When finishing up our work on plane subspace codes in $\mathrm{PG}(5, \mathbb{F}_q)$, we have discovered that one of the five isomorphism types of optimal binary subspace codes of size 77 can be constructed from a set of 48 binary 3×3 matrices that is not extendable to an MRD code.

References

[1] L. Teirlinck, Non-trivial t-designs without repeated blocks exist for all t, *Discrete Mathematics*. **65**(3), 301–311 (1987).

[2] P. Keevash. The existence of designs. Preprint arXiv:1401.3665 [math.CO] (Jan., 2014).

[3] P. J. Cameron. Generalisation of Fisher's inequality to fields with more than one element. In *Combinatorics (Proc. British Combinatorial Conf., Univ. Coll. Wales, Aberystwyth, 1973)*, 9–13. London Math. Soc. Lecture Note Ser., No. 13. Cambridge Univ. Press, London (1974).

[4] S. Thomas, Designs over finite fields, *Geometriae Dedicata*. **24**, 237–242 (1987).

[5] A. Fazeli, S. Lovett, and A. Vardy, Nontrivial t-designs over finite fields exist for all t, *Journal of Combinatorial Theory, Series A*. **127**, 149–160 (2014). Preprint arXiv:1306.2088 [math.CO].

[6] R. Koetter and F. Kschischang, Coding for errors and erasures in random network coding, *IEEE Transactions on Information Theory*. **54**(8), 3579–3591 (Aug., 2008).

[7] A. Beutelspacher, Partial spreads in finite projective spaces and partial designs, *Mathematische Zeitschrift*. **145**, 211–230 (1975). Corrigendum, ibid. 147:303 (1976).

[8] S. El-Zanati, H. Jordon, G. Seelinger, P. Sissokho, and L. Spence, The maximum size of a partial 3-spread in a finite vector space over GF(2), *Designs, Codes and Cryptography*. **54**(2), 101–107 (2010).

[9] A. Kohnert and S. Kurz. Construction of large constant dimension codes with a prescribed minimum distance. In eds. J. Calmet, W. Geiselmann, and J. Müller-Quade, *Mathematical Methods in Computer Science. Essays in Memory of Thomas Beth*, number 5393 in Lecture Notes in Computer Science, 31–42. Springer-Verlag (2008).

[10] M. Braun and J. Reichelt, q-analogs of packing designs, *Journal of Combinatorial Designs*. **22**(7), 306–321 (July, 2014). Preprint arXiv:1212.4614 [math.CO].

[11] T. Honold, M. Kiermaier, and S. Kurz. Optimal binary subspace codes of length 6, constant dimension 3 and minimum subspace distance 4. In eds. G. Kyureghyan, G. L. Mullen, and A. Pott, *Topics in Finite Fields. 11th International Conference on Finite Fields and their Applications, July 22–26, 2013, Magdeburg, Germany*, vol. 632, *Contemporary Mathematics*, 157–176. American Mathematical Society (2015). Preprint arXiv:1311.0464 [math.CO].

[12] H. Liu and T. Honold. Poster: A new approach to the main problem of subspace coding. In *9th International Conference on Communications and Networking in China (ChinaCom 2014, Maoming, China, Aug. 14–16)*, 676–677 (2014). Full paper available as arXiv:1408.1181 [math.CO].

[13] A.-L. Trautmann and J. Rosenthal. New improvements on the Echelon-Ferrers construction. In ed. A. Edelmayer, *Proceedings of the 19th International Symposium on Mathematical Theory of Networks and Systems (MTNS 2010)*, 405–408, Budapest, Hungary (5–9 July, 2010). Reprint arXiv:1110.2417 [cs.IT].

[14] P. J. Cameron, Note on large sets of infinite Steiner systems, *Journal of Combinatorial Designs*. **3**(4), 307–311 (1995).

[15] M. Kiermaier and M. O. Pavčević. Intersection numbers for subspace designs. Journal of Combinatorial Designs, **23**(11), 463–480 (Nov., 2015).

[16] M. Braun, T. Etzion, P. R. J. Östergård, A. Vardy, and A. Wassermann. Existence of q-analogs of Steiner systems. Preprint arXiv:1304.1462 [math.CO] (Apr., 2013).

[17] D. Silva, F. Kschischang, and R. Koetter, A rank-metric approach to error control in random network coding, *IEEE Transactions on Information Theory*. **54**(9), 3951–3967 (Sept., 2008).

[18] P. Delsarte, Bilinear forms over a finite field, with applications to coding theory, *Journal of Combinatorial Theory, Series A.* **25**, 226–241 (1978).

[19] E. M. Gabidulin, Theory of codes with maximum rank distance, *Problems of Information Transmission.* **21**(1), 1–12 (1985).

[20] R. M. Roth, Maximum-rank array codes and their application to crisscross error correction, *IEEE Transactions on Information Theory.* **37**(2), 328–336 (Mar., 1991). Comments by Ernst M. Gabidulin and Author's Reply, ibid. 38(3):1183 (1992).

[21] T. Etzion and N. Silberstein, Error-correcting codes in projective spaces via rank-metric codes and Ferrers diagrams, *IEEE Transactions on Information Theory.* **55**(7), 2909–2919 (July, 2009).

[22] A. Beutelspacher, On parallelisms in finite projective spaces, *Geometriae Dedicata.* **3**(1), 35–40 (1974).

[23] R. H. F. Denniston. Packings of PG(3, q). In ed. A. Barlotti, *Finite Geometric Structures and their Applications*, number 60 in CIME Summer Schools, 195–199. Springer-Verlag (2011). Reprint of the 1st ed. C.I.M.E., Ed. Cremonese, Roma, 1973.

[24] G. L. Ebert, Paritioning projective geometries into caps, *Canadian Journal of Mathematics.* **37**(6), 1163–1175 (1985).

[25] J. W. P. Hirschfeld, *Projective Geometries over Finite Fields.* (Oxford University Press, 1998), 2nd edition.

[26] A. Beutelspacher and U. Rosenbaum, *Projektive Geometrie.* Number 41 in Vieweg Studium (Vieweg, 1992).

[27] P. Dembowski, *Finite Geometries.* (Springer-Verlag, 1968). Classics in Mathematics Series (1997).

[28] D. G. Glynn, On a set of lines of PG(3, q) corresponding to a maximal cap contained in the Klein quadric of PG(5, q), *Geometriae Dedicata.* **26**(3), 273–280 (1988).

[29] T. Feulner. Canonical forms and automorphisms in the projective space. Preprint arXiv:1305.1193 [cs.IT] (May, 2013).

[30] T. Feulner, The automorphism groups of linear codes and canonical representatives of their semilinear isometry classes, *Advances in Mathematics of Communications.* **3**(4), 363–383 (Nov., 2009).

[31] T. Feulner. *Eine kanonische Form zur Darstellung äquivalenter Codes – Computergestützte Berechnung und ihre Anwendung in der Codierungstheorie, Kryptographie und Geometrie.* Phd thesis, Universität Bayreuth (2014).

[32] M. Kiermaier and R. Laue, Derived and residual subspace designs, *Advances in Mathematics of Communications.* **9**(1), 105–115 (Feb., 2015).

Chapter 9

Integral Orthogonal Groups

Aloys Krieg

Lehrstuhl A für Mathematik, RWTH Aachen University, D-52056 Aachen
E-mail: `krieg@rwth-aachen.de`

We describe integral orthogonal groups of signature $(2, n)$ and results on their generators. Special emphasis is devoted to the cases $n = 2$ and $n = 3$ as the latter case corresponds to the paramodular group of degree 2.

Dedicated to Armin Leutbecher on the occasion of his
80th birthday

Contents

1. Introduction

Considering automorphic forms, the integral groups $SL(2; \mathbb{Z})$ and in higher ranks $Sp(n; \mathbb{Z})$ have played an important role for more than a century. In particular the work of Borcherds on product expansions of automorphic forms brought more attention to the orthogonal group $\mathcal{O}(2, n)$ (cf. [1]). Moreover orthogonal modular varieties have played an important role in algebraic geometry, in particular in the context of moduli spaces of polarized $K3$ surfaces. For details the reader is referred to the joint papers of Gritsenko and Hulek (e.g. [9, 10]).

The attached half-space is a Hermitian symmetric space of rank 2. If $n = 2$ we obtain two copies of the upper complex half-plane. The case $n = 3$ yields the Siegel half-space of degree 2. We have a closer look at the integral groups $\mathcal{O}(2, n)$

and determine generators as well as precise descriptions of isomorphisms with well-known other groups. Results can partly be found in the literature.

2. Real orthogonal groups

Throughout the chapter let $S = S^{(n)}$ be a positive definite $n \times n$ matrix,

$$S_0 = \begin{pmatrix} 0 & 0 & 1 \\ 0 & -S & 0 \\ 1 & 0 & 0 \end{pmatrix} \in \mathrm{Sym}(n+2; \mathbb{R}) \quad \text{of signature} \quad (1, n+1),$$

$$S_1 = \begin{pmatrix} 0 & 0 & 1 \\ 0 & S_0 & 0 \\ 1 & 0 & 0 \end{pmatrix} \in \mathrm{Sym}(n+4; \mathbb{R}) \quad \text{of signature} \quad (2, n+2).$$

The orthogonal group under consideration is given by

$$\mathcal{O}(S_1; \mathbb{R}) := \{M \in M(n+4; \mathbb{R}); \ S_1[M] = S_1\},$$

where the prime denotes the transpose and $A[B] := B'AB$ for matrices of suitable size. We will mostly use a decomposition

$$M = \begin{pmatrix} \alpha & a' & \beta \\ b & K & c \\ \gamma & d' & \delta \end{pmatrix} \in M(n+4; \mathbb{R}), \tag{1}$$

$$\alpha, \beta, \gamma, \delta \in \mathbb{R}, \ a, b, c, d \in \mathbb{R}^{n+2}, \ K \in M(n+2; \mathbb{R}).$$

Then a straightforward calculation yields

$$S_1[M] = \begin{pmatrix} 2\alpha\gamma + S_0[b] & \alpha d' + \gamma a' + b' S_0 K & \alpha\delta + \beta\gamma + b' S_0 c \\ \alpha d + \gamma a + K' S_0 b & a d' + d a' + S_0[K] & \delta a + \beta d + K' S_0 c \\ \alpha\delta + \beta\gamma + c' S_0 b & \delta a' + \beta d' + c' S_0 K & 2\beta\delta + S_0[c] \end{pmatrix}. \tag{2}$$

Here one can read off the fundamental relations, i.e. the identity $S_1[M] = S_1$ in terms of coefficients. Given $M \in \mathcal{O}(S_1; \mathbb{R})$ one calculates

$$M^{-1} = S_1^{-1} M' S_1 = \begin{pmatrix} \delta & c' S_0 & \beta \\ S_0^{-1} d & S_0^{-1} K' S_0 & S_0^{-1} a \\ \gamma & b' S_0 & \alpha \end{pmatrix}. \tag{3}$$

Using (2) a simple calculation leads to

Proposition 1. *The following matrices belong to* $\mathcal{O}(S_1; \mathbb{R})$:

a) $\quad M_0 := \begin{pmatrix} 0 & 0 & 1 \\ 0 & E & 0 \\ 1 & 0 & 0 \end{pmatrix}, \quad \begin{pmatrix} \alpha & 0 & 0 \\ 0 & E & 0 \\ 0 & 0 & 1/\alpha \end{pmatrix}, \quad 0 \neq \alpha \in \mathbb{R},$

b) $\quad J := \begin{pmatrix} 0 & 0 & -1 \\ 0 & V & 0 \\ -1 & 0 & 0 \end{pmatrix}, \quad V := \begin{pmatrix} 0 & 0 & -1 \\ 0 & E & 0 \\ -1 & 0 & 0 \end{pmatrix},$

c) $\quad M_\lambda := \begin{pmatrix} 1 & -\lambda' S_0 & -\frac{1}{2} S_0[\lambda] \\ 0 & E & \lambda \\ 0 & 0 & 1 \end{pmatrix}, \quad \lambda \in \mathbb{R}^{n+2},$

d) $\quad \widetilde{M}_\lambda := M_0 M_\lambda M_0 = J M_{V\lambda} J = \begin{pmatrix} 1 & 0 & 0 \\ \lambda & E & 0 \\ -\frac{1}{2} S_0[\lambda] & -\lambda' S_0 & 1 \end{pmatrix}, \quad \lambda \in \mathbb{R}^{n+2},$

e) $\quad R_K := \begin{pmatrix} 1 & 0 & 0 \\ 0 & K & 0 \\ 0 & 0 & 1 \end{pmatrix}, \quad K \in \mathcal{O}(S_0; \mathbb{R}),$

in particular $R_{(-V)}$ and for $\mu \in \mathbb{R}^n$

$$R_{K_\mu}, \ R_{\widetilde{K}_\mu}, \ K_\mu = \begin{pmatrix} 1 & \mu' S & \frac{1}{2} S[\mu] \\ 0 & E & \mu \\ 0 & 0 & 1 \end{pmatrix}, \quad \widetilde{K}_\mu = \begin{pmatrix} 1 & 0 & 0 \\ \mu & E & 0 \\ \frac{1}{2} S[\mu] & \mu' S & 1 \end{pmatrix},$$

f) $\quad M_D := \begin{pmatrix} D & 0 & 0 \\ 0 & E & 0 \\ 0 & 0 & D^* \end{pmatrix}, \quad D = \begin{pmatrix} \alpha & \beta \\ \gamma & \delta \end{pmatrix} \in \mathrm{SL}(2; \mathbb{R}),$

$$D^* = \begin{pmatrix} \alpha & -\beta \\ -\gamma & \delta \end{pmatrix} = D[I], \quad I = \begin{pmatrix} -1 & 0 \\ 0 & 1 \end{pmatrix},$$

g) $\quad \widetilde{M}_D := R_V M_{D[I]} R_V = \begin{pmatrix} \alpha E^{(2)} & 0 & \beta I \\ 0 & E^{(n)} & 0 \\ \gamma I & 0 & \delta E^{(2)} \end{pmatrix}, \quad D = \begin{pmatrix} \alpha & \beta \\ \gamma & \delta \end{pmatrix} \in \mathrm{SL}(2; \mathbb{R}).$

Here E stands for the identity matrix of suitable size.

Let $\mathcal{H}$ denote the upper half-plane in $\mathbb{C}$ and let $\mathcal{H}_S$ denote the half-space attached to S, i.e. (cf. [13])

$$\mathcal{H}_S := \{ w = u + iv = \begin{pmatrix} \tau \\ z \\ \omega \end{pmatrix} \in \mathbb{C}^{n+2}; \ S_0[v] > 0, \ \tau, \omega \in \mathcal{H} \}.$$

Then $\mathcal{O}(S_1; \mathbb{R})$ acts on $\mathcal{H}_S \cup (-\mathcal{H}_S)$ via

$$w \mapsto M\langle w \rangle := \left(-\tfrac{1}{2}S_0[w] \cdot b + Kw + c\right) \cdot (M\{w\})^{-1},$$
$$M\{w\} := -\tfrac{1}{2}\gamma S_0[w] + d'w + \delta.$$

In view of Proposition 1 we have

$$
\begin{cases}
J\langle w \rangle = \dfrac{-1}{\frac{1}{2}S_0[w]}\begin{pmatrix} \omega \\ -z \\ \tau \end{pmatrix}, \\[2em]
M_\lambda \langle w \rangle = w + \lambda, \\[1em]
R_K \langle w \rangle = Kw, \\[1em]
M_D \langle w \rangle = \begin{pmatrix} \tau - \frac{\gamma}{2(\gamma\omega+\delta)}S[z] \\ \frac{1}{\gamma\omega+\delta}z \\ \frac{\alpha\omega+\beta}{\gamma\omega+\delta} \end{pmatrix}, \quad
\widetilde{M}_D \langle w \rangle = \begin{pmatrix} \frac{\alpha\tau+\beta}{\gamma\tau+\delta} \\ \frac{1}{\gamma\tau+\delta}z \\ \omega - \frac{\gamma}{2(\gamma\tau+\delta)}S[z] \end{pmatrix}.
\end{cases}
\tag{4}
$$

Let $\mathcal{O}_0(S_1; \mathbb{R})$ denote the subgroup of matrices in $\mathcal{O}(S_1; \mathbb{R})$, which map $\mathcal{H}_S$ onto $\mathcal{H}_S$. This is the connected component of E in $\mathcal{O}(S_1; \mathbb{R})$. Given

$$M = \begin{pmatrix} * & * & * \\ C & * & D \end{pmatrix} \in \mathcal{O}(S_1; \mathbb{R}), \quad C, D \in M(2; \mathbb{R}),$$

one has

$$M \in \mathcal{O}_0(S_1; \mathbb{R}) \iff \det(CP + D) > 0, \quad P = \begin{pmatrix} 0 & 1 \\ 1 & 0 \end{pmatrix}, \tag{5}$$

because of $ie = i(1, 0, \ldots, 1)' \in \mathcal{H}_S$. Let $\mathcal{O}_0(S_0; \mathbb{R})$ denote the subgroup of matrices $K \in \mathcal{O}(S_0, \mathbb{R})$ such that R_K belongs to $\mathcal{O}_0(S_1; \mathbb{R})$. Given

$$K = \begin{pmatrix} * & * & * \\ \gamma & * & \delta \end{pmatrix} \in \mathcal{O}(S_0; \mathbb{R}), \quad \gamma, \delta \in \mathbb{R},$$

one has

$$K \in \mathcal{O}_0(S_0; \mathbb{R}) \iff \gamma + \delta > 0. \tag{6}$$

If $K \in \mathcal{O}(S_0; \mathbb{R})$ thus either K or $-K$ belongs to $\mathcal{O}_0(S_0; \mathbb{R})$.

3. The integral orthogonal groups

For the remaining part let S be positive definite and even, i.e. $S[\lambda] \in 2\mathbb{Z}$ for all $\lambda \in \mathbb{Z}^n$. Denote the associated *modular group* by

$$\Gamma_S := \mathcal{O}_0(S_1; \mathbb{Z}) := \mathcal{O}_0(S_1; \mathbb{R}) \cap M(n + 4; \mathbb{Z}).$$

Using Proposition 1 and (5) we obtain

$$J \in \Gamma_S, \quad M_\lambda \in \Gamma_S \quad \text{for all} \quad \lambda \in \mathbb{Z}^{n+2}. \tag{7}$$

Proposition 2. *The subgroup of Γ_S generated by J and $M_\lambda, \lambda \in \mathbb{Z}^{n+2}$, contains the following matrices*

a) $\widetilde{M}_\lambda, \quad \lambda \in \mathbb{Z}^{n+2}$,

b) $\begin{pmatrix} \varepsilon & 0 & 0 \\ 0 & W & 0 \\ 0 & 0 & \varepsilon \end{pmatrix}$, $\quad W = (E - \varepsilon \lambda \lambda' S_0) V$,

for $\lambda \in \mathbb{Z}^{n+2}$ such that $\varepsilon = \frac{1}{2} S_0[\lambda] = \pm 1$, *inparticular*

$$M_1 = \begin{pmatrix} -E^{(2)} & 0 & 0 \\ 0 & E & 0 \\ 0 & 0 & -E^{(2)} \end{pmatrix},$$

c) $M_D, \widetilde{M}_D, \quad D \in \mathrm{SL}(2; \mathbb{Z})$,

d) $R_K, K = K_\mu = \begin{pmatrix} 1 & \mu'S & \frac{1}{2}S[\mu] \\ 0 & E & \mu \\ 0 & 0 & 1 \end{pmatrix}$, $\quad K = \widetilde{K}_\mu = \begin{pmatrix} 1 & 0 & 0 \\ \mu & E & 0 \\ \frac{1}{2}S[\mu] & \mu'S & 1 \end{pmatrix}$, $\quad \mu \in \mathbb{Z}^n$.

Proof. a) Use Proposition 1 c).
b) Use the Proposition in [13] and take $\lambda = (1, 0, \dots, 0, -1)'$ in order to get M_1.
c) If $D = \begin{pmatrix} 1 & 1 \\ 0 & 1 \end{pmatrix}$ resp. $D = \begin{pmatrix} 1 & 0 \\ 1 & 1 \end{pmatrix}$ one has

$$M_D = M_{-e_{n+2}}, \quad \widetilde{M}_D = M_{e_1} \quad \text{resp.} \quad M_D = \widetilde{M}_{e_1}, \quad \widetilde{M}_D = \widetilde{M}_{-e_{n+2}}, \tag{8}$$

if the $e_i, i = 1, \dots, n+2$, denote the canonical basis of $\mathbb{R}^{n+2}$. As these matrices D generate $\mathrm{SL}(2; \mathbb{Z})$ according to [16] the claim follows.
d) One has

$$R_{K_\mu} = M_{\begin{pmatrix} 0 & -1 \\ 1 & 0 \end{pmatrix}} M_\lambda M_{\begin{pmatrix} 0 & -1 \\ 1 & 0 \end{pmatrix}}^{-1}, \quad \lambda = \begin{pmatrix} 0 \\ \mu \\ 0 \end{pmatrix},$$

$$R_{\widetilde{K}_\mu} = J^{-1} R_{K_{-\mu}} J.$$

□

We describe certain parabolic subgroups of Γ_S. We start with the parabolic subgroup $\Gamma_{S,\infty}$, which plays a crucial role in the compactification. It is characterized by the fact that the functional determinant of the associated transformation does not depend on the variable $w \in \mathcal{H}_S$.

Proposition 3. *One has*

a) $\quad \Gamma_{S,\infty} := \left\{ \begin{pmatrix} * & * \\ 0 & * \end{pmatrix} \in \Gamma_S;\ 0 = 0^{(n+3,1)} \right\} = \left\{ \begin{pmatrix} * & * \\ 0 & * \end{pmatrix} \in \Gamma_S;\ 0 = 0^{(1,n+3)} \right\}$

$$= \left\{ \pm \begin{pmatrix} 1 & * & \cdots & \cdots & * \\ 0 & * & \cdots & \cdots & \vdots \\ \vdots & \vdots & & & \\ \vdots & * & \cdots & * & * \\ 0 & \cdots & \cdots & 0 & 1 \end{pmatrix} \in \Gamma_S \right\}$$

is a subgroup of Γ_S. Each $M \in \Gamma_{S,\infty}$ has a unique representation of the form

$$M = \pm R_K M_\lambda = \pm M_{K\lambda} R_K, \quad K \in \mathcal{O}_0(S_0; \mathbb{Z}),\ \lambda \in \mathbb{Z}^{n+2}.$$

and

b) $\quad \Gamma_S^J = \left\{ \begin{pmatrix} * & * \\ 0 & * \end{pmatrix} \in \Gamma_S,\ 0 = 0^{(n+2,2)} \right\} = \left\{ \begin{pmatrix} * & * \\ 0 & * \end{pmatrix} \in \Gamma_S;\ 0 = 0^{(2,n+2)} \right\}$

is a subgroup of Γ_S. Each $M \in \Gamma_S^J$ has a unique representation of the form

$$M = R_K R_{K_\mu} M_\lambda M_D,$$

$$K = \begin{pmatrix} 1 & 0 & 0 \\ 0 & U & 0 \\ 0 & 0 & 1 \end{pmatrix},\ U \in \mathcal{O}(S; \mathbb{Z}),\ \mu \in \mathbb{Z}^n,\ \lambda = \begin{pmatrix} \lambda^* \\ 0 \end{pmatrix},\ \lambda^* \in \mathbb{Z}^{n+1}.$$

Note that $\mathcal{O}(S; \mathbb{Z})$ is a finite group as S is positive definite. We call Γ_S^J the *Jacobi group*. It is fundamental for additive lifts because Gritsenko [6] lifted Jacobi forms, i.e. automorphic forms with respect to Γ_S^J, to modular forms with respect to Γ_S.

Proof. The subset of block triangular matrices is always a subgroup. The additional zeros follow from (2).

a) In view of $b = c = 0,\ \gamma = 0$ in (1) we get $\alpha\delta = 1$ from (2), hence

$$M = \pm \begin{pmatrix} 1 & a' & \beta \\ 0 & K & c \\ 0 & 0 & 1 \end{pmatrix} = \pm M_c R_K = \pm R_K M_\lambda,\ \lambda = K^{-1} c \in \mathbb{Z}^{n+2},\ K \in \mathcal{O}_0(S_0; \mathbb{Z}).$$

b) We have

$$M = \begin{pmatrix} A & * & * \\ 0 & U & * \\ 0 & 0 & D \end{pmatrix},\quad A, D \in M(2; \mathbb{Z}),\ U \in M(n; \mathbb{Z}).$$

As M is unimodular we get $D \in \mathrm{GL}(2; \mathbb{Z})$ and then $D \in \mathrm{SL}(2; \mathbb{Z})$ by (5). Moreover (2) yields $D = A^* = A[I]$ as well as $U \in \mathcal{O}(S; \mathbb{Z})$.

Then (2) yields $K = \begin{pmatrix} 1 & 0 & 0 \\ 0 & U & 0 \\ 0 & 0 & 1 \end{pmatrix} \in \mathcal{O}_0(S_0; \mathbb{Z})$ and

$$R_K^{-1} M M_A{}^1 = \begin{pmatrix} E & * & * & * \\ 0 & E & \mu^* & \lambda^* \\ 0 & 0 & 1 & 0 \\ 0 & 0 & 0 & 1 \end{pmatrix}, \ \lambda^* \in \mathbb{Z}^{n+1}, \ \mu^* = \begin{pmatrix} \frac{1}{2} S[\mu] \\ \mu \end{pmatrix}, \ \mu \in \mathbb{Z}^n.$$

Thus (2) implies that this matrix is equal to

$$R_{K_\mu} M_\lambda, \quad \lambda = \begin{pmatrix} \lambda^* \\ 0 \end{pmatrix}.$$

$\square$

Following [2] we obtain

Theorem 1. Γ_S *is generated by the matrices*

$$J, \quad M_\lambda, \ \lambda \in \mathbb{Z}^{n+2}, \quad R_K, \ K \in \mathcal{O}_0(S_0; \mathbb{Z}).$$

Proof. Let Δ denote the subgroup of Γ_S generated by the matrices quoted above. Then the matrices quoted in Proposition 2 belong to Δ. Proposition 3 yields $\Gamma_{S,\infty} \subset \Delta$.

Now let $M \in \Gamma_S$ be arbitrary of the form (1). Multiplying by a matrix M_D, $D \in \mathrm{SL}(2; \mathbb{Z})$, from the right we may assume

$$\gamma = 0$$

without restriction. If $d = 0$ we have $\delta = \pm 1$ and $\delta = 1$ by appropriate choice of D. Otherwise we may multiply by R_{K_μ} in Proposition 2 d) from the right in order to assume

$$d_{n+2} \neq 0.$$

We have

$$gcd(d_1, \ldots, d_{n+2}, \delta) = 1$$

in view of $\Gamma_S \subset \mathrm{GL}(n + 4; \mathbb{Z})$. Let

$$P_1 = \emptyset, \quad P_j := \{p \in \mathbb{P}; \exists l < j \text{ such that } p \nmid d_l, \ p \mid d_{n+2}\}, \ 2 \leq j \leq n+1,$$

$$u := \prod_{p \in \mathbb{P}, \ p \nmid \delta, \ p \mid d_{n+2}} p, \ \lambda_j := \prod_{p \in P_j} p, \ 1 \leq j \leq n+1, \ \lambda_{n+2} = 0,$$

$$\lambda = (\lambda_1, \ldots, \lambda_{n+1})' \in \mathbb{Z}^{n+2}.$$

Multiplication by $M_{u\lambda}$ from the right yields a form

$$\begin{pmatrix} * & * & * \\ * & d_{n+2} & \delta^* \end{pmatrix}, \quad \delta^* = \delta + u \sum_{j=1}^{n+1} \lambda_j d_j.$$

We get

$$gcd(d_{n+2}, \delta^*) = 1$$

from the construction. Multiplying by an appropriate M_F, $F \in SL(2; \mathbb{Z})$, from the right we get a matrix

$$\begin{pmatrix} * & \ldots & * & * \\ * & \ldots & * & 1 \end{pmatrix}.$$

Thus $S_0^{-1} d = \mu \in \mathbb{Z}^n$ follows from (3). Now multiply by $\widetilde{M}_\mu$ from the right in order to obtain

$$M^* = \begin{pmatrix} * & \ldots & * & * \\ 0 & \ldots & 0 & 1 \end{pmatrix}.$$

As this matrix belongs to Δ, we are done.   □

Define

$$\widehat{\Gamma}_S := \{M \in \Gamma_S; \ M \in E + M(n+4; \mathbb{Z}) \cdot S_1\}.$$

Let

$$M = \begin{pmatrix} A & F & B \\ * & U & * \\ C & G & D \end{pmatrix} \in \Gamma_S, \quad A, B, C, D \in M(2; \mathbb{Z}). \tag{9}$$

As M^{-1} is integral again, we conclude from the description of M^{-1} in (3) that FS^{-1} and GS^{-1} are integral, hence

$$M \in \widehat{\Gamma}_S \iff U \in E + M(n; \mathbb{Z})S.$$

We call $\widehat{\Gamma}_S$ the *discriminant kernel* of Γ_S.

A verification using (3) yields

Lemma 1. Γ_S *acts as a group of permutations on* $\mathbb{Z}^{n+4}/S_1\mathbb{Z}^{n+4}$ *via*

$$\phi_M : \lambda + S_1\mathbb{Z}^{n+4} \mapsto M'^{-1}\lambda + S_1\mathbb{Z}^{n+4}, \quad M \in \Gamma_S. \tag{10}$$

whose kernel is $\widehat{\Gamma}_S$. *Thus* $\widehat{\Gamma}_S$ *is a normal subgroup of* Γ_S *of finite index containing the matrices*

$$J, \quad M_\lambda, \ \lambda \in \mathbb{Z}^{n+2}.$$

Using Lemma 1 and Proposition 1, the same arguments as above lead to

Corollary 1. $\widehat{\Gamma}_S$ *is generated by the matrices*

$$J, \ M_\lambda, \ \lambda \in \mathbb{Z}^{n+2}, \ R_K, \ K \in \widehat{O}_0(S_0; \mathbb{Z}), \ i.e. \ K \in E + M(n+2; \mathbb{Z}) \cdot S_0.$$

As the matrices F, G in (9) belong to

$$M(2 \times n; \mathbb{Z})S$$

a straightforward verification yields

Corollary 2. *a) If $G \in M(n; \mathbb{Z})$, $\det G \neq 0$, the mapping*

$$\widehat{\Gamma}_{G'SG} \to \widehat{\Gamma}_S, \quad M \mapsto \begin{pmatrix} E & 0 & 0 \\ 0 & G & 0 \\ 0 & 0 & E \end{pmatrix} M \begin{pmatrix} E & 0 & 0 \\ 0 & G & 0 \\ 0 & 0 & E \end{pmatrix}^{-1},$$

is an injective homomorphism of the groups.
b) If $r \in \mathbb{N}$, the mapping

$$\Gamma_{r^2 S} \to \Gamma_S, \quad M \mapsto \begin{pmatrix} E & 0 & 0 \\ 0 & rE & 0 \\ 0 & 0 & E \end{pmatrix} M \begin{pmatrix} E & 0 & 0 \\ 0 & rE & 0 \\ 0 & 0 & E \end{pmatrix}^{-1},$$

is an injective homomorphism of the groups.

4. Euclidean matrices

Similar to Leutbecher's approach [14] we consider those matrices S, which allow us to proceed with a kind of Euclidean algorithm. We call S *Euclidean* if for all $x \in \mathbb{Q}^n$ there exists $\mu \in \mathbb{Z}^n$ such that

$$\tfrac{1}{2}S[x - \mu] < 1. \tag{11}$$

Examples of Euclidean matrices S are given by

$$(2), (4), (6), \begin{pmatrix} 2 & 0 \\ 0 & 2 \end{pmatrix}, \begin{pmatrix} 2 & 1 \\ 1 & 2 \end{pmatrix}, \begin{pmatrix} 2 & 0 \\ 0 & 4 \end{pmatrix}, \begin{pmatrix} 2 & 1 \\ 1 & 4 \end{pmatrix}, \begin{pmatrix} 2 & 1 \\ 1 & 6 \end{pmatrix},$$

$$\begin{pmatrix} 2 & 0 & 0 \\ 0 & 2 & 0 \\ 0 & 0 & 2 \end{pmatrix}, \begin{pmatrix} 2 & 0 & 0 & 1 \\ 0 & 2 & 0 & 1 \\ 0 & 0 & 2 & 1 \\ 1 & 1 & 1 & 2 \end{pmatrix}$$

or if S is 8×8 and additionally unimodular, which is the Gram matrix of the E_8-lattice (cf. [3]) or of integral Cayley numbers (cf. [4]).

Theorem 2. *If S is euclidean the group Γ_S is generated by*

$$J, M_\lambda, \lambda \in \mathbb{Z}^{n+2}, R_K, K = \begin{pmatrix} 0 & 0 & 1 \\ 0 & E & 0 \\ 1 & 0 & 0 \end{pmatrix}, \begin{pmatrix} 1 & 0 & 0 \\ 0 & U & 0 \\ 0 & 0 & 1 \end{pmatrix}, U \in \mathcal{O}(S; \mathbb{Z}).$$

In particular Γ_S is finitely generated.

Proof. In view of Theorem 1 it suffices to show that the matrices R_K, $K \in \mathcal{O}_0(S_0; \mathbb{Z})$, are generated by the matrices quoted above. Let $K \in \mathcal{O}_0(S_0; \mathbb{Z})$ with g, $g' = (\alpha, b', \gamma)$, $b \in \mathbb{Z}^n$, as its first column. We obtain

$$2\alpha\gamma = S[b]$$

from (2). If $\gamma \neq 0$ we use (11) in order to find $\mu \in \mathbb{Z}^n$ such that

$$\tfrac{1}{2}S[b + \gamma\mu] < \gamma^2.$$

Hence we have

$$K_\mu g = \begin{pmatrix} \alpha^* \\ b + \gamma\mu \\ \gamma \end{pmatrix}, \quad |\alpha^*| < \gamma.$$

If $\alpha \neq 0$ we proceed in the same way and multiply by $\widetilde{K}_\mu$ from the left. As S is positive definite we obtain

$$\pm(1, 0, \ldots, 0)' \quad \text{or} \quad \pm (0, \ldots, 0, 1)'$$

as the first column after finitely many steps. In the latter case multiply by $K_0 = \begin{pmatrix} 0 & 0 & 1 \\ 0 & E & 0 \\ 1 & 0 & 0 \end{pmatrix}$. Thus we get

$$K = \pm \begin{pmatrix} 1 & * & * \\ 0 & U & * \\ 0 & 0 & 1 \end{pmatrix} \in \mathcal{O}_0(S_0; \mathbb{Z}).$$

As R_K belongs to Γ_S^J the claim follows from Proposition 3. $\qquad\square$

Remark 1. a) A refinement of the argument shows that the same result is true for matrices $S \equiv 0 \bmod 2$ satisfying

$$\tfrac{1}{2}S[x - \mu] \leq 1 \tag{11'}$$

instead of (11). The main reason is that α, γ cannot both be even because of

$$\alpha\delta + \beta\gamma - c'Sb = 1$$

just as in the proof of Theorem 2. Examples of this type are

$$S = (8), \quad \begin{pmatrix} 4 & 0 \\ 0 & 4 \end{pmatrix}, \quad \begin{pmatrix} 2 & 0 & 0 & 0 \\ 0 & 2 & 0 & 0 \\ 0 & 0 & 2 & 0 \\ 0 & 0 & 0 & 2 \end{pmatrix}.$$

b) Other simple set of generators of Γ_S are known whenever S_1 satisfies the so-called Kneser condition (cf. [7]) which in our situation says

$$S \not\equiv 0 \bmod 2 \quad \text{and} \quad S \not\equiv 0 \bmod 3.$$

5. The case $n = 0$, $S_0 = \left(\begin{smallmatrix} 0 & 1 \\ 1 & 0 \end{smallmatrix}\right)$

If $n = 0$, we have

$$\widetilde{M}_D = R_{-V} M_{D[I]} R_{-V}, \quad R_{-V} = \begin{pmatrix} 1 & 0 & 0 & 0 \\ 0 & 0 & 1 & 0 \\ 0 & 1 & 0 & 0 \\ 0 & 0 & 0 & 1 \end{pmatrix},$$

due to Proposition 1. If $C, D = \left(\begin{smallmatrix} \alpha & \beta \\ \gamma & \delta \end{smallmatrix}\right) \in \mathrm{SL}(2; \mathbb{Z})$, a verification yields

$$M_C \cdot \widetilde{M}_D = \widetilde{M}_D \cdot M_C = \begin{pmatrix} \alpha C & \beta C I \\ \gamma I C & \delta I C I \end{pmatrix}, \quad I = \begin{pmatrix} -1 & 0 \\ 0 & 1 \end{pmatrix},$$

$$M_\lambda = \widetilde{M}_{\left(\begin{smallmatrix} 1 & \alpha \\ 0 & 1 \end{smallmatrix}\right)} \cdot M_{\left(\begin{smallmatrix} 1 & -\beta \\ 0 & 1 \end{smallmatrix}\right)}, \quad \lambda = \begin{pmatrix} \alpha \\ \beta \end{pmatrix},$$

$$J = M_{-D} \cdot \widetilde{M}_D, \quad D = \begin{pmatrix} 0 & -1 \\ 1 & 0 \end{pmatrix}.$$

It follows from Theorem 1 that Γ_S is generated by

$$M_C, \quad \widetilde{M}_D, \quad C, D \in \mathrm{SL}(2; \mathbb{Z}), \quad R_K, \quad K \in \mathcal{O}_0(S_0; \mathbb{Z}).$$

One easily verifies

$$\mathcal{O}(S_0; \mathbb{Z}) = \left\{ \pm \begin{pmatrix} 1 & 0 \\ 0 & 1 \end{pmatrix}, \pm \begin{pmatrix} 0 & 1 \\ 1 & 0 \end{pmatrix} \right\}, \quad \mathcal{O}_0(S_0; \mathbb{Z}) = \left\{ \begin{pmatrix} 1 & 0 \\ 0 & 1 \end{pmatrix}, \begin{pmatrix} 0 & 1 \\ 1 & 0 \end{pmatrix} \right\}.$$

Hence Theorem 1 and the calculations above yield

Theorem 3. *Let $n = 0$.*
a) Γ_S is generated by the matrices

$$M_C, \quad \widetilde{M}_D, \quad C, D \in \mathrm{SL}(2; \mathbb{Z}), \quad R_{-V}.$$

b) One has

$$\Gamma_S = \{ M = M_C \cdot \widetilde{M}_D \cdot R_{-V}^r; \; C, D \in \mathrm{SL}(2; \mathbb{Z}), r = 0, 1 \}.$$

In the representation $M = M_C \cdot \widetilde{M}_D \cdot R_{-V}^r$ now $\det M = (-1)^r$ holds and $\pm(C, D)$ are uniquely determined, i.e.

$$\Gamma_S / \{\pm E\} \cong \mathrm{PSL}(2; \mathbb{Z})^2 \ltimes \mathbb{Z}/2\mathbb{Z}.$$

From the knowledge of the abelian characters of $\mathrm{SL}(2; \mathbb{Z})$ we can calculate the abelian characters of Γ_S, because $M_{D[I]}$ and $\widetilde{M}_D$ are conjugate in Γ_S.

Corollary 3. *Let $n = 0$. Then the abelian characters of Γ_S are given by*

$$M = M_C \cdot \widetilde{M}_D \cdot R_{-V}^r \mapsto (-1)^{r\alpha} \cdot \kappa(C \cdot D),$$

where $\alpha \in \{0, 1\}$ and κ runs through the abelian characters of $\mathrm{SL}(2; \mathbb{Z})$.

The abelian characters of $\mathrm{SL}(2; \mathbb{Z})$ are described in [16].

6. The case $n = 1$, $S = (2t)$, $t \in \mathbb{N}$

Let $H_2(\mathbb{R})$ denote the Siegel half space of degree 2, consisting of all

$$Z = X + iY, \quad X, Y \in \mathrm{Sym}(2; \mathbb{R}), \quad Y \text{ positive definite.}$$

Given $t \in \mathbb{N}$ the mapping

$$\phi_t : \mathcal{H}_S \to H_2(\mathbb{R}), \quad w = \begin{pmatrix} \tau \\ z \\ \omega \end{pmatrix} \mapsto Z = \begin{pmatrix} \tau & z \\ z & \omega/t \end{pmatrix},$$

is a bijection satisfying

$$S_0[w] = 2(\tau\omega - tz^2) = 2t \cdot \det Z.$$

Given $U = \begin{pmatrix} \alpha & \beta \\ \gamma & \delta \end{pmatrix} \in \mathrm{GL}(2; \mathbb{R})$ we calculate

$$UZU' = \begin{pmatrix} \alpha^2\tau + 2\alpha\beta z + \omega\beta^2/t & \alpha\gamma\tau + (\alpha\delta + \beta\gamma)z + \omega\beta\delta/t \\ * & (\gamma^2 t\tau + 2\gamma\delta tz + \delta^2\omega)/t \end{pmatrix}.$$

Hence we have got

$$\phi_t^{-1}(U\phi_t(w)U') = K_U \cdot w, \quad \text{where} \quad K_U = \begin{pmatrix} \alpha^2 & 2\alpha\beta & \beta^2/t \\ \alpha\gamma & \alpha\delta + \beta\gamma & \beta\delta/t \\ \gamma^2 t & 2\gamma\delta t & \delta^2 \end{pmatrix}. \qquad (12)$$

A simple calculation yields

$$S_0[K_U] = (\det U)^2 \cdot S_0, \quad \det K_U = (\det U)^3. \qquad (13)$$

Given $A = \begin{pmatrix} \alpha & a & \beta \\ b & m & c \\ \gamma & d & \delta \end{pmatrix} \in M(3; \mathbb{R})$ a verification analogous to (1) shows that

$$S_0[A] = \begin{pmatrix} 2\alpha\gamma - 2tb^2 & \alpha d + \gamma a - 2tmb & \alpha\delta + \beta\gamma - 2tbc \\ \alpha d + \gamma a - 2tmb & 2ad - 2tm^2 & \delta a + \beta d - 2tmc \\ \alpha\delta + \beta\gamma - 2tbc & \delta a + \beta d - 2tmc & 2\beta\delta - 2tc^2 \end{pmatrix}. \qquad (14)$$

Theorem 4. *Let $n = 1$, $S = (2t)$, $t \in \mathbb{N}$. Then the mapping*

$$\mathrm{PSL}(2; \mathbb{R}) \to S\mathcal{O}_0(S_0; \mathbb{R}), \quad \pm U \mapsto K_U,$$

is an isomorphism of the groups.

Proof. As $(Z, U) \mapsto UZU'$ is a group action, the mapping is a homomorphism of the groups due to (12) and (13). As $UZU' = Z$ for all $Z \in H_2(\mathbb{R})$ holds if and only if $U = \pm E$, the mapping is injective. Now let $K \in S\mathcal{O}_0(S_0; \mathbb{R})$ be given in the form above. Then (14) yields

$$2\alpha\gamma - 2tb^2 = 0.$$

Hence there exist $\varepsilon = \pm 1$, $x, y \in \mathbb{R}$ such that

$$\alpha = \varepsilon x^2, \quad b = \varepsilon xy, \quad \gamma = \varepsilon y^2 t.$$

Since K is invertible we conclude $(x, y) \neq (0, 0)$. Hence there exists

$$U = \begin{pmatrix} x & * \\ y & * \end{pmatrix} \in \mathrm{SL}(2; \mathbb{R}).$$

As the first column of K_U coincides with the first column of εK, we get

$$B := K_U^{-1} \cdot K = \begin{pmatrix} \varepsilon & * & * \\ 0 & * & * \\ 0 & * & * \end{pmatrix} \in S\mathcal{O}_0(S_0; \mathbb{R}).$$

Using (14) we conclude

$$B = \begin{pmatrix} \varepsilon & 0 & 0 \\ 0 & \delta & 0 \\ 0 & 0 & \varepsilon \end{pmatrix} \begin{pmatrix} 1 & 2t\lambda & t\lambda^2 \\ 0 & 1 & \lambda \\ 0 & 0 & 1 \end{pmatrix} = \begin{pmatrix} \varepsilon & 0 & 0 \\ 0 & \delta & 0 \\ 0 & 0 & \varepsilon \end{pmatrix} \cdot K_{\left(\begin{smallmatrix} 1 & \lambda t \\ 0 & 1 \end{smallmatrix} \right)}, \quad \lambda \in \mathbb{R}, \ \delta = \pm 1.$$

This yields

$$\begin{pmatrix} \varepsilon & 0 & 0 \\ 0 & \delta & 0 \\ 0 & 0 & \varepsilon \end{pmatrix} \in S\mathcal{O}_0(S_0; \mathbb{R}), \quad \text{hence} \quad \varepsilon = \delta = 1 \quad \text{and} \quad K = K_{U \left(\begin{smallmatrix} 1 & \lambda t \\ 0 & 1 \end{smallmatrix} \right)}.$$

$\square$

Now we assume that t is squarefree. We consider

$$\Gamma^0(t) := \left\{ \begin{pmatrix} \alpha & \beta \\ \gamma & \delta \end{pmatrix} \in \mathrm{SL}(2; \mathbb{Z}); \ \beta \equiv 0 \ \mathrm{mod} \ t \right\}$$

and its maximal discrete extension

$$\Gamma^0(t)^* := \langle \Gamma^0(t), V_d; \ d \in \mathbb{N}, \ d \mid t \rangle,$$

where

$$V_d = \frac{1}{\sqrt{d}} \begin{pmatrix} rd & t \\ s & d \end{pmatrix}, \quad r, s \in \mathbb{Z}, \quad rd - st/d = 1.$$

The cosets

$$V_d \cdot \Gamma^0(t) = \Gamma^0(t) \cdot V_d = \left\{ U = \frac{1}{\sqrt{d}} \begin{pmatrix} \alpha d & \beta t \\ \gamma & \delta d \end{pmatrix}; \ \alpha, \beta, \gamma, \delta \in \mathbb{Z}, \ \det U = 1 \right\}$$

are independent of the choice of r, s. Such a U satisfies

$$K_U = \begin{pmatrix} \alpha^2 d & 2\alpha\beta t & \beta^2 t/d \\ \alpha\gamma & \alpha\delta d + \beta\gamma t/d & \beta\delta \\ \gamma^2 t/d & 2\gamma\delta t & \delta^2 d \end{pmatrix} \in S\mathcal{O}_0(S_0; \mathbb{Z}). \tag{15}$$

Corollary 4. *Let $n = 1$, $S = (2t)$, $t \in \mathbb{N}$ squarefree. Then the mapping*

$$\Gamma^0(t)^* / \{\pm E\} \to S\mathcal{O}_0(S_0; \mathbb{Z}), \quad \pm U \mapsto K_U,$$

is an isomorphism of the groups.

Proof. Due to Theorem 2 and (15) it remains to be proved that the mapping is surjective. Let $(\alpha, b, \gamma)'$ be the first column of a matrix $K \in S\mathcal{O}_0(S_0; \mathbb{Z})$. Thus (14) shows that

$$\alpha\gamma = tb^2 \quad \text{and} \quad gcd(\alpha, b, \gamma) = 1.$$

As t is squarefree $gcd(\alpha, \gamma) = 1$ holds. Hence there exist

$$d, r, s \in \mathbb{Z} \quad \text{such that} \quad d \mid t, \quad gcd(rd, st/d) = 1$$

and

$$\alpha = r^2 d, \quad b = rs, \quad \gamma = s^2 t/d.$$

Theorem 4 yields $d \in \mathbb{N}$ and there are $u, v \in \mathbb{Z}$ satisfying $urd - vst/d = 1$, hence

$$U = \frac{1}{\sqrt{d}} \begin{pmatrix} rd & vt \\ s & ud \end{pmatrix} \in V_d \cdot \Gamma^0(t).$$

Just as in the proof of Theorem 4 we conclude

$$B = K_U^{-1} K = \begin{pmatrix} 1 & * & * \\ 0 & * & * \\ 0 & * & * \end{pmatrix} \in S\mathcal{O}_0(S_0; \mathbb{Z}).$$

Using (14) and $\det B = 1$ we get

$$B = \begin{pmatrix} 1 & 2\lambda t & \lambda^2 t \\ 0 & 1 & \lambda \\ 0 & 0 & 1 \end{pmatrix} = K_{\left(\begin{smallmatrix} 1 & \lambda t \\ 0 & 1 \end{smallmatrix}\right)}, \quad \lambda \in \mathbb{Z},$$

hence

$$K = K_{U\left(\begin{smallmatrix} 1 & \lambda t \\ 0 & 1 \end{smallmatrix}\right)}, \quad U \begin{pmatrix} 1 & \lambda t \\ 0 & 1 \end{pmatrix} \in V_d \cdot \Gamma^0(t).$$

$$\square$$

In (15) we have $\alpha\delta d - \beta\gamma t/d = 1$, hence

$$gcd(\alpha\delta d + \beta\gamma t/d - 1, 2t) = gcd(2\beta\gamma t/d, 2t) = 2t/d$$

in view $gcd(\beta\gamma, d) = 1$. Thus we have

$$K_U \in \widehat{S\mathcal{O}}_0(S_0; \mathbb{Z}), \quad \text{i.e.} \quad A \in E + M(3; \mathbb{Z})S_0 \iff d = 1.$$

The result is

Corollary 5. *Let* $n = 1$, $S = (2t)$, $t \in \mathbb{N}$ *squarefree. Then the mapping*

$$\Gamma^0(t)/\{\pm E\} \to \widehat{S\mathcal{O}}_0(S_0; \mathbb{Z}), \quad \pm U \mapsto K_U,$$

is an isomorphism of the groups.

The results of this and the following section can partly be found in the literature, cf. in particular [8].

7. The paramodular group

Given $t \in \mathbb{N}$ we consider the paramodular group Σ_t of degree 2 and level t which consists of the matrices

$$M = \begin{pmatrix} * & *t & * & * \\ * & * & * & */l \\ * & *t & * & * \\ *t & *t & *t & * \end{pmatrix} \in \mathrm{Sp}(2; \mathbb{Q}),$$

where $*$ always stands for an integer. It is well-known (or easily verified) that the following matrices

$$J_t = \begin{pmatrix} 0 & 0 & 1 & 0 \\ 0 & 0 & 0 & 1/t \\ -1 & 0 & 0 & 0 \\ 0 & -t & 0 & 0 \end{pmatrix}, \quad \begin{pmatrix} E & S \\ 0 & E \end{pmatrix}, \quad S = \begin{pmatrix} \alpha & \beta \\ \beta & \gamma/t \end{pmatrix}, \quad \alpha, \beta, \gamma \in \mathbb{Z},$$

$$\begin{pmatrix} U & 0 \\ 0 & U'^{-1} \end{pmatrix}, \quad U \in \Gamma^0(t)$$

belong to Σ_t. There are two canonical embeddings, i.e. injective group homomorphisms

$$\mathrm{SL}(2; \mathbb{Z}) \to \Sigma_t, \quad \begin{pmatrix} \alpha & \beta \\ \gamma & \delta \end{pmatrix} \mapsto \begin{pmatrix} \alpha & \beta \\ \gamma & \delta \end{pmatrix}^{\uparrow} := \begin{pmatrix} \alpha & 0 & \beta & 0 \\ 0 & 1 & 0 & 0 \\ \gamma & 0 & \delta & 0 \\ 0 & 0 & 0 & 1 \end{pmatrix}, \tag{16}$$

$$\mathrm{SL}(2; \mathbb{Z}) \to \Sigma_t, \quad \begin{pmatrix} \alpha & \beta \\ \gamma & \delta \end{pmatrix} \mapsto \begin{pmatrix} \alpha & \beta \\ \gamma & \delta \end{pmatrix}^{\downarrow} := \begin{pmatrix} 1 & 0 & 0 & 0 \\ 0 & \alpha & 0 & \beta/t \\ 0 & 0 & 1 & 0 \\ 0 & \gamma t & 0 & \delta \end{pmatrix}. \tag{17}$$

We recall from [11] and [12] the following

Theorem 5. *Given $t \in \mathbb{N}$ then Σ_t is generated by the matrices*

$$J_t, \quad \begin{pmatrix} E & S \\ 0 & E \end{pmatrix}, \quad S = \begin{pmatrix} 1 & 0 \\ 0 & 0 \end{pmatrix}, \quad \begin{pmatrix} 0 & 1 \\ 1 & 0 \end{pmatrix}, \quad \begin{pmatrix} 0 & 0 \\ 0 & 1/t \end{pmatrix}.$$

Proof. The result is well-known for $t = 1$ in view of $\Sigma_1 = \mathrm{Sp}(2; \mathbb{Z})$ (cf. [15]). Therefore let $t > 1$ and let G denote the subgroup generated by the 4 matrices above. Then G contains the matrices

$$\begin{pmatrix} E & S \\ 0 & E \end{pmatrix}, \quad \begin{pmatrix} E & 0 \\ S^* & E \end{pmatrix} = J_t \begin{pmatrix} E & -S \\ 0 & E \end{pmatrix} J_t^{-1},$$

$$\tag{18}$$

$$S = \begin{pmatrix} \alpha & \beta \\ \beta & \gamma/t \end{pmatrix}, \quad S^* = \begin{pmatrix} \alpha & \beta t \\ \beta t & \gamma t \end{pmatrix}, \quad \alpha, \beta, \gamma \in \mathbb{Z}.$$

As $SL(2;\mathbb{Z})$ is generated by $\left(\begin{smallmatrix}1&1\\0&1\end{smallmatrix}\right)$ and $\left(\begin{smallmatrix}1&0\\1&1\end{smallmatrix}\right)$ (cf. [16]) and the embedded matrices of them in (16) and (17) are of the form (18) we conclude

$$U^\uparrow, U^\downarrow \in G \quad \text{for all } U \in SL(2;\mathbb{Z}). \tag{19}$$

Let

$$S := \begin{pmatrix} 0 & 1 \\ 1 & 0 \end{pmatrix}, \quad U := \begin{pmatrix} 1 & 1 \\ 0 & 1 \end{pmatrix}.$$

Then we calculate

$$\begin{pmatrix} E & 0 \\ tE_{22} & E \end{pmatrix} \begin{pmatrix} E & -S \\ 0 & E \end{pmatrix} \begin{pmatrix} E & 0 \\ -tE_{22} & E \end{pmatrix} \begin{pmatrix} E & S \\ 0 & E \end{pmatrix} \begin{pmatrix} E & -tE_{11} \\ 0 & E \end{pmatrix}$$

$$= \begin{pmatrix} U^t & 0 \\ 0 & U'^{-t} \end{pmatrix} \in G,$$

$$\begin{pmatrix} E & 0 \\ E_{11} & E \end{pmatrix} \begin{pmatrix} E & -S \\ 0 & E \end{pmatrix} \begin{pmatrix} E & 0 \\ -E_{11} & E \end{pmatrix} \begin{pmatrix} E & S \\ 0 & E \end{pmatrix} \begin{pmatrix} E & -E_{22} \\ 0 & E \end{pmatrix}$$

$$= \begin{pmatrix} U' & 0 \\ 0 & U^{-1} \end{pmatrix} \in G.$$

Given $V = \left(\begin{smallmatrix} \alpha & \beta t \\ \gamma & \delta \end{smallmatrix}\right) \in \Gamma^0(t)$ a straightforward calculation yields

$$\begin{pmatrix} U' & 0 \\ 0 & U^{-1} \end{pmatrix}^{\alpha\gamma} \begin{pmatrix} \delta & \beta^2\gamma t^2 \\ -\gamma & \alpha(1-\beta\gamma t) \end{pmatrix}^\uparrow \begin{pmatrix} U^t & 0 \\ 0 & U'^{-t} \end{pmatrix}^{-\beta} \begin{pmatrix} \alpha(1-\beta\gamma t) & -\gamma t \\ \beta^2\gamma t & \delta \end{pmatrix}^\downarrow.$$

$$\begin{pmatrix} E & 0 \\ T & E \end{pmatrix} \begin{pmatrix} V & 0 \\ 0 & V'^{-1} \end{pmatrix} = \begin{pmatrix} E & * \\ 0 & E \end{pmatrix} \in G, \quad T = \begin{pmatrix} \gamma\delta & -\beta\gamma t \\ -\beta\gamma t & 0 \end{pmatrix},$$

thus

$$\begin{pmatrix} V & 0 \\ 0 & V'^{-1} \end{pmatrix} \in G \quad \text{for all } V \in \Gamma^0(t). \tag{20}$$

Now we start with $M \in \Sigma_t$ with $(a_1, a_3, c_1, tc_3)'$ as its first column. We find $U, V \in SL(2;\mathbb{Z})$ such that

$$U \begin{pmatrix} a_1 \\ c_1 \end{pmatrix} = \begin{pmatrix} \bar{a}_1 \\ 0 \end{pmatrix}, \quad V \begin{pmatrix} a_3 \\ c_3 \end{pmatrix} = \begin{pmatrix} \bar{a}_3 \\ 0 \end{pmatrix}.$$

Hence $(\bar{a}_1, \bar{a}_3, 0, 0)'$ is the first column of

$$\overline{M} = V^\uparrow U^\downarrow M \in \Sigma_t.$$

We conclude $\bar{a}_1 \neq 0$ from $\det \overline{M} = 1$ and $t > 1$. The same argument leads to a matrix $W \in SL(2;\mathbb{Z})$ such that $(0, 0, *, *)$ is the last row of $\overline{M} \cdot W^\downarrow$. As $A'C$ is a symmetric matrix and $\bar{a}_1 \neq 0$ we conclude that

$$M_0 := V^\downarrow U^\downarrow M W^\downarrow = \begin{pmatrix} A & B \\ 0 & D \end{pmatrix} \in \Sigma_t.$$

If we replace W by $-W$ if necessary, we may assume $A \in \Gamma^0(t)$, hence

$$M_0 = \begin{pmatrix} A & 0 \\ 0 & A'^{-1} \end{pmatrix} \begin{pmatrix} E & * \\ 0 & E \end{pmatrix} \in G.$$

Thus $G = \Sigma_t$ follows. □

The maximal discrete extension Σ_t^* for squarefree t is known to be generated by Σ_t and

$$\begin{pmatrix} V_d & 0 \\ 0 & V_d'^{-1} \end{pmatrix}, \quad d \in \mathbb{N}, \, d \mid t.$$

Looking at (4) we see that

$$J_t \langle Z \rangle = - \begin{pmatrix} 1 & 0 \\ 0 & 1/t \end{pmatrix} \begin{pmatrix} \tau & z \\ tz & \omega \end{pmatrix}^{-1}$$

$$= \frac{1}{\tau\omega - tz^2} \begin{pmatrix} \omega & -z \\ -z & \tau/t \end{pmatrix} = \phi_t(J\langle w \rangle),$$

$$\begin{pmatrix} E & S \\ 0 & E \end{pmatrix} \langle Z \rangle = Z + S = \begin{pmatrix} \tau + \alpha & z + \beta \\ z + \beta & (\omega + \gamma)/t \end{pmatrix} = \phi_t(M_\lambda \langle w \rangle), \qquad (21)$$

$$S = \begin{pmatrix} \alpha & \beta \\ \beta & \gamma/t \end{pmatrix}, \quad \lambda = \begin{pmatrix} \alpha \\ \beta \\ \gamma \end{pmatrix}.$$

Since the generators correspond to each other a consequence of Corollaries 4, 5 and Theorem 5 is

Corollary 6. *Let $n = 1$, $S = (2t)$, $t \in \mathbb{N}$ squarefree.*
a) The groups $\Sigma_t/\{\pm E\}$ and $\widehat{\Gamma}_S/\{\pm E\}$ as well as $\widehat{\Gamma}_S \cap \mathrm{SL}(5; \mathbb{Z})$ are isomorphic.
b) The groups $\Sigma_t^*/\{\pm E\}$ and $\Gamma_S/\{\pm E\}$ as well as $\Gamma_S \cap \mathrm{SL}(5; \mathbb{Z})$ are isomorphic.

An application of Theorem 5 and (21) moreover yields

Corollary 7. *Set $S = (2t)$, $t \in \mathbb{N}$ be squarefree. Then $\widehat{\Gamma}_S$ is generated by*

$$J, \, M_\lambda, \, \lambda = \begin{pmatrix} 1 \\ 0 \\ 0 \end{pmatrix}, \begin{pmatrix} 0 \\ 1 \\ 0 \end{pmatrix}, \begin{pmatrix} 0 \\ 0 \\ 1 \end{pmatrix}.$$

If t is not squarefree, we can refer to Corollary 2.

8. Abelian characters

In this section we construct some abelian characters for orthogonal groups.
Let $S \in \mathrm{Sym}(n; \mathbb{Z})$ be positive definite and even. Considering Lemma 1 there are

two obvious abelian characters of Γ_S, namely

$$M \mapsto \det M \quad \text{and} \quad M \mapsto \operatorname{sgn} \phi_M.$$

Note that the second one may be trivial. Let again $P = \left(\begin{smallmatrix} 0 & 1 \\ 1 & 0 \end{smallmatrix}\right)$.

Lemma 2. *Let* $S \equiv 0 \,(\mathrm{mod}\ 2)$. *Let* $M = \left(\begin{smallmatrix} A & * & B \\ * & * & * \\ C & * & D \end{smallmatrix}\right)$, $A, B, C, D \in M(2; \mathbb{Z})$. *Then the mapping*

$$F : \Gamma_S \to \operatorname{Sp}(2; \mathbb{Z}/2\mathbb{Z}), \quad M \mapsto F_M := \begin{pmatrix} E & 0 \\ 0 & P \end{pmatrix} \begin{pmatrix} A & B \\ C & D \end{pmatrix} \begin{pmatrix} E & 0 \\ 0 & P \end{pmatrix},$$

is a surjective homomorphism of the groups.

Proof. Clearly F is a homomorphism of the groups. It is easily verified from Proposition 1 that generators of $\operatorname{Sp}(2; \mathbb{Z}/2\mathbb{Z})$ belong to the image of F. $\square$

If $S \equiv 0 \,(\mathrm{mod}\ 2)$ clearly

$$M \mapsto \chi(F_M)$$

becomes a non-trivial abelian character of Γ_S, if χ denotes the non-trivial abelian character of $\operatorname{Sp}(2; \mathbb{Z}/2\mathbb{Z}) \cong S_6$. The abelian characters of the paramodular groups were determined in [7] and more generally in [5].

References

[1] Borcherds, R.E.: Automorphic forms with singularities on Grassmannians. Invent. Math **132** (1998), 267-280.

[2] Bühler, F.: Modulformen zur orthogonalen Gruppe. Diplomarbeit RWTH Aachen 1996.

[3] Conway, J., Sloane, N.J.A.: Sphere packings, lattices and groups. 3rd ed., Springer-Verlag, New York 1999.

[4] Conway, J., Smith, D.A.: On quaternions and octonions: their geometry, arithmetic, and symmetry. A.K. Peters, Natick, MA 2003.

[5] Dern, T., Marschner, A.: Characters of paramodular groups and some extensions. *Comm. Algebra* **30** (2002), 4589-4604.

[6] Gritsenko, V.: Fourier-Jacobi functions of n (variables). J. Sov. Math. **53** (1991), 243-252.

[7] Gritsenko, V., Hulek, K.: Commutator coverings of Siegel modular threefolds. Duke Math. J. **94** (1998), 509-542.

[8] Gritsenko, V., Hulek, K.: Minimal Siegel modular threefolds. Math. Proc. Cambridge Phil. Soc. **123** (1998), 461-485.

[9] Gritsenko, V., Hulek, K.: Uniruledness of orthogonal modular varieties. J. Algebr. Geom. **23** (2014), 711-725.

[10] Gritsenko, V., Hulek, K., Sankaran, G.K.: Abelianisation of orthogonal groups and the fundamental group of modular varieties. J. Algebra **322** (2009), 463-478.

[11] Grosche, J.: Über die Erzeugbarkeit paramodularer Gruppen durch Elemente mit Fixpunkten. *J. Reine Angew. Math.* **293/294** (1977), 86-98.

[12] Kappler, F.: Über die Charaktere Siegelscher Stufengruppen. PhD thesis, Freiburg 1977.

[13] Krieg, A.: Jacobi Forms of Several Variables and the Maaß Space. *J. Number Th.* **56** (1996), 242-255.

[14] Leutbecher, A.: Euklidischer Algorithmus und die Gruppe GL₂. Math. Ann. **231** (1978), 269-285.

[15] Maaß, H.: *Siegel's modular forms and Dirichlet series.* Lect. Notes Math. **216**, Springer-Verlag, Berlin 1971.

[16] Newman, M.: Integral Matrices. Academic Press, New York–London 1972.

Chapter 10

The Role of Fourier Analysis in X-Ray Crystallography

Florian Rupp & Jürgen Scheurle

*Department of Mathematics and Science, German University of Technology
in Oman, Halban Campus, Muscat, Sultanate of Oman*
florian.rupp@gutech.edu.om

*Lehrstuhl für Höhere Mathematik und Analytische Mechanik,
Technische Universität München, Fakultät für Mathematik,
D-85747 Garching, Germany*
scheurle@ma.tum.de

One of the most important applications and actual drivers for new insights in the structure of mathematical groups is crystallography. It is also one of the areas were mathematics predicted the physical reality long before experiments could really contribute. This changed about a hundred years ago with the emergence of X-ray diffraction which however turned out to still heavily rely on Mathematics that was now needed to solve the phase problem. New challenges in the interpretation came to light when quasi-crystals were discovered and thus opened the scene for further and deeper mathematical insights.

This chapter informally surveys the essential mathematics of X-ray crystallography emphasizing the role of Fourier analysis in this context and, based on several highly recommended sources, intends to give a straight introduction to this subject which is often only found in scattered locations dispersed in the literature.

*Dedicated to Professor Dr. Armin Leutbecher on the occasion of his 80-th birthday
who experienced the modern crystallographic insights as contemporary witness.*

Contents

1. Introduction

Traditionally, crystals are considered to be three-periodic arrangements of unit cells containing atoms. These unit cells form a tiling of space and are the corner stone for the mathematical analysis of crystals by means of symmetries and corresponding space groups. It took until the very beginning of the 20th century before atomic structure was rigorously understood by Max von Laue, William Lawrence Bragg and William Henry Bragg utilizing a refined X-ray diffraction method. This method worked remarkably well to verify the group theoretic concepts developed in order to explain the periodic inner structure of crystals. Generalizations of diffraction methods to metal and organic crystals soon appeared, for instance, diffraction revealed the spatial structure of DNA by James D. Watson and Francis Crick and based on the work of Rosalind Franklin. At this stage the mathematics of tilings and experiments were in alignment. Though, in the 1980s Dan Schechtman discovered the unexpected phenomenon of a metallic crystal with five-fold symmetry leading to the development of the field of quasi-crystals. Thus, the typical picture of a symmetric packing of unit cells could no longer persist. On the other hand, Schechtman's experiments were based on diffraction such that a new combined definition of crystals necessarily had to include the mathematical description of diffraction and scattering.

In this regard the International Union of Crystallography (IUC) refined in 1991 the term 'crystal' to mean 'any solid having an essentially discrete diffraction diagram', and introduced the term 'aperiodic crystal' to mean 'any crystal in which three-dimensional periodicity can be considered to be absent' (see [2], p. 928). As diffraction patterns can mathematically be interpreted in terms of Fourier theory, the IUCs new interpretation shifts the essential attribute of crystals from position space into Fourier space. As [6], p. 1, outlines this 'broader definition reflects our current understanding that microscopic periodicity is a sufficient but not a necessary condition for crystallinity'.

The nature of this definition is rather interesting in a second, philosophical aspect as it shifts the focus from what the (Platonic) idea of a crystal should be (a part of an ordered three-periodic pattern) to the tangible nature of the outcome of an (Aristotelian) observation (a somehow discrete point pattern on a screen). Though, being broad and unifying and having the strength of excluding amorphous materials form the definition of a crystal this definition does only indirectly and opaquely for the layman answer the question of the inner structure of a crystal. Only when we know how the diffraction diagram translates to an atomic structure do we really understand the nature of the crystal we are looking at. This translation is essentially made possible by Fourier analysis.

The connections between Fourier series and diffraction form the core of the present article together with essential remarks about the mathematical description of symmetries in Fourier space and group theory. Thus, we intend to make this translation between the International Union of Crystallography (IUC) definition of a crystal and its inherent atomic structure decodable for classical periodic crystals. A thorough mathematical investigation of aperiodic crystals is beyond the scope of this article. In respect of our aim it is worth to note, that there are essentially two representations of crystalline structures (cf. [7], p. 108):

(1) by means of discrete atoms, and equivalently
(2) by means of a continuous electron density where atoms are localized as concentrations or peaks of the density function. The focus of our exposition will be on this view of structures.

These representations give raise to specific methods for the detection of unknown crystalline structures. In the classical way, where the crystal is viewed as a three-periodic arrangement of distinct atoms, structure identification is closely connected to diffraction and long-range scattering.

By purpose, in our introduction to Fourier Crystallography, the quantum mechanics of diffraction have been omitted as well as the effects of temperature on scattering and impurities in the structure of the original crystal. Our notation is aligned with the main source [7] of this contribution.

The remainder of this survey is structured as follows: First, in Section 2 we give an overview of the processes that occur during a typical diffraction experiment and link the diffraction diagram with the structure of the crystal. Next, Section 3 discusses the basics of unmodified scattering at an electron in order to motivate the essential concept of the electronic scattering factor by means of physical descriptions. Section 4 then focuses on the derivation of certain essential formulas that connect the diffraction diagram with Fourier series. It is there were we identify the Fourier coefficients with the electronic scattering factors. Finally, Section 5 gives a short outlook and summary.

2. How does the Diffraction Pattern Emerge?

For simplicity of the discussion, let us assume that we are dealing with a perfect monocrystal that is crystallographically aligned at axes with units a, b and c, called a-, b- and c-axes. I.e., the unit cells of the crystal are aligned to this coordinate frame. As sketched in Fig. 1 such a monocrystal is then, for instance[1], irradiated by a monochromatic laser beam with wavelength λ and amplitude E_0. A sequence of lenses focuses the beam and aligns (i.e. polarizes) the directions of its waves such that a common arrival angle at the crystal results. Parts of the beam are transmitted instantaneously whereas other parts are diffracted/ scattered at the electron cloud of the crystals atoms. As the beam consist of a multitude of parallel light waves such diffraction events simultaneously take place at neighboring atoms as well. The standard assumption of crystallography postulates interference of the

---

[1]We refer to [4] or [7] for a survey discussion of other types of experimental diffraction set-ups; like the powder diffraction or rotation diffraction with respect to the experimental set-up or like neutron-diffraction with respect to the nature of the beam.

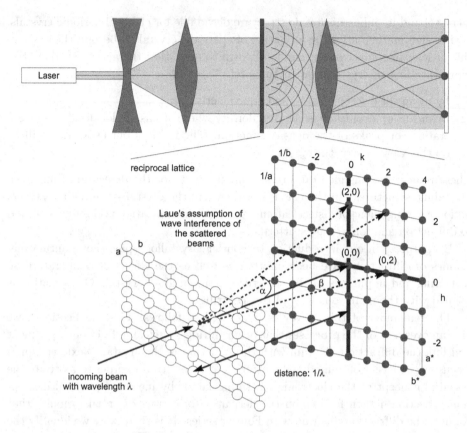

Fig. 1. Sketch of an experimental set-up to generate diffraction patterns from a monocrystal. During the diffraction event wave interference generates a lattice reciprocal to the crystal lattice that is then detected on a photo plate. See the text for more information.

thus scattered beams. Then, at a distance of $1/\lambda$ from the layer of atoms that the photons hit the first layer of interference points emerges. The more waves interfere at a point the more intensity (i.e. energy) this point will display.

Next, of course, the third dimension of the crystal has to be considered and those interference waves will have to pass the successively piled atomic lattice planes and finally form the reciprocal lattice with a^*-, b^*- and c^*-axes (cf. [1], [4] or [7] for the construction of the reciprocal lattice). Here, the optically inspired insight of William Lawrence Bragg reveals the occurring dynamics, cf. [3]: a diffraction spot only appears if what is nowadays called the **Bragg condition** is satisfied

$$m \cdot \lambda = 2 \cdot d \cdot \sin(\vartheta) ,$$

where m is an integer, d is the spacing between the atomic lattice planes and ϑ is the angle of incidence. During diffraction experiments the angle ϑ is altered to gain optimal knowledge about the inner structure of the crystal.

To be precise, we have to state that there are two types of scattering: unmodified scattering where the incoming and outgoing waves have the same wavelength and Compton-modified scattering where the outgoing wave has a longer wave length. In diffraction by a crystal it is thus only the unmodified scattering that gives rise to the Bragg reflection. The modified scattering by different electrons is completely incoherent and in general produces just a noisy background that may be suppressed by appropriate filters later on for the reconstruction of the crystallographic structure[2].

In the atomistic representation, these resulting interactions of the entering X-rays and the electron clouds of the crystal's atoms are described in two steps: First, the scattering effect of all the electrons in a single atom n are combined in terms of the **electronic scattering factor**

$$f_n := \int_0^\infty 4\pi \cdot r^2 \cdot \varrho_n(r) \cdot \frac{\sin(\kappa \cdot r)}{\kappa \cdot r}\, dr\,, \tag{1}$$

where ϱ_n is a radially symmetric electron density function, r is the distance function with respect to the center of atom n, and $\kappa := 4\pi\lambda^{-1}\sin(\vartheta)$. In Section 3 this electronic scattering factor is motivated.

Next, the scattering effect of all the atoms in a unit cell are combined in terms of the **structure factor**

$$F_{h,k,\ell} = \sum_n f_n \cdot e^{2\pi i \cdot (h \cdot x_n/a + k \cdot y_n/b + \ell \cdot z_n/c)}\,, \tag{2}$$

where f_n is the structure factor of the single atom n, and x_n/a, y_n/b, and z_n/c are (scaled) fractional coordinates of the atom n in the unit cell and h, k, and ℓ are the **Miller indices**, i.e. the coordinates of the outward normal vector corresponding to the atomic lattice planes.

Though, physically the structure factor $F_{h,k,\ell}$ can be interpreted as the effective electronic scattering factor of each unit cell. In view of Fourier theory, on the other hand, this essentially is the Fourier coefficient $C_{h,k,\ell}$ of the electron density. In fact, the determination of the electron density function $\varrho(x, y, z)$ is the ultimate goal. It completely encodes the crystal structure and it is mathematically given by a Fourier sum, where the correspondence between the complex Fourier coefficients $C_{h,k,\ell}$ and the complex structure factor $F_{h,k,\ell}$ is

$$C_{h,k,\ell} = \text{volume of the unit cell} \cdot F_{h,k,\ell}\,.$$

Finally, as the last step in the experimental set-up shown in Fig. 1 a collection of lenses streamlines the interference waves such that a clear photographic image of the inference pattern can be taken. This image is the 'diffraction diagram' the International Union of Crystallography is referring to.

---

[2]In alignment with Thomson's formula the intensity of the Compton radiation is weak for closely bound electrons independent of the angle ϑ. For weakly bound electrons, like in the outer shells of very heavy atoms, and especially larger angles this incoherent scattering becomes more prominent. In the limit case of a free electron, though, Compton diffraction is the only scattering observed, cf. [4], p. 14.

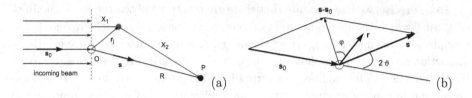

Fig. 2.   (a) Scattering by a group of electrons located at $\mathbf{r}_j$, the vector $\mathbf{s}_0$ points in the direction of the entering beam and $\mathbf{s}$ in the direction to the point of observation. (b) Relations between the vectors $\mathbf{s} - \mathbf{s}_0$, $\mathbf{r}$ and the angles φ and ϑ for scattering at a single atom, cf. [7] p. 7.

Unfortunately, this interpretation has a serious drawback: The diffraction diagram just displays the intensities (amplitudes) of the interference waves, i.e. the absolute values $|F_{h,k,\ell}|$, while the phases of the structure factors are still missing.

Typically, the missing phases are detected by an iterative trail and error procedure where assumptions on the structure of the crystal (e.g. based on its chemical composition or mathematical auxiliary functions like the Patterson function, cf. [7], p. 112) are used to get a first approximation of the structure that is then refined in order to match the diffraction results. It is here were still the traditional information from group theory and symmetry proves to be essential for crystallography as it is this kind of information that is used to obtain good initial guesses to determine the phases of the structure factors. Hence, crystallography remains a subject were a huge and specific experience is required and that cannot be computerized easily.

3. How is the Electronic Scattering Factor Motivated?

The electronic scattering factor is easily motivated by means of unmodified classical scattering as illustrated in Fig. 2. Let us follow the exposition given in [7], p. 7, and assume that the monochromatic beam with an amplitude E_0 is polarized such that it travels in the direction of the unit vector $\mathbf{s}_0$. At an atom in the crystallographic unit cell m electrons are presumed to be clustered about the point O and the position of each is represented by cell vectors $\mathbf{r}_1, \mathbf{r}_2, \ldots, \mathbf{r}_m$ originating at O. Finally, we detect the scattered wave at a point P at a large distance R from the electrons in the direction given by the unit vector $\mathbf{s}$.

In this setting the incoming wave's intensity ϵ_0 acting on the electron j ($j = 1, 2, \ldots, m$) is given by

$$\epsilon_0 = E_0 \cdot \cos\left(2\pi\nu^i t - \frac{2\pi X_1}{\lambda}\right),$$

where ν^i is the frequency of the wave and X_1 its travel distance to the electron j. To represent the absolute value of the amplitude and the phase of the wave scattered at this electron at the point of observation P, we multiply by the familiar factor for an electron and take into account the total travel distance to P as $X_1 + X_2$:

$$\epsilon_j = \frac{E_0 e^2}{m_e c^2 X_2} \cdot \cos\left(2\pi\nu^i t - \frac{2\pi(X_1 + X_2)}{\lambda}\right),$$

where e and m_e are the charge and mass of the electron respectively and c denotes the speed of light. Note, that this approach neglects a jump of $180°$ in the phase during scattering. This can be neglected as it is the same for all electrons involved. By virtue of our setting the source of the beam and the point of observation P are both at distances that are very large compared to $\|\mathbf{r}_j\|$ such that the usual classical plane wave approximation is justified: $X_2 \to R$ (X_2 is replaced by R), where R is the distance between O and P, such that the distinct location of the electron looses relevance. Also

$$X_1 + X_2 \to \langle \mathbf{s}_0, \mathbf{r}_j \rangle + R - \langle \mathbf{s}, \mathbf{r}_j \rangle = R - \langle (\mathbf{s} - \mathbf{s}_0), \mathbf{r}_j \rangle.$$

Taking into account all electrons of atom n, the overall intensity ϵ_n of the unmodifiedly scattered wave as observed at P is

$$\epsilon_n = \frac{E_0 e^2}{m_e c^2 R} \cdot e^{2\pi i (\nu^i t - (R/\lambda))} \cdot \sum_{j=1}^{m} e^{(2\pi i/\lambda) \cdot \langle (\mathbf{s} - \mathbf{s}_0), \mathbf{r}_j \rangle}.$$

Considering density elements $\varrho_n dV$ at positions $\mathbf{r}$ rather than distinct electrons at positions $\mathbf{r}_j$, and by replacing the sum by an integral, leads in the continuum limit to

$$\epsilon_n = \frac{E_0 e^2}{m_e c^2 R} \cdot e^{2\pi i (\nu^i t - (R/\lambda))} \cdot \underbrace{\int e^{(2\pi i/\lambda) \cdot \langle (\mathbf{s} - \mathbf{s}_0), \mathbf{r} \rangle} \varrho_n dV}_{= f_n}.$$

Next, we make the simplification that ϱ_n is spherically symmetric: Closed groups of inner electrons possess spherical symmetry (as we have outlined in particular these electrons contribute heavily to unmodified scattering) and it is a common simplification to work with the combined electron distribution for the electrons of a closed group. Hence, it is not unreasonable to work with spherical symmetric electron densities $\varrho_n = \varrho_n(r)$ concerning a single atom n where the origin of the coordinate frame is located at the center of this atom. Thus f_n introduced in the previous formula simplifies to

$$f_n = \int_{r=0}^{\infty} \int_{\varphi=0}^{\pi} 2\pi \cdot e^{i\kappa r \cos(\varphi)} \cdot \varrho_n(r) \cdot r^2 \sin(\varphi) \, d\phi dr,$$

where $\langle (\mathbf{s} - \mathbf{s}_0), \mathbf{r} \rangle = 2\sin(\vartheta) r \cos(\varphi)$ as sketched in Fig. 2 (b) and $\kappa := 4\pi\lambda^{-1} \sin(\vartheta)$ as above. Integration with respect to φ finally leads to

$$f_n = \int_0^{\infty} 4\pi \cdot r^2 \cdot \varrho_n(r) \cdot \frac{\sin(\kappa \cdot r)}{\kappa \cdot r} \, dr,$$

which indeed is the electronic scattering factor f_n as introduced in (1).

4. Fourier Analysis

As mentioned before we are aiming at reconstructing in a consistent and slow paced manner the arrangement of three-periodic crystals from diffraction information obtained through Fourier coefficients. The key for this method is representing a crystalline structure by introducing a continuous electron density function $\varrho(x, y, z)$ expressed in electrons per unit volume and including all the electrons in a unit cell. We assume that the reader is aware of the mathematical methods of Fourier analysis as outlined, e.g., in [5].

Let us assume a perfect crystal with a three-periodic arrangement of distinct atoms. Since the complete electron density $\varrho(x, y, z)$ of the crystal then is triply periodic, it can be represented by a triple Fourier series in the x-, y-, and z-coordinates parallel to the a-, b- and c-axes to which the unit cell of the crystal is assumed to be aligned. To obtain the general form of this series we follow a three step approach. Hereby, our exposition will follow the lines of argumentation presented in [7], p. 108.

Let us moreover assume that the occurring functions are sufficiently smooth then we obtain the following representation for the triply periodic electron density:

$$\varrho(x, y, z) = \sum_{p=-\infty}^{\infty} \sum_{q=-\infty}^{\infty} \sum_{r=-\infty}^{\infty} C_{p,q,r} \cdot e^{-2\pi i \cdot (p \cdot x/a + q \cdot y/b + r \cdot z/c)}, \tag{3}$$

with some period $T_z = c$ in the z-direction, some period $T_y = b$ in the y-direction (that may be independent of T_z) and some period $T_x = a$ in the x-direction. Here the three periods may be rationally independent.

The complex Fourier coefficients $C_{p,q,r}$ are obtained in the usual way by multiplying both sides of Eq. (3) by

$$e^{2\pi i \cdot (h \cdot x/a + k \cdot y/b + \ell \cdot z/c)},$$

where h, k, ℓ are fixed integers, and integrating over one period cell

$$\int_0^a \int_0^b \int_0^c \varrho(x, y, z) \cdot e^{2\pi i \cdot (h \cdot x/a + k \cdot y/b + \ell \cdot z/c)} \, dz \, dy \, dx$$

$$= \sum_{p=-\infty}^{\infty} \sum_{q=-\infty}^{\infty} \sum_{r=-\infty}^{\infty} C_{p,q,r} \int_0^a \int_0^b \int_0^c e^{-2\pi i \cdot ((p-h) \cdot x/a + (q-k) \cdot y/b + (r-\ell) \cdot z/c)} \, dz \, dy \, dx.$$

$$\tag{4}$$

Since

$$\int_0^c e^{-2\pi i \cdot ((r-\ell) \cdot z/c)} \, dz = \begin{cases} c & \text{if } r = \ell, \\ 0 & \text{if } r \neq \ell, \end{cases}$$

all the terms on the right-hand side of Eq. (4) are zero except the one term for which simultaneously $p = h$, $q = k$, and $r = \ell$. Hence, the Fourier coefficients $C_{h,k,\ell}$

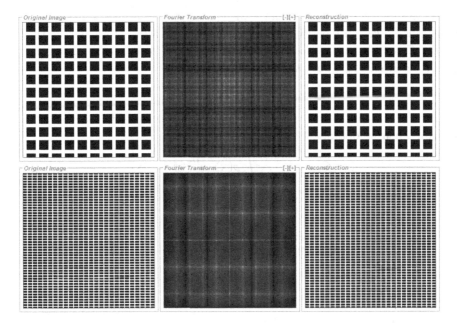

Fig. 3.   Two examples of planar structures (left-hand sides), their Fourier transform (middle column) and the reconstruction of the structure by means of an inverted Fourier transform (right-hand sides).   Images obtained and generated with the software *Fourier Transform Lab* (www.jcrsytal.com).

are given by

$$\int_0^a \int_0^b \int_0^c \varrho(x, y, z) \cdot e^{2\pi i \cdot (h \cdot x/a + k \cdot y/b + \ell \cdot z/c)} \, \mathrm{d}z \, \mathrm{d}y \, \mathrm{d}x \; = \; a \cdot b \cdot c \cdot C_{h,k,\ell}, \quad (5)$$

where $a \cdot b \cdot c$ is the volume V of the corresponding unit cell relative to the x-, y- and z-coordinates.

This establishes an isomorphism between the state space (position coordinates) and the Fourier space (amplitudes and phases) that, in particular, allows the reconstruction of a structure in state space from the data provided in Fourier space.

Figures 3 and 4 illustrate the application of the Fourier transform to some periodic and aperiodic images to highlight the symmetries and discreteness of the resulting structures in phase space. Note, that symmetries of a structure in state space with respect to a group carry over to the image in Fourier space: Let us consider a periodic crystal, and let G be an orthogonal linear representation of its point group on both spaces such that $\varrho(\Gamma \mathbf{x}) = \varrho(\mathbf{x})$ for all $\Gamma \in G$ with $\mathbf{x} = (x, y, z)^T \in \mathbb{R}^3$ (where T means the transposed vector). So, we assume that the density function ϱ is invariant under the group action G. Also, denote the unit cell by Z and set $\xi_{h,k,\ell} := 2\pi \, (h/a, k/b, \ell/c)^T$ for any integers h, k, ℓ. Note, with respect to the coordinates chosen, $\Gamma \in G$ is a matrix with integer-valued entries, i.e. $\Gamma \xi_{h,k,\ell} = \xi_{\tilde{h},\tilde{k},\tilde{\ell}}$ for some integers $\tilde{h}, \tilde{k}$, and $\tilde{\ell}$ depending on h, k, and ℓ. Moreover, in the present context this matrix commutes with the diagonal matrix $\text{diag}\,(1/a, 1/b, 1/c)$. Finally, by periodicity we can identify ΓZ and Z. For simplicity, now we assume that the unit cell Z is rectangular. Then the matrix Γ is orthogonal, and it follows that the Fourier coefficients of ϱ obey the following relations

$$C_{\tilde{h},\tilde{k},\tilde{\ell}} \;=\; C_{h,k,\ell}\,.$$

To show this, we consider the representation of the Fourier coefficients according to Eq. (5). Then we obtain

$$C_{h,k,\ell} \;=\; \frac{1}{V}\int_Z \varrho(\mathbf{x})\cdot e^{i\langle \mathbf{x}, \xi_{h,k,\ell}\rangle}\,d\mathbf{x}$$

which immediately leads to

$$C_{\tilde{h},\tilde{k},\tilde{\ell}} = \frac{1}{V}\int_Z \varrho(\mathbf{x})\cdot \underbrace{e^{i\langle \mathbf{x}, \Gamma\xi_{h,k,\ell}\rangle}}_{=\,e^{i\langle \Gamma^T\mathbf{x}, \xi_{h,k,\ell}\rangle}}\,d\mathbf{x}$$

$$= \frac{1}{V}\int_Z \varrho(\Gamma^{-1}\tilde{\mathbf{x}})\cdot e^{i\langle \tilde{\mathbf{x}}, \xi_{h,k,\ell}\rangle}\cdot \underbrace{\left|\det(\Gamma^{-T})\right|}_{=\,1}\,d\tilde{\mathbf{x}} \;=\; \frac{1}{V}\int_Z \varrho(\tilde{\mathbf{x}})\cdot e^{i\langle \tilde{\mathbf{x}}, \xi_{h,k,\ell}\rangle}\,d\tilde{\mathbf{x}}$$

$$= C_{h,k,\ell}$$

for all integers $h, k,$ and ℓ.

If no manipulations in Fourier space are performed (like the mentioned filtering to get rid of background noise in the diffraction diagram) and thus the complete information is available for the back-transformation from Fourier space to state space then the original image and the reconstructed one are indistinguishable. In experiments, though, the data measured and interpreted as elements in Fourier space are often a bit blurred which consequently leads to artifacts or ambiguities in the actual reconstruction that have to be corrected using additional information about the original structure.

Next, let us link this to the actual electron density arrangement inside the crystal. As usual, we consider a crystallographic unit cell Z with volume V and (scaled) fractional coordinates x_n/a, y_n/b, and z_n/c. As discussed, in the atomic representation these cells are equipped with cell vectors $\mathbf{r}_n$. The structure factor $F_{h,k,\ell}$ corresponding to such a unit cell is as given in Eq. (2):

$$F_{h,k,\ell} \;=\; \sum_n f_n \cdot e^{2\pi i\cdot(h\cdot x_n/a + k\cdot y_n/b + \ell\cdot z_n/c)}\,.$$

The above discrete expression for the structure factor may be replaced by one represented in terms of a continuous electron density ϱ. As above in the discussion of scattering in Section 3, by considering that instead of electrons at position $\mathbf{r}_n$ we have charge elements $\varrho\,dV$ at position $\mathbf{r}$, and by replacing the sum by an integral the structure factor becomes

$$F_{h,k,\ell} \;=\; \int_0^a\int_0^b\int_0^c \varrho(x,y,z)\cdot e^{2\pi i\cdot(h\cdot x_n/a + k\cdot y_n/b + \ell\cdot z_n/c)}\,dV\,. \tag{6}$$

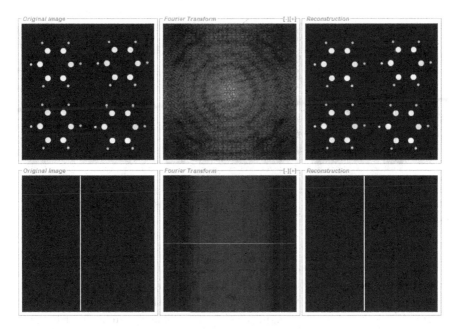

Fig. 4. Two further examples of planar structures (left-hand sides), their Fourier transform (middle column) and the reconstruction of the structure by means of an inverted Fourier transform (right-hand sides). Images obtained and generated with the software *Fourier Transform Lab* (www.jcrsytal.com).

Comparing Eq. (6) with Eq. (5), we get

$$V \cdot C_{h,k,\ell} = F_{h,k,\ell}.$$

Thus, the electron density expressed in electrons per unit volume is given by a triple Fourier series in which the Fourier coefficients are the structure factors $F_{h,k,\ell}$:

$$\varrho(x,y,z) = \frac{1}{V} \sum_{h=-\infty}^{\infty} \sum_{k=-\infty}^{\infty} \sum_{\ell=-\infty}^{\infty} F_{h,k,\ell} \cdot \mathrm{e}^{-2\pi i \cdot (h \cdot x/a + k \cdot y/b + \ell \cdot z/c)}. \qquad (7)$$

Each h, k, ℓ-reflection from a crystal corresponds to a term in the Fourier representation of the electron density, a strong h, k, ℓ-reflection signifying a strong h, k, ℓ-harmonic mode in the electron density. From this point of view, X-ray diffraction in a crystal amounts to a Fourier analysis of the electron density in the crystal. Each coefficient in the series for $\varrho(x,y,z)$ corresponds to a point (h, k, ℓ) in the reciprocal lattice. As [7], p. 110, describes, it was pointed out by Paul P. Ewald that the reciprocal lattice represents completely the electron density function (cf. [1]), if to each point in the reciprocal lattice we attach a weighting function $F_{h,k,\ell}$.

At first sight it would appear that Eq. (7) offers a straightforward way for structure determination, since the positions of the atoms will be given by the peaks in the electron density function $\varrho(x,y,z)$. The trouble, pointed out already, comes

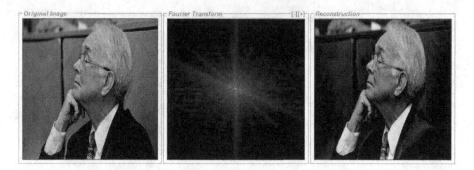

Fig. 5. Fourier-transformation of a photo Prof. A. Leutbecher taken at the colloquium in his honor. In analogy to Fig. 3. Images generated with the software *Fourier Transform Lab* (www.jcrsytal.com).

from the fact that what we measure is some sort of an integrated intensity, and this is proportional to the square of the absolute value of $F_{h,k,\ell}$, i.e. to $F_{h,k,\ell} \cdot \overline{F_{h,k,\ell}}$. Hence, in terms of an experimental setting, Eq. (7) represents the electron density $\varrho(x, y, z)$ by a Fourier series in which the magnitudes of all coefficients can be determined from measurements, but for which the phases are unknown. Hence, in order to construct an unknown electron density from experiments additional structural information is required to be included to gain knowledge (or good estimates) of the phases.

5. Resume

From the point of view of a representation of a periodic or aperiodic crystalline structure by its electron density function the connection between the diffraction diagram and the Fourier transform is quite straightforward, with the (electronic) structure factor as the connecting physical quantity. Of course, several simplifications have to be taken into account for such a classical description. It turns out that they work extremely well for structures that are composed of lighter atoms with tightly bounded shells and not too large angles of the entering beam.

Though, the reconstruction of the inner structure from a diffraction diagram is not that obvious as the diagram measures only the integrated intensities of the diffracted beams. Thus it just provides a part of the complete representation of the structure in Fourier space. The required information about the phases has to be derived in a different way that is typically related to group theoretic guessing of the possible internal composition and then an iterative approach can be invoked to distill the correct combinations of amplitudes and phases.

References

[1] Ewald, P. P. (1923). *Kristalle und Röntgenstrahlen* (Springer).

[2] International Union of Crystallography (1992). *Report of the Executive Committee for 1991 Acta Cryst.* **A48**, pp. 922–946.

[3] Chalmers, M. (2014). *A structural revolution, ESRFnews* 66, pp. 16/17.

[4] Guinier, A. (1994). *X-Ray Diffraction in Crystals, Imperfect Crystals, and Amorphous Bodies*, New York: Dover.

[5] Leutbecher, A. (1999). *Analysis IV*, Lecture Notes at the TU München.

[6] Lifshitz, R. (1995). *Introduction to Fourier-Space Crystallography*, Lecture Notes for the International School on Quasicrystals, May 13-20, 1995, Balatonfüred, Hungary.

[7] Warren, B. E. (1969). *X-Ray Diffraction*, New York: Dover.

Chapter 11

An Elementary Proof for Joint Continuity of Semiflows

Stephan Schmitz

Lehrstuhl für Höhere Mathematik und Analytische Mechanik,
Technische Universität München, Fakultät für Mathematik,
D-85747 Garching, Germany
E-mail: `schmitz@ma.tum.de`

Given a componentwise continuous semiflow $S = S(t, x)$ on a metric space X, it will be shown that S is continuous at every point (t, x) with $t > 0$. Furthermore, if X is locally compact or the time of the system is $\mathbb{N}_0, \mathbb{Z}$ or $\mathbb{R}$ then S is continuous. For this we give a new, completely elementary proof.

Contents

1. Introduction

It is well-known, that in general a componentwise continuous function of several variables is not necessarily jointly continuous unless some uniformity assumptions are met. This situation is however very different for semiflows, or semidynamical systems. The main objective in this article is to answer the question, where and when the dynamics of a topological semidynamical system is actually continuous. The main result (theorem 1 below) will be that a componentwise continuous semiflow is continuous, if $\mathbb{T}$ is one of the sets $\mathbb{N}_0, \mathbb{Z}, \mathbb{R}$. In the case $\mathbb{T} = \mathbb{R}_+$ discontinuities may arise, but only for time $t = 0$, so that the semiflow is 'instantly' continuous. To this end, a careful analysis of discontinuities will be made (to show later that they - mostly - do not occur).

All of these results have been proved before [1–3, 5], although they seem to be far from being well-known in the field of dynamical systems theory. In contrast to previous proofs, the approach in this paper is completely elementary and the only requirement is a basic understanding of metric spaces, with the Baire category theorem as the deepest topological result needed in advance. The price to pay for this is a certain loss of generality: The assumptions in the paper at hand, namely the requirement of a componentwise continuous semiflow on a metric space, while still being absolutely reasonable for dynamical systems theory, can be considerably weakend with the main result remaining valid. For example a general second countable topological space is sufficient as phase space, and the assumption of componentwise continuity can be replaced by Baire measurability for the time component. See the paper of Ball [1] for details.

On the other hand there also is considerable gain. For one, there is no more need to leave the framework of metric spaces, which is quite common for much of semidynamical system theory, see for example the book by Sell and You [5]. So the results presented here become 'generic' in the sense, that they do not require the use of some 'deeper' theory. And then those results are quite useful to have at an early stage of, for example, attractor theory, because they widen the range of applications to componentwise continuous semiflows rather than jointly continuous ones.

2. Basic definitions and notation

Let $\mathbb{R}_+ := \{t \in \mathbb{R} \mid t \geq 0\} = [0; \infty)$. In a metric space X, $B_r(x_0)$ will denote the open ball with center $x_0 \in X$ and radius $r > 0$.

A semidynamical system is a triple $(X, \mathbb{T}, S)$, where X is a metric space, $\mathbb{T}$ is one of the sets $\mathbb{N}_0, \mathbb{Z}, \mathbb{R}_+, \mathbb{R}$, and $S : \mathbb{T} \times X \longrightarrow X$ satisfies the following conditions:

(1) $S(0, x) = x$ for all $x \in X$ (identity property).
(2) $S(t_1, S(t_2, x)) = S(t_1 + t_2, x)$ for all $t_1, t_2 \in \mathbb{T}$, $x \in X$ (flow property).

X is called the *phase space*, $\mathbb{T}$ the *time* and S the *dynamics* or the *semiflow* of the system. A semidynamical system $(X, \mathbb{T}, S)$ is a *topological semidynamical system* if the map S is componentwise continuous. That is, if for fixed $t \in \mathbb{T}$ the mapping $S(t, \cdot) : x \mapsto S(t, x)$ is continuous, and for fixed $x \in X$ the mapping $S(\cdot, x) : t \mapsto S(t, x)$ is continuous. The system $(X, \mathbb{T}, S)$ is a *continuous semidynamical system* if the mapping S is continuous (or *jointly* continuous to stress the difference from componentwise continuity).

Before attacking the question of joint continuity, some general remarks on continuity and discontinuity will be given in the next section. Section 4 then focuses on semiflows and semidynamical systems.

3. Elementary discontinuity

The main technical tool for the investigation and – later – the preclusion of discontinuities is the modulus of discontinuity.

Definition 1 (Modulus of discontinuity). Let X, Y be metric spaces, $f : X \longrightarrow Y$, $x_0 \in X$. Then

$$\varepsilon_0(f; x_0) := \sup\{\varrho \geq 0 \mid \operatorname{diam}(f(U)) \geq \varrho \text{ for every open } U \ni x_0\}$$

is called the *modulus of discontinuity* of f at x_0. Obviously $\varepsilon_0(f; x_0) \geq 0$.

The modulus of discontinuity is very similar to the notorious ε_0 in the negation of the ε-δ-definition of continuity. So the close connection to continuity of f, which is stated in point 2 of the following lemma does not come as a surprise. In general, the modulus of discontinuity $\varepsilon_0(f; x_0)$ can be regarded as a means of measuring how much a function f jumps at x_0. Of course, the value $\varepsilon_0(f; x_0) = \infty$ is possible.

Lemma 1. *Let X, Y be metric spaces, $f : X \longrightarrow Y$.*

(1) If U is a neighborhood of $x_0 \in X$, then $\operatorname{diam}(f(U)) \geq \varepsilon_0(f, x_0)$.
(2) f is continuous at $x_0 \in X$ if and only if $\varepsilon_0(f; x_0) = 0$.
(3) The function $\varepsilon_0(f; \cdot) : X \longrightarrow \mathbb{R}$ is upper semi-continuous. That is, if $x_n \to x_0$ in X, then $\varepsilon_0(f; x_0) \geq \limsup_{n \to \infty} \varepsilon_0(f; x_n)$. In particular, if $X_0 \subseteq X$ and $r \geq 0$ then the set

$$\{x \in X_0 \mid \varepsilon_0(f; x) \geq r\}$$

is closed in X_0.

Proof. We start with (1): It is immediately clear from the supremum in the definition of $\varepsilon_0(f; x_0)$, that

$$\operatorname{diam}(f(U)) \geq \varepsilon_0(f; x_0) - \delta \text{ if } \delta > 0.$$

But then $\operatorname{diam}(f(U)) \geq \varepsilon_0(f; x_0)$ since $\delta > 0$ was arbitrary.

Next, to (2): Let $y_0 := f(x_0)$. If $r := \varepsilon_0(f; x_0) > 0$, then

$$\operatorname{diam}(f(f^{-1}(B_{r/3}(y_0)))) \leq \operatorname{diam}(B_{r/3}(y_0)) < r$$

since $f(f^{-1}(B_{r/3}(y_0))) \subseteq B_{r/3}(y_0)$. But then $f^{-1}(B_{r/3}(y_0))$ is not a neighborhood of x_0, because otherwise $\operatorname{diam}(f(f^{-1}(B_{r/3}(y_0))) \geq r$ by the preceding point, a contradiction. Thus f is not continuous at x_0.

If, on the other hand, $\varepsilon_0(f; x_0) = 0$, then

$$\{\varrho \geq 0 \mid \operatorname{diam}(f(U)) \geq \varrho \text{ for every open } U \ni x_0\} = \{0\}.$$

Thus, for every $\varepsilon > 0$, there is an open neighborhood U of x_0, such that $\operatorname{diam}(f(U)) < \varepsilon$. In particular $d(f(x), f(x_0)) < \varepsilon$ whenever $x \in U$, where d is the metric on Y. This shows that f is continuous at x_0.

Finally to (3): Let $r := \limsup_{n \to \infty} \varepsilon_0(f; x_n)$. If $r = 0$ there is nothing to show, since $\varepsilon_0(f; x_0) \geq 0$ is clear. Now consider the case $r > 0$. Let $0 \leq r' < r$. Then $\varepsilon_0(f; x_n) \geq r'$ for infinitely many $n \in \mathbb{N}$. If then $U \ni x_0$ is open, U also contains some x_{n_0} ($n_0 \in \mathbb{N}$) with $\varepsilon_0(f; x_{n_0}) \geq r'$, and therefore $\operatorname{diam}(f(U)) \geq \varepsilon_0(f; x_{n_0}) \geq r'$. Since U was arbitrary, one has

$$r' \in \{\varrho \geq 0 \mid \operatorname{diam}(f(U)) \geq \varrho \text{ for every open } U \ni x_0\}.$$

Therefore $\varepsilon_0(f; x_0) \geq r'$. Since $r' < r$ was arbitrary, $\varepsilon_0(f; x_0) \geq r$ follows.   □

If $(X, \mathbb{T}, S)$ is a topological semidynamical system, then the domain of S, $\mathbb{T} \times X$, is a metric space with metric

$$d'((t_1, x_1), (t_2, x_2)) := |t_1 - t_2| + d(x_1, x_2), \quad \text{for } t_1, t_2 \in \mathbb{T}, x_1, x_2 \in X,$$

where d is the metric on X. Hence, for $(t_0, x_0) \in \mathbb{T} \times X$, the modulus of discontinuity, $\varepsilon_0(S; (t_0, x_0))$ is well-defined. We use the shorthand notation

$$\varepsilon_0(t, x) := \varepsilon_0(S; (t, x)) \quad \text{for } (t, x) \in \mathbb{T} \times X,$$

for a given system. This will be useful in the proof of the following lemma about some rather elementary facts concerning the (dis)continuity of a semiflow.

Lemma 2 (Continuity and discontinuity: elementary facts). *Let* $(X, \mathbb{T}, S)$ *be a topological semidynamical system on a metric space X with metric d.*

(1) If $\mathbb{T} = \mathbb{N}_0$ or $\mathbb{T} = \mathbb{Z}$, then S is continuous.

(2) If $\mathbb{T} \in \{\mathbb{R}_+, \mathbb{R}\}$ and S is continuous at some point $(t_0, x_0) \in \mathbb{T} \times X$ with $t_0 > 0$, then S is continuous at every point $(t_0 + \tau, x_0)$ with $\tau \in \mathbb{T}$. If $\mathbb{T} = \mathbb{R}$, this implies that S is continuous on $\mathbb{T} \times \{x_0\}$.

(3) Let $(t_0, x_0) \in \mathbb{T} \times X$ such that $\varepsilon_0(t_0, x_0) > 0$. Then for every open $U \ni x_0$ and every $\delta > 0$, there are $t \in \mathbb{T}$, $x \in U$ such that $|t - t_0| < \delta$ and $d(S(t, x), S(t, x_0)) > \frac{1}{4}\varepsilon_0(t_0, x_0)$.

Proof. First to (1): Let $(t_0, x_0) \in \mathbb{T} \times X$. Let $\varepsilon > 0$. Since $S(t_0, \cdot)$ is continuous, there is for every $\varepsilon > 0$ some open U_ε containing x_0 such that

$$\operatorname{diam}(S(\{t_0\} \times U_\varepsilon)) = \operatorname{diam}(S(t_0, U_\varepsilon)) < \varepsilon.$$

$\mathbb{T}$ is discrete and therefore the set $\{t_0\} \times U_\varepsilon$ is open in $\mathbb{T} \times X$. Thus

$$0 \leq \varepsilon_0(t_0, x_0) \leq \operatorname{diam}(S(\{t_0\} \times U_\varepsilon)) < \varepsilon.$$

This is valid for every $\varepsilon > 0$, which implies $\varepsilon_0(t_0, x_0) = 0$. Thus S is continuous at every point (t_0, x_0) by lemma 1.

For (2), we let $\tau \in \mathbb{T}$. The mapping $g : \mathbb{T} \times X \longrightarrow X$, $g(t, x) := S(t + \tau, x)$ is continuous at (t_0, x_0) since there $g = S(\tau, \cdot) \circ S$ is a composition of continuous functions. Now let $t_n \to t_0 + \tau$ in $\mathbb{T}$ and $x_n \to x_0$ in X. Since $t_0 > 0$, one may assume without loss of generality, that $t_n \geq \tau$ for all n. Thus $t_n - \tau \to t_0$ even for $\mathbb{T} = \mathbb{R}_+$, and thus

$$S(t_n, x_n) = S(\tau, S(t_n - \tau, x_n)) = g(t_n - \tau, x_n) \longrightarrow g(t_0, x_0) = S(t_0 + \tau, x_0).$$

This shows that S is continuous at $(t_0 + \tau, x_0)$.

Finally, to (3): Let $U \ni x_0$ be open and $\delta > 0$. Define $J := (t_0 - \delta; t_0 + \delta) \cap \mathbb{T}$. Without loss of generality, δ is so small, that $\operatorname{diam}(S(J, x_0)) \leq \frac{1}{4}\varepsilon_0(t_0, x_0)$. Choose $(t_1, x_1), (t_2, x_2) \in J \times U$ such that

$$d(S(t_1, x_1), S(t_2, x_2)) > \operatorname{diam}(S(J \times U)) - \frac{1}{4}\varepsilon_0(t_0, x_0).$$

Since

$$d(S(t_1, x_1), S(t_2, x_2))$$
$$\leq d(S(t_1, x_1), S(t_1, x_0)) + \underbrace{d(S(t_1, x_0), S(t_2, x_0))}_{\leq \operatorname{diam}(S(J, x_0))} + d(S(t_2, x_0), S(t_2, x_2))$$
$$\leq d(S(t_1, x_1), S(t_1, x_0)) + d(S(t_2, x_0), S(t_2, x_2)) + \frac{1}{4}\varepsilon_0(t_0, x_0),$$

one has

$$d(S(t_1, x_1), S(t_1, x_0)) + d(S(t_2, x_0), S(t_2, x_2))$$
$$> \operatorname{diam}(S(J \times U)) - 2 \cdot \frac{1}{4}\varepsilon_0(t_0, x_0)$$
$$\geq \frac{1}{2}\varepsilon_0(t_0, x_0).$$

But this is only possible if one of the summands on the left is strictly greater than $\frac{1}{4}\varepsilon_0(t_0, x_0)$. This concludes the proof with $t = t_1$ or $t = t_2$. $\qquad\square$

Next, the tools discussed so far are applied to the question of joint continuity of semiflows.

4. Joint continuity of semiflows

The following lemma looks, at first glance, like some interesting fact about discontinuities of semiflows. Which it is, but only in a matter of speaking. The lemma is the crucial ingredient for the proof by contradiction of a later theorem (theorem 1), which implies that the requirements here are actually never fulfilled. Thus the somewhat funny name.

Lemma 3 (A lemma which applies only once). *Let $(X, \mathbb{T}, S)$ be a topological semidynamical system with time $\mathbb{T} \in \{\mathbb{R}, \mathbb{R}_+\}$. If there are $t_0 > 0$ and $x_0 \in X$, such that S is not continuous at (t_0, x_0), then there exists an open interval $J \subseteq \left[\frac{t_0}{2}; t_0\right]$ and a constant $\varepsilon_J > 0$, such that*

$$\varepsilon_0(t, x_0) \geq \varepsilon_J \text{ for all } t \in J.$$

Proof. Let $I := \left[\frac{t_0}{2}; t_0\right]$. For $n \in \mathbb{N}$ let

$$I_n := \left\{ t \in I \,\middle|\, \varepsilon_0(t, x_0) \geq \frac{1}{n} \right\}.$$

Then every I_n is closed. To see this, note that $I_n = S(\,\cdot\,, x_0)^{-1}(K_n)$ where

$$K_n := \left\{ (t,x) \in I \times \{x_0\} \;\Big|\; \varepsilon_0(t,x) \geq \frac{1}{n} \right\} \quad \text{for } n \in \mathbb{N}.$$

Due to point 3 of lemma 1, every K_n is closed in $I \times \{x_0\}$, and since $I \times \{x_0\}$ is compact, every K_n is closed in X. Hence every I_n is closed as the continuous counterimage of a closed set.

By point 2 of lemma 2, S is discontinuous at every point of $I \times \{x_0\}$, because otherwise it would be continuous at (t_0, x_0). Therefore $\varepsilon_0(t, x_0) > 0$ for every $t \in I$, and consequently

$$I = \bigcup_{n \in \mathbb{N}} I_n.$$

The compact interval I is also a complete metric space. Since all I_n are closed (also in I), the Baire category theorem (see e.g. Dunford and Schwartz [4], theorem I.6.9) applies. Hence there is some $n_0 \in \mathbb{N}$, such that I_{n_0} contains a subset which is open in I and thus contains some open interval J. Letting $\varepsilon_J := \frac{1}{n_0}$ this proves the lemma. $\qquad\square$

Now the main theorem will be stated and proven, namely that discontinuities of semiflows exist only for $t = 0$. This is remarkable, since it is very well-known that this is in general false for arbitrary componentwise continuous functions of two arguments. So, vaguely, being a semiflow makes a function continuous.

Theorem 1 (Continuity I). *Let $(X, \mathbb{T}, S)$ be a topological semidynamical system. If the time $\mathbb{T}$ is one of the sets $\mathbb{N}_0, \mathbb{Z}, \mathbb{R}$, then S is continuous. If $\mathbb{T} = \mathbb{R}_+$, then S is continuous on $(0; \infty) \times X$, that is at every (t, x) with $t > 0$, $x \in X$.*

Proof. Let d denote the metric on X.

For discrete time, that is $\mathbb{T} = \mathbb{N}_0$ or $\mathbb{T} = \mathbb{Z}$, the assertion was already proven as point 1 of lemma 2. Therefore let $\mathbb{T} = \mathbb{R}$ or $\mathbb{T} = \mathbb{R}_+$. It will be shown that S is continuous at every point (t, x_0) with $t > 0$, $x_0 \in X$. If $\mathbb{T} = \mathbb{R}$, this implies the continuity of S via point 2 of lemma 2.

Assume that the assertion is wrong, that is, S is discontinuous at some point $(\tau_0, x_0) \in \mathbb{T} \times X$ with $\tau_0 > 0$. Then by lemma 3 there is an open interval $J \subseteq \mathbb{T}$ and a constant $\varepsilon_J > 0$ such that

$$\varepsilon_0(t, x_0) \geq \varepsilon_J \quad \text{for all } t \in J.$$

Now a sequence $(x_n)_{n \in \mathbb{N}}$ of points in X, and a sequence $(J_n)_{n \in \mathbb{N}}$ of intervals in $\mathbb{T}$ will be constructed recursively.

For n=1: Let $T \in J$. By point 3 of lemma 2 there is $x_1 \in B_1(x_0)$ and $t_1 \in J$ (near T), such that

$$d(S(t_1, x_1), S(t_1, x_0)) > \frac{\varepsilon_0(T, x_0)}{4} \geq \frac{\varepsilon_J}{4}.$$

Since the map $t \mapsto d(S(t, x_1), S(t, x_0))$ is continuous, there is some $\delta_1 > 0$, such that $J_1 := [t_1; t_1 + \delta_1] \subseteq J$ and

$$d(S(t, x_1), S(t, x_0)) > \frac{\varepsilon_J}{4} \text{ for all } t \in J_1.$$

Then $J_1 \subseteq J$ is a compact interval with nonempty interior.

For $n \geq 2$: Given a compact interval $J_{n-1} \subseteq J$ with nonempty interior. Let $T \in \overset{\circ}{J}_{n-1}$. By point 3 of lemma 2 there is $x_n \in B_{1/n}(x_0)$, and $t_n \in J_{n-1}$ (near T), such that

$$d(S(t_n, x_n), S(t_n, x_0)) > \frac{\varepsilon_0(T, x_0)}{4} \geq \frac{\varepsilon_J}{4}.$$

Since $t \mapsto d(S(t, x_n), S(t, x_0))$ is continuous, there is some $\delta_n > 0$, such that $J_n := [t_n; t_n + \delta_n] \subseteq J_{n-1}$ and

$$d(S(t, x_n), S(t, x_0)) > \frac{\varepsilon_J}{4} \text{ for all } t \in J_n.$$

Then $J_n \subseteq J_{n-1} \subseteq J$ is a compact interval with nonempty interior.

This construction yields a sequence (x_n) in X with $x_n \to x_0$. Furthermore, it yields nested intervals $I_1 \supseteq I_2 \supseteq I_3 \supseteq \ldots$ with

$$d(S(t, x_n), S(t, x_0)) > \frac{\varepsilon_J}{4} \text{ for all } t \in J_k, n \leq k.$$

This is clear by construction for $k = n$, and for $k \geq n$ it follows immediately from $J_n \subseteq J_k$.

Since all J_k are nonempty and compact, $\bigcap_{k \in \mathbb{N}} J_k$ is nonempty. Thus, there is some t_0, which is contained in all J_k, implying

$$d(S(t_0, x_n), S(t_0, x_0)) > \frac{\varepsilon_J}{4} \text{ for all } n \in \mathbb{N}.$$

But $S(t_0, x_n) \to S(t_0, x_0)$, since S is componentwise continuous and $x_n \to x_0$, a contradiction. Thus the assumption was false, and the assertion is true. $\square$

Remark 1. Note that up to this point, the flow property of the semiflow S has been used only once, namely in the proof of point 2 of lemma 2, while the identity properties has not been applied at all. Thus, if point 2 of lemma 2 can be substituted somehow, an equivalent of theorem 1 will hold for a more general class of maps than just semiflows. For example, other kinds of functional identities might yield the same result.

As for the case $t = 0$ and $\mathbb{T} = \mathbb{R}_+$, a semiflow might actually be discontinuous, but a simple example in $\mathbb{R}^d$ does not exist.[1] This follows immediately from the following theorem.

---

[1] An example will be presented later, because the oncoming theorem is quite helpful for the proof of its properties.

Theorem 2 (Continuity II). *Let $(X, \mathbb{T}, S)$ be a topological semidynamical system, and let $x_0 \in X$ have a compact neighborhood. Then S is continuous on $\mathbb{T} \times \{x_0\}$. In particular, if the metric space X is locally compact, then S is continuous.*

Proof. Let d denote the metric of X. Assume that the assertion is false. Then theorem 1 states $\mathbb{T} = \mathbb{R}_+$, and there are sequences $(t_n)_{n \in \mathbb{N}}$ in $\mathbb{R}_+$ and $(x_n)_{n \in \mathbb{N}}$ in X such that $t_n \downarrow 0$, $x_n \to x_0$, but (cf. point 3 of lemma 2)

$$d(S(t_n, x_n), x_0) \geq \frac{\varepsilon_0(0, x_0)}{4} > 0.$$

Let $0 < r < \frac{1}{4}\varepsilon_0(0, x_0)$ such that $K := \{x \in X \mid d(x, x_0) = r\}$ is compact or empty. This is possible because x_0 has a compact neighborhood. Since $x_n \to x_0$, there is some $N \in \mathbb{N}$, such that $d(x_n, x_0) < r$ for all $n \geq N$. Thus, by the intermediate value theorem (applied to $t \mapsto d(S(t, x_n), x_0)$ with n fixed), for $n \geq N$ there is some $\tau_n \in (0; t_n)$ such that $S(\tau_n, x_n) \in K$. Hence K is nonempty, and the sequence $S(\tau_n, x_n)$ has an accumulation point in K. For simplicity, let $S(\tau_n, x_n)$ without loss of generality be convergent with limit $\hat{x}_0 \in K$.

Since S is componentwise continuous and $d(x_0, \hat{x}_0) = r \neq 0$, there is a $\tau_0 > 0$ such that

$$d(S(\tau_0, x_0), S(\tau_0, \hat{x}_0)) \neq 0.$$

Note that $\tau_m \to 0$, since $t_n \to 0$. Thus $\tau_n + \tau_0 \to \tau_0$, and since S is continuous at (τ_0, x_0), due to theorem 1, one has

$$S(t_n + \tau_0, x_n) \longrightarrow S(\tau_0, x_0).$$

But

$$S(\tau_n + \tau_0, x_n) = S(\tau_0, S(\tau_n, x_n)) \longrightarrow S(\tau_0, \hat{x}_0) \neq S(\tau_0, x_0),$$

a contradiction since limits are unique. Thus the assumption was false. $\square$

Since $\mathbb{R}^d$ is locally compact, a componentwise continuous semiflow on $\mathbb{R}^d$ is jointly continuous. This includes, of course, flows generated by ordinary differential equations.

Finally, the following example demonstrates, that a semiflow for time $\mathbb{T} = \mathbb{R}_+$ on a metric space might actually be discontinuous at least at one point. Since the phase space of such a system cannot be locally compact, the construction is somewhat complicated.

Example 1 (A not entirely continuous system). The system will be constructed by glueing together a countable number of systems $(X_n, \mathbb{R}_+, S_n)$ at a common fixed point x_0.

First step: The systems $(X_n, \mathbb{R}_+, S_n)$, $n \in \mathbb{N}$. Here X_n is a metric space with metric d, containing two points x_0 and x_n and satisfying the following properties (please note the picture).

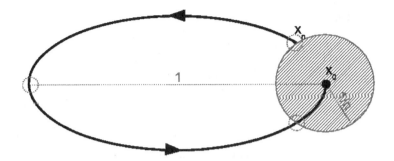

- $X_n = \{x_0\} \cup S_n(\mathbb{R}_+ \times \{x_n\})$ is a compact metric space.
- x_0 is a fixed point.
- $d(x_n, x_0) = \frac{1}{n}$.
- One has $d(S_n(\frac{1}{2n}, x_n), x_0) = 1$.
- For every $t > \frac{1}{n}$ one has $d(S_n(t, x_n), x_0) < \frac{1}{n}$.
- If $d(x, x_0) < \frac{1}{n}$, then $d(S_n(t, x), x_0) \leq d(x, x_0)$ for all $t \geq 0$.
- S_n is continuous.

Looking at (the black part of) the picture, the system $(X_n, \mathbb{R}_+, S_n)$ is easily constructed, for example as a subset of the unit circle in $\mathbb{C}$. The given properties imply: For $(t, x) \in \mathbb{R}_+ \times X_n$ with $t > \frac{1}{n}$ or $d(x, x_0) < \frac{1}{n}$ one has $d(S_n(t, x), x_0) < \frac{1}{n}$.

Second step: Construction of the system $(X, \mathbb{R}_+, S)$. To get X, one glues all X_n together at the common fixed point x_0. (Take the topological sum $\sum_n X_n$ and identify all $x_0 \in X_n$.) As a metric d on X one takes $d(x, y)$ if both x and y are elements of the same X_n. Else, let $d(x, y) := d(x, x_0) + d(y, x_0)$. It is easily proven that d is really a metric on X. The mapping $S : \mathbb{R}_+ \times X \longrightarrow X$ is defined by

$$S(t, x) := S_n(t, x) \text{ for } x \in X_n.$$

Then $(X, \mathbb{R}_+, S)$ is obviously a semidynamical system. Furthermore:

- S is componentwise continuous, that is $(X, \mathbb{R}_+, S)$ is a topological semidynamical system.
- S is continuous at every point of $(\mathbb{R}_+ \times X) \setminus \{(0, x_0)\}$. But S is not continuous at $(0, x_0)$. In particular, the semiflow S is *not* continuous.

Proof. For fixed x, the mapping $t \mapsto S(t, x)$ is continuous, because there is some $n \in \mathbb{N}$ such that x in x_n. Therefore $S(t, x) = S_n(t, x)$ and S_n is continuous. Thus S is continuous with respect to t.

Now let t be fixed and consider the mapping $\varphi : x \mapsto S(t, x)$. If $t = 0$ then $\varphi = \text{id}$ is continuous. Now let $t > 0$. If $x \neq x_0$ then φ is continuous at x, because x has a neighborhood contained in one X_n, and on this neighborhood $\varphi(x) = S_n(t, x)$ where S_n is continuous. Now consider $x = x_0$. Let $\varepsilon > 0$. There is $N \in \mathbb{N}$ such that $\frac{1}{N} < \varepsilon$ and $\frac{1}{N} < t$. Let $\delta := \frac{1}{N}$. Let $x \in X$ with $d(x, x_0) < \delta$. Then there is $n \in \mathbb{N}$

with $x \in X_n$. If $n \geq N$, then $t > \frac{1}{N} \geq \frac{1}{n}$ and therefore

$$d(\varphi(x), \varphi(x_0)) = d(S_n(t,x), x_0) < \frac{1}{n} \leq \frac{1}{N} < \varepsilon.$$

If $n < N$ then $d(x, x_0) < \delta = \frac{1}{N} \leq \frac{1}{n}$ and therefore

$$d(\varphi(x), \varphi(x_0)) = d(S_n(t,x), x_0) \leq d(x, x_0) < \frac{1}{N} < \varepsilon.$$

This shows, that φ is continuous at x_0. In total, S is componentwise continuous, and thus $(X, \mathbb{R}_+, S)$ is a topological semidynamical system.

By theorem 1 the semiflow S is continuous at every point (t, x) with $t > 0$. Furthermore if $x \in X$, $x \neq x_0$ then x has a compact neighborhood in one of the X_n. Therefore S is continuous at $(x, 0)$ by theorem 2. Thus S is continuous at every point except (x_0, t_0). But here $x_n \to x_0$, $\frac{1}{2n} \longrightarrow 0$ but

$$d(S(\tfrac{1}{2n}, x_n), S(0, x_0)) = d(S(\tfrac{1}{n}, x_n), x_0) = 1$$

and thus $S(\frac{1}{2n}, x_n)$ does not converge to x_0. Hence S is not continuous at (x_0, t_0).

$\square$

The above example is minimal in the sense, that it consists solely of ingredients necessary for the occurrence of a discontinuity at $t = 0$. An example in Hilbert space can be found in an article by Chernoff [3]. As an example with practical impact, the two-dimensional Navier-Stokes equation is mentioned in a book by Sell and You [5] to exhibit discontinuities at $t = 0$.

Acknowledgements

I would like to thank the referees for their comments and suggestions. They really helped clearing some points up. And thanks to the editor for his patience during several delays (caused by me) of the review process.

References

[1] J.M. Ball (1974): *Continuity Properties of Nonlinear Semigroups*, J Funct Ana, 17: 91-103.
[2] P.R. Chernoff & J. Marsden (1970): *On Continuity and Smoothness of Group Actions*, Bull Am Math Soc, 76: 1044-1049. (In a later paper from 1975 Chernoff admits that one example given here does not work.)
[3] P.R. Chernoff (1975): A Note on Continuity of Semigroups of Maps, Proc AMS, 53(2): 318-320.
[4] N. Dunford & J.T. Schwartz (1958): Linear Operators, Part I: General Theory, Interscience Publishers, Inc., New York.
[5] G.R. Sell & Y. You (2002): *Dynamics of Evolutionary Equations*, Applied Mathematical Sciences, 143, Springer Verlag, New York Berlin Heidelberg.

Chapter 12

A Convergent String Method: Existence and Approximation for the Hamiltonian Boundary-Value Problem

Hartmut Schwetlick & Johannes Zimmer

both Department of Mathematical Sciences, University of Bath, Bath BA2 7AY, United Kingdom,

E-mail: {schwetlick|zimmer}@maths.bath.ac.uk

This chapter studies the existence of long-time solutions to the Hamiltonian boundary value problem, and their consistent numerical approximation. Such a boundary value problem is, for example, common in Molecular Dynamics, where one aims at finding a dynamic trajectory that joins a given initial state with a final one, with the evolution being governed by classical (Hamiltonian) dynamics. The setting considered here is sufficiently general so that long time transition trajectories connecting two configurations can be included, provided the total energy E is chosen suitably. In particular, the formulation presented here can be used to detect transition paths between two stable basins and thus to prove the existence of long-time trajectories. The starting point is the formulation of the equation of motion of classical mechanics in the framework of Jacobi's principle; a curve shortening procedure inspired by Birkhoff's method is then developed to find geodesic solutions. This approach can be viewed as a string method.

Dedicated to Professor Armin Leutbecher on the occasion of his 80th birthday

Contents

1. Introduction

The aim of this chapter is to study the existence and give a consistent approximation procedure of the boundary value problem for the conservative dynamical system

$$\frac{d^2 q(t)}{dt^2} = -\nabla V(q), \tag{1}$$

where V is a smooth potential on Q. We assume that Q is an open subset of $\mathbb{R}^n$ as this is the relevant case for the applications we have in mind; extensions to a more general setting are possible but not discussed here.

For the boundary conditions, we write

$$q(0) = q_a \text{ and } q(T_0) = q_b \tag{2}$$

with $q_a, q_b \in Q$ and $T_0 > 0$. Here, T_0 is part of the problem and has to be determined (however, the total energy E, defined as the sum of kinetic and potential energy, is fixed). The focus on the boundary-value problem is motivated by applications, as discussed below.

1.1. *Hamiltonian systems, rare events and path sampling*

Equation (1) (furnished with various initial or boundary conditions) can be reformulated as the classic Hamiltonian problem

$$\dot{p} = -\frac{\partial H}{\partial q}(p, q),$$
$$\dot{q} = \frac{\partial H}{\partial q}(p, q) \tag{3}$$

for $p, q \in \mathbb{R}^n$, where H is the *Hamiltonian*

$$H = \frac{1}{2}p^2 + V(q). \tag{4}$$

Mathematically, the existence of solutions to (3), often more succinctly written as

$$\dot{z} = \begin{pmatrix} 0 & -\text{Id} \\ \text{Id} & 0 \end{pmatrix} H_z(z), \tag{5}$$

with $z := (p, q)$, is a classical problem. Periodic solutions have been a particular focus, and existence results obtained until the early 1980s are discussed in the beautiful survey article [19]. Already for periodic solutions, a clear distinction has to be made for local results (that is, short time solutions) and global solutions describing solutions in the large. Apparently the first global global result was obtained by Seifert [22] for a Hamiltonian which is slightly more general than the one in (4). The key idea of his proof is based on differential geometry, using an equivalent reformulation of (3) in which solutions can be found as a geodesic in a (degenerate) Riemannian metric, the so-called Jacobi metric. A curve shortening procedure proposed by G. D. Birkhoff [3, Section V.7] can then be applied to show the existence of a geodesic. This result has later been extended by Weinstein, and a more general result based on a different variational approach was given by Rabinowitz [18].

In the Sciences, the interest in non-periodic long time solutions has recently been rejuvenated by various applications. Namely, complex systems in physics, chemistry or biology can often be described by a potential energy landscape with many wells, separated by barriers. A common problem is then to find a trajectory joining a given initial point (configuration) with a given final point. We study this problem in the situation where the dynamics is determined by (1), and the points given in (2) are potentially far apart. In particular, the two configurations will generically be located in different wells of the energy landscape. Rare events are an example of these transitions between two wells. Typically, thermally activated reactions have many deep wells separated by large energy barriers. Reactants will then spend most of the time jostling around in one well before a rare spontaneous fluctuation occurs that lifts the atoms of the reactant over the barrier into the next (product) valley. Information on rare events is crucial since they represent important changes in the system, such as chemical reactions or conformational modifications of molecules. A major challenge in Molecular Dynamics (MD) is that these hopping events take place so rarely that the computational limits of MD simulations can be easily exceeded. Since the problem (1)–(2) is central in MD, a number of solution strategies have been proposed; see [20] for a brief review of some methods. Further, for practitioners of MD, the question arises whether any numerical approximation shadows a physical one [12] (and if so, whether it shadows a generic physical trajectory). The lack of hyperbolicity rules out standard tools to prove shadowing (e.g., [15, Theorem 18.1.3]). Thus, for MD, computations are "based on trust" [10, Section 4.3].

One further difficulty for the computation of Hamiltonian trajectories, in particular for MD, is that these trajectories are often chaotic, and one has to restrict oneself to averaged statistical information. However, of particular interest in the analysis of rare events are trajectories going directly from one well of a potential to another one; such transition paths can be used to define so-called reaction coordinates. Often the most efficient algorithms using for example path sampling assume a knowledge of these coordinates and thus these non-chaotic trajectories are of significant practical importance. The method presented in this chapter is concerned with the calculation of such non-chaotic transition trajectories. Even these relatively "simple" trajectories are in practise very hard to compute, since they correspond to rare events and take place on very long time scales.

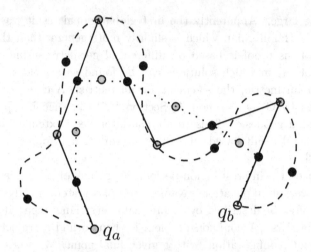

Fig. 1. Birkhoff's algorithm, for the toy example of the Euclidean metric in $\mathbb{R}^2$ and for $i = 5$. The initial curve is plotted as a dashed line. Points with odd index are marked by black dots, points with even index by grey dots. In the first step, the points with even indices are kept fixed, and joined by a geodesic. New positions for the points with odd indices on the new curve are determined (solid curve). In a next step, these points are joined by geodesics, which determines new positions for the points with even indices (dotted curve). The curves (slowly) converge to the geodesic line segment connecting q_0 and q_{2i}.

1.2. *Jacobi metric and Birkhoff curve shortening*

Rather working with the Hamiltonian boundary value problem (1)–(2) directly, we use an equivalent variational formulation, namely the Maupertuis principle, according to which trajectories to (1) with total energy E are suitably re-parametrised geodesics with respect to the *Jacobi metric*

$$g_{ij}^{\text{Jac}}(q) := (E - V(q))\,\delta_{ij}(q) \tag{6}$$

(more generally, if Q is equipped with a Riemannian metric g_{ij}, then $g_{ij}^{\text{Jac}}(q) := (E - V(q))\,g_{ij}(q)$). So Hamiltonian trajectories are critical points of the length functional associated with (6). While the equivalence has been known for centuries, it seems that little advantage has been taken of the fact that this variational formulation has a very convenient mathematical structure. Note that Hamiltonian problems such as (1) are commonly indefinite, while a geodesic problem is elliptic and thus bounded from below.

So the problem reduces to that of finding geodesics in the Jacobi metric. This problem has been addressed by Birkhoff, who described a curve shortening procedure to find global geodesics under the assumption that local (sufficiently short) geodesics can be found explicitly. This assumption is not met here since local geodesics have to be approximated; the main task is to show that nevertheless global convergence can be obtained for a suitably devised local approximation scheme. Since the local scheme we propose also relies on a Birkhoff curve shortening idea, we present his idea in the global setting first to keep the presentation self-contained.

Initially, one joins the given initial and final point q_a and q_b by an arbitrary curve. Then sufficiently many points are marked on the curve so that a local geodesic can be computed between next to nearest neighbours (see Figure 1). Note that the argument assumes that local geodesics can be computed with sufficient accuracy. In the first step, the points with even indices are kept fixed, and joined by a geodesic. Now positions for the points with odd indices on the new curve are determined (solid curve in Figure 1). The procedure of joining next to nearest neighbouring points by local geodesics is then repeated for the points with odd indices (dotted curve in Figure 1). It is not hard to show that this iterative procedure decreases the length. Under suitable assumptions (e.g., [14]), this method can be shown to converge to a global geodesic; however, there are situations such as for degenerating metrics where convergence will not take place.

The central result of this chapter is an analogous local result, introducing a sequence of approximating sequences converging to a sufficiently short geodesic. The trade-off is that the result is local (applicable only to sufficient short geodesics), but does not assume that the approximating sequences consist of exact geodesic segments. We rely on the observation that the global Birkhoff argument localises the geodesic problem in a geometrically tractable way. That is, one can restrict the local analysis to points which are sufficiently close. We then show that they can be joined by a geodesic which in addition can be represented as a graph and prove consistency and convergence of the proposed local approximation.

1.3. *Results*

The main result of this chapter is a consistent approximation of what we call local geodesics; important aspects are that the proof is constructive and yields bounds on the allowed distance between points to be joined by local geodesics. The bounds are not in terms of the usual estimates from differential geometry (such as the injectivity radius, that is, the radius for which there is a unique geodesic starting at the centre, with arbitrary velocity), but are expressed explicitly in terms of the total and potential energy of the molecular Hamiltonian system (1). We point out that locality of the geodesics does not necessarily require that the two end points are very close; an important aspect of the proof is that we may consider local geodesics which can be represented as graphs. Global geodesics can violate this assumption, while suitably small segments of geodesics remain geodesics and can be represented as graphs. Birkhoff's idea to segment an original connecting curve allows us to confine our algorithm to such local geodesics. The efficiency and applicability of a global Birkhoff method as in [21] will depend on the chosen parametrisation (number and location of points in Figure 1). For example, it is possible that refinements in a numerical implementation are required. However, the proof shows that in the setting studied in this chapter, no refinement or reparametrisation is required.

As a by-product, we show the existence of a continuous (physical) trajectory for suitable points q_a, q_b, using the Jacobi formulation as Seifert [22], but replacing the periodic boundary conditions considered there by Equation (2). There are some related existence results [2, 13], which also rely on the Jacobi formulation. The novelty of the results presented here are twofold: (i) while a formulation using the Jacobi metric is natural, a difficulty is that the metric degenerates at the boundary ∂Q of the configuration space, where kinetic energy $\int \frac{1}{2}\dot{q}^2 \, dx$ and potential energy $\int V(q) \, dx$ agree. We provide *a priori* bounds to ensure that the geodesic stays away from ∂Q; the bounds depend on the location of the boundary points q_a and q_b or on the total energy E. Bounds could be obtained along the lines of thought presented in this paper, but in a simpler fashion. The reason why we give a more complicated argument is that the approximation we give here is constructive, giving existence of a solution and at the same time a consistent approximation procedure. Thus, we obtain an approximation procedure which may not necessarily be the most efficient but one for which can dispense with the need for trust. Since the algorithm we develop is consistent, the issue of shadowing is answered in an affirmative way for the procedure we propose, under the assumptions made on the potential energy V and the bounds in term of total energy E made on the end points.

The existence of geodesics joining a given initial and final point in open domains Q, which is trivial within the radius of injectivity, is not obvious if the two points are further away from each other. We point out that the argument of this paper automatically proves the existence of such a geodesics for the Jacobi metric with sufficiently large total energy E. Other existence results, proceeding along quite different lines, can be found elsewhere [2, 13]. We remark that we will not address the question of how to choose E; this choice typically requires insight in the physics, chemistry or biology of the problem in question and thus cannot be answered in the general mathematical framework considered here. However, the existence result given here can be interpreted in two ways: given the initial point q_a and the total energy E, the arguments provide estimates on possible locations of the final point q_b so that q_a and q_b can be joined by a trajectory with total energy E. Alternatively, given q_a and q_b, the analysis provides lower bounds on E such that q_a and q_b can be joined with this total energy. It is easy to see that no general existence theorem can hold if q_a, q_b and E are unrestricted (E determines the configuration manifold Q, and in particular for small E the configuration manifold may be disconnected, so q_a and q_b could be in different components).

1.4. *Applications and limitations*

Approximations which are proven to be consistent, such as Godunov's scheme for hyperbolic equations, are often less efficient than algorithms for which consistency cannot be shown. The approximation introduced in this chapter is no exception. Yet, it is often possible to take inspiration from a consistent approximation and deduce efficient (but not provably consistent) formulations. This is the case for the formulation introduced here; a related flow model approximation has been shown to be able to detect different trajectories joining points in different wells of the energy landscape of the Müller potential and the collinear reaction $H_2 + H \rightarrow H + H_2$ [20]. While our formulation relies both on the choice of Maupertuis' formulation and Birkhoff's curve shortening, which seems to be new to the field of numerical methods for Molecular Dynamics, the curve-shortening procedures resembles other rubber-band algorithms [12]. For the isothermal case, a nudged elastic band method has been proposed by E, Ren and Vanden-Eijnden [6]. There, the aim is to find minimal energy paths, which are defined as paths along which the orthogonal component of the deterministic vector field vanishes. The approach in [6] to reduce the orthogonal contributions iteratively bears many similarities with the method presented here. As examples of string methods with temperature, we refer to the pioneering work by E, Ren and Vanden-Eijnden [7, 9].

Maupertuis' principle has been used before in MD [1], but without the connection to Birkhoff's curve shortening method. We also refer the reader to the recent work by Cameron, Kohn and Vanden-Eijnden which gives an analysis and in particular convergence results for a steepest descent string method [5].

The usual numerical approach for a boundary value problem is a shooting method; there, the existence of a solution is assumed as well as the closeness of an initial guess to the solution. (There are abstract existence results available, see for example [16]; however, as noted by Stoer and Bulirsch [23, 7.3.3], the abstract formulation of the boundary conditions to be imposed rules out the condition (2), for the first-order system considered there, even for the case $n = 2$.) The method discussed in this paper provides both an explicit existence proof and estimates on the closeness required.

We point out that the Birkhoff algorithm also provides a strategy for gluing together local geodesics to obtain global paths, which converge to a global geodesic. Note that this latter aspect of the Birkhoff approach defines a natural tool to localise the computation of geodesics which can be exploited for parallelisation. This aspect is discussed in more detail in [21]. An advantage of Birkhoff's method is that these local steps are intrinsically parallelisable.

Finally, we mention connections to the Onsager-Machlup / Freidlin-Wentzell theory. There, an action functional is derived, with the minimal action path describing the most likely trajectory. The connection between that theory, Hamilton-Jacobi theory and the Maupertuis principle as discussed here is a topic with many open questions; we refer to [8] for results in this direction together with applications to rare events.

1.5. *Notation*

Throughout the presentation, Q is the *configuration manifold* of a system and thus describes all possible states the system can occupy. The coordinates of the *phase space* (cotangent bundle) T^*Q are (q^j, p_j), position and momentum. Analogously, the coordinates of the tangent bundle TQ are $(q^j, \dot{q}^j)$, where $\dot{q}^j$ denotes the velocity. We assume that the system dynamics is conservative with $3N$ degrees of freedom.

Then, the *Hamiltonian* $H: T^*Q \to \mathbb{R}$ is defined as $H := E := T + V$. Here, the kinetic energy $T = T(p)$ is a function of the momenta only and $V = V(q)$ is the potential energy, depending on the coordinates q alone. The *Lagrangian* of the system is a function $L: TQ \to \mathbb{R}$, namely $L(q, \dot{q}) = T - V$. For a wide class of applications, it is sufficient to consider $\langle \dot{q}, \dot{q} \rangle = \sum_{j=1}^{3N} m_j \dot{q}_j \dot{q}_j$, the inner product for a system with N particles with mass m_j, and to assume that the Lagrangian is of the standard form

$$L(q, \dot{q}) := \frac{1}{2} \langle \dot{q}, \dot{q} \rangle - V(q). \tag{7}$$

This chapter is organised as follows: In Section 2, we review the Maupertuis principle and the Birkhoff curve shortening algorithm, both for the continuous setting. Section 3 describes the analogous discrete setting, contains the relevant *a priori* estimates and introduces the discrete Birkhoff procedure for a fixed discretisation. Section 4 describes the Birkhoff refinement and convergence to the continuous limit. Numerical simulations and numerical convergence rates for a model problem are the content of Section 5.

2. The continuous setting

Our construction will be guided by a variational formulation, equivalent to (1), where convergent approximations can be obtained with relative ease. This continuous setting is sketched in the present section.

It is a well known fact that solutions to (1) with pre-assigned total energy E are re-parametrised geodesics in *Jacobi's metric* (6). This formulation is sometimes denoted *Maupertuis' principle*, or *Jacobi's least action principle*. For the special Lagrangian (7), the *Routhian* associated with Jacobi's principle is

$$R(q, q') = 2(E - V(q)) \langle q', q' \rangle. \tag{8}$$

Obviously, $R(\gamma, \gamma')$ is a metric in those regions of Q where $V(q) < E$. The action functional

$$J[\gamma] := \int_a^b R(\gamma, \gamma')\, d\tau \tag{9}$$

is the measure of the length of γ in this metric. For a given curve γ, the value $J[\gamma]$ is often called the *energy* of γ; the *length* of the curve is then

$$L[\gamma] := \int_a^b \sqrt{R(\gamma, \gamma')}\, d\tau. \tag{10}$$

It is trivial to verify that critical points of the energy functional are critical points of the length functional; the converse is true for curves parameterised by arc length. The length functional is invariant under re-parametrisations, while a minimiser of the energy functional is automatically parameterised by arc length.

The Maupertuis' principle seeks geodesics, that is, stationary solutions of the functional (9) with the metric (8). Maupertuis' principle has been employed in a number of computational approaches and is regarded as a very accurate method for the verification of other algorithmic formulations [17].

We already mentioned in passing that solutions of (1) are re-parametrisations of solutions of (9) (or, equivalently, (10)). The re-parametrisation is such that the physical time for a solution of (1) can be recovered via the explicit formula

$$t = \int_0^\tau \sqrt{\frac{\langle q', q' \rangle}{2(E - V)}} \, ds. \tag{11}$$

With the exception of [20], numerical methods for (1) seem, to the best of our knowledge, not have taken advantage of the geodesic formulation (8) (see [20] for a recent survey and a method that relies on observations similar to, but simpler than, those made in this chapter). This is somewhat surprising, since the *Birkhoff curve shortening algorithm* is a classic method for the convergent approximation of geodesics.

2.1. *Existence of extended geodesics*

Birkhoff's curve shortening method [3] is a constructive way to find extended geodesics, based on the assumption that local (short) geodesics (that is, geodesics joining points within the radius of injectivity) can be computed exactly. In this subsection, we recall the classic Birkhoff method; the main part of the chapter then addresses the question of how to find the local geodesics constructively, for the case of the Jacobi metric. A straightforward implementation of the Birkhoff method relies on an approximation of local geodesics within the radius of injectivity and thus requires knowledge of the radius of injectivity. For the complex energy land-scapes we have in mind, this radius is not easily computable in a quantitative way. Thus, we have here two aims. Firstly, we present an algorithm for the numerical approximation of local geodesics in an explicitly given neighbourhood. Secondly, we obtain a quantitative description of the size of this neighbourhood. A Birkhoff method then glues together these local geodesics to obtain an extended piecewise geodesic curve.

The aim of this chapter is to develop a discrete framework that mimics the Birkhoff procedure. Besides the usual difficulty of discretisation errors inherent in any numerical approach, we face the challenge that even the computation of local geodesics, as in Birkhoff's algorithm, is time-consuming and difficult to control for non-Euclidean metrics. We propose an approximation of Jacobi's metric (8) by the trapezoidal rule. The key observation is that the difference between the Jacobi metric and the Euclidean metric occurs on a fine scale, described in greater detail below. The analysis of Section 3 will show that crucial bounds on the discrete curvature in a Birkhoff procedure can be obtained since it is possible to show that in some (quantitatively characterised) situations the Birkhoff procedure for the Jacobi metric is locally identical to that for the Euclidean metric. The analysis will also reveal that for other configurations, that is, other geometric configurations, the two approximations differ, which results in the Jacobi procedure making steps which seem counter-intuitive if regarded within a Euclidean picture. Obviously, such a disagreement of the two approximations is necessary as we need to compute a geodesic in the Jacobi metric and thus have to differ at some point from the Euclidean picture.

3. A local discretised Birkhoff method

This section mimics the continuous framework laid out in Section 2 in a discrete setting.

3.1. *The discrete setting*

Throughout this section, we assume that the total energy E is sufficiently large, as described in the next paragraph. We point out that the choice of E determines the configuration manifold $Q \subset \mathbb{R}^n$, which we take as the set of points q where $E - V(q) > 0$. Let q_a and $q_b \in Q$ be given; define $\ell = \frac{|q_a - q_b|}{2}$, where $|\cdot|$ is the Euclidean distance on Q (not the Jacobi distance). We choose an orthonormal basis for Q such that $q_b - q_a = 2\ell e_1$. For $q \in Q$, we write $q = (X, Y) \in \mathbb{R} \times \mathbb{R}^{n-1}$, and in particular $e_1 = (1, 0)$.

Let us write the Jacobi metric in the form

$$g_{ij}^{\text{Jac}}(q) = e^{2h}(q)\delta_{ij}. \tag{12}$$

We require E to be sufficiently large so that the line segment joining q_a and q_b is contained in Q as well. In fact, we will work in a framework where either E is chosen large enough, depending on the given points q_a and q_b, or, given E, we choose points q_a and q_b in Q with sufficiently small distance ℓ, such that $q_b \in B_\ell(q_a) \subset Q$.

We consider a convex set Q_H such that Q_H is compactly contained in Q, $Q_H \Subset Q$; then the Jacobi metric is not degenerate, and hence Riemannian, on Q_H. It is bounded from above and from below by Euclidean metrics, but we do not use this fact. Then, if V is sufficiently smooth, the metric factor h is $C^{1,\alpha}(Q_H)$ and there is a finite $H \in \mathbb{R}$ such that the estimates

$$\left| h(X + x, Y + y) - \left(h(X, Y) + x\frac{\mathrm{d}}{\mathrm{d}x}h(X, Y) + y \cdot \nabla_N h(X, Y) \right) \right| \leq H \left(|x| + |y| \right)^{1+\alpha}, \tag{13}$$

$$|\nabla_N h(X + x, Y + y) - \nabla_N h(X, Y)| \leq H \left(|x| + |y| \right)^{\alpha} \tag{14}$$

hold.

Applying Equation (10) to the length of a straight line segment $[q_1, q_2]$ between two points,

$$\gamma_{[q_1, q_2]}(t) := q_1 + t(q_2 - q_1) \text{ for } t \in (0, 1),$$

we obtain

$$L[\gamma_{[q_1,q_2]}] = \int_0^1 e^h(\gamma_{[q_1,q_2]}(t)) \|\dot{\gamma}_{[q_1,q_2]}(t)\| \, dt = \int_0^1 e^h(\gamma_{[q_1,q_2]}(t)) \, dt \, \|q_1 - q_2\|. \quad (15)$$

We now introduce the discretised setting. We first define an equidistant Cartesian grid on $Q_H \subset \mathbb{R}^n$. We discretise the integral for the length of a straight line segment as follows.

Definition 1 (Discretised length of segment). For the straight line segment $[q_1, q_2]$ between two points on the grid, we define the *discretised length* by applying the 2-point trapezoidal rule to (15)

$$\bar{L}_{[q_1,q_2]} := \frac{e^h(\gamma_{[q_1,q_2]}(0)) + e^h(\gamma_{[q_1,q_2]}(1))}{2} \|q_1 - q_2\| = \frac{e^h(q_1) + e^h(q_2)}{2} \|q_1 - q_2\|.$$

Below we will introduce a discrete Birkhoff procedure which chooses to move points of polygonal curves in a direction normal to e_1 in order to decrease the length. It is thus necessary to estimate changes in the length as normal variations of a curve are considered. This is the content of the following lemma.

Lemma 1. *Let Q_H be convex set such that $Q_H \Subset Q$ and assume the metric factor satisfies the Hölder estimates (13) and (14) for $H > 0$. Let $q = (X, Y) \in Q_H$ be a grid point.*

Assume further that $\varepsilon \neq 0$ and $\delta, \Delta \in \mathbb{R}^{n-1}$ are such that $\delta \neq 0$ and $q_\varepsilon = q + (\varepsilon, \Delta)$ and $q_\delta = q + (0, \delta)$ are also grid points in Q_H satisfying the bounds

$$\left| \frac{\Delta}{\varepsilon} \right| \leq 1 \quad and \quad \left| \frac{\delta}{\varepsilon} \right| \leq 1. \quad (16)$$

Then the difference quotient

$$\bar{D}(\varepsilon, \delta, \Delta) := \frac{\bar{L}_{[q_\delta, q_\varepsilon]} - \bar{L}_{[q, q_\varepsilon]}}{|\varepsilon \delta|} \quad (17)$$

satisfies

$$\bar{D}(\varepsilon, \delta, \Delta) = \frac{\delta}{|\delta|} \cdot \left(-\frac{e^h(q_{\frac{\varepsilon}{2}}) F\left(\frac{\Delta}{\varepsilon}\right)}{\varepsilon} + \frac{1}{2} \nabla_N e^h(q) P_F\left(\frac{\Delta}{\varepsilon}\right) \right) + O\left(\left| \frac{\delta}{\varepsilon^2} \right| \right) + O(|\varepsilon|^\alpha),$$

where

$$q_{\frac{\varepsilon}{2}} = \frac{q + q_\varepsilon}{2} = q + \tfrac{1}{2}(\varepsilon, \Delta) \quad and \quad P_F(\eta) := \sqrt{1 + |\eta|^2}, \quad F(\eta) := \nabla P_F(\eta) = \frac{\eta}{P_F(\eta)}. \quad (18)$$

Note that $|F(\eta)| \leq 1$ for all $\eta \in \mathbb{R}$.

Proof. We can write

$$\frac{1}{|\varepsilon|} \bar{L}_{[q_\delta, q_\varepsilon]} = \frac{e^h(q_\delta) + e^h(q_\varepsilon)}{2} \cdot P_F\left(\frac{\Delta - \delta}{\varepsilon} \right).$$

Then it follows directly that

$$D(\varepsilon, \delta, \Delta) = D_1 S_2 + S_1 D_2$$

with

$$D_1 = \frac{1}{|\delta|} \left[\frac{e^h(q_\delta) + e^h(q_\varepsilon)}{2} - \frac{e^h(q) + e^h(q_\varepsilon)}{2} \right] = \frac{1}{2|\delta|} \left[e^h(q_\delta) - e^h(q) \right],$$

$$S_1 = \frac{1}{2} \left[\frac{e^h(q_\delta) + e^h(q_\varepsilon)}{2} + \frac{e^h(q) + e^h(q_\varepsilon)}{2} \right] = \frac{1}{2} \left[e^h(q) + e^h(q_\varepsilon) \right] + \frac{|\delta|}{2} \cdot D_1,$$

$$D_2 = \frac{1}{|\delta|} \left(P_F \left(\frac{\Delta - \delta}{\varepsilon} \right) - P_F \left(\frac{\Delta}{\varepsilon} \right) \right),$$

$$S_2 = \frac{1}{2} \left(P_F \left(\frac{\Delta - \delta}{\varepsilon} \right) + P_F \left(\frac{\Delta}{\varepsilon} \right) \right) = P_F \left(\frac{\Delta}{\varepsilon} \right) + \frac{|\delta|}{2} D_2.$$

Let us define D_{10} by writing

$$D_1 = \frac{1}{2} \frac{\delta}{|\delta|} \cdot \nabla_N e^h(q) + D_{10},$$

where we estimate the error term D_{10}, using the Hölder bounds for the metric factor h

$$|D_{10}| = \frac{1}{2|\delta|} \left| e^h(q_\delta) - e^h(q) - \delta \cdot \nabla_N e^h(q) \right| \le |\delta|^\alpha \frac{H}{2}.$$

Secondly, we consider S_1 and write for the first term

$$\frac{1}{2} \left[e^h(q) + e^h(q_\varepsilon) \right] = e^h(q_{\frac{\varepsilon}{2}}) + S_{10}$$

with $q_{\frac{\varepsilon}{2}}$ defined in (18) being the mid-point on the segment $[q, q_\varepsilon]$. By symmetry, the Hölder bounds imply

$$|S_{10}| \le H \left(|\varepsilon| + |\Delta| \right)^{1+\alpha} = H \left(1 + \left| \frac{\Delta}{\varepsilon} \right|^{1+\alpha} \right) |\varepsilon|^{1+\alpha} \le 2H |\varepsilon|^{1+\alpha}.$$

Furthermore, we deduce

$$D_2 = \frac{1}{|\delta|} \left(P_F \left(\frac{\Delta - \delta}{\varepsilon} \right) - P_F \left(\frac{\Delta}{\varepsilon} \right) \right) = \frac{1}{|\delta|} \int_0^1 F \left(\frac{\Delta - \delta s}{\varepsilon} \right) \cdot \frac{-\delta}{\varepsilon} \, ds,$$

where F is defined in (18). Thus, we can write

$$D_2 = -\frac{\delta}{|\delta|} \cdot \frac{1}{\varepsilon} F + D_{20},$$

where, using $|DF(\eta)| \le C$ for all $\eta \in \mathbb{R}$,

$$|D_{20}| = \frac{1}{|\delta|} \left| \int_0^1 \int_0^s \left(\frac{-\delta}{\varepsilon} \right)^t \cdot DF \left(\frac{\Delta - \delta \tau}{\varepsilon} \right) \cdot \left(\frac{-\delta}{\varepsilon} \right) \, d\tau \, ds \right| \le \left| \frac{\delta}{\varepsilon^2} \right| C.$$

We summarise

$$
\begin{aligned}
D_1 S_2 + S_1 D_2 &= \left(\frac{1}{2}\frac{\delta}{|\delta|} \cdot \nabla_N e^h(q) + D_{10}\right)\left(P_F\left(\frac{\Delta}{\varepsilon}\right) + \frac{|\delta|}{2} D_2\right) \\
&\quad + \left(e^h(q_{\frac{\varepsilon}{2}}) + S_{10} + \frac{|\delta|}{2} n_1\right)\left(-\frac{\delta}{|\delta|} \cdot \frac{1}{\varepsilon} F\left(\frac{\Delta}{\varepsilon}\right) + D_{20}\right) \\
&= \frac{\delta}{|\delta|} \cdot \left(\frac{1}{2}\nabla_N e^h(q) P_F\left(\frac{\Delta}{\varepsilon}\right) - e^h(q_{\frac{\varepsilon}{2}})\frac{1}{\varepsilon} F\left(\frac{\Delta}{\varepsilon}\right)\right) + C\left|\frac{\delta}{\varepsilon^2}\right| + O(|\varepsilon|^\alpha),
\end{aligned}
$$

with the functions F and P_F from (18).   □

We now prepare a crucial quantitative upper bound for the discrete bend of a polygon (Proposition 1). To this aim, consider three neighbouring points along the polygon. We assume that for this triplet $(q_{-\varepsilon}, q, q_{+\varepsilon})$, the X co-ordinates are distributed equidistantly for an $\varepsilon > 0$,

$$
q = (X, Y), \quad q_{\pm\varepsilon} = q + (\pm\varepsilon, \Delta_\pm).
$$

For a given $\delta \neq 0$, we want to estimate the *centred differences*

$$
\Delta\bar{L}(\varepsilon, \delta) := \left(\bar{L}_{[q_{-\varepsilon}, q+(0,\delta)]} + \bar{L}_{[q+(0,\delta), q_{+\varepsilon}]}\right) - \left(\bar{L}_{[q_{-\varepsilon}, q]} + \bar{L}_{[q, q_{+\varepsilon}]}\right).
$$

Using centred coordinates, we can rewrite this as

$$
\Delta\bar{L}(\varepsilon, \delta) = \varepsilon\,|\delta|\left(\bar{D}^- + \bar{D}^+\right),
$$

where

$$
\bar{D}^\pm := \bar{D}(\pm\varepsilon, \delta, \Delta_\pm)
$$

with $\bar{D}(\pm\varepsilon, \delta, \Delta_\pm)$ defined in (17).

We will now combine the length calculation of Lemma 1 on both sides of the centre point q. We define, in analogy to (18),

$$
q_{\frac{\pm\varepsilon}{2}} = q + \tfrac{1}{2}(\pm\varepsilon, \Delta_\pm) \tag{19}
$$

$$
F_\pm := F\left(\frac{\Delta_\pm}{\pm\varepsilon}\right) = \frac{\frac{\Delta_\pm}{\pm\varepsilon}}{\sqrt{1 + \left|\frac{\Delta_\pm}{\pm\varepsilon}\right|^2}} \tag{20}
$$

and obtain

$$\frac{\Delta \bar{L}(\varepsilon, \delta)}{\varepsilon |\delta|} = \bar{D}^- + \bar{D}^+$$

$$= \frac{\delta}{|\delta|} \cdot \left(-\frac{e^h(q_{\frac{-\varepsilon}{2}})F_-}{-\varepsilon} + \frac{1}{2}\nabla_N e^h(q)\sqrt{1 + \left|\frac{\Delta_-}{-\varepsilon}\right|^2} \right.$$

$$\left. -\frac{e^h(q_{\frac{\varepsilon}{2}})F_+}{\varepsilon} + \frac{1}{2}\nabla_N e^h(q)\sqrt{1 + \left|\frac{\Delta_+}{\varepsilon}\right|^2} \right)$$

$$+ O\left(\left|\frac{\delta}{\varepsilon^2}\right|\right) + O(|\varepsilon|^\alpha)$$

$$= \frac{\delta}{|\delta|} \cdot \left(-\frac{e^h(q_{\frac{\varepsilon}{2}})F_+ - e^h(q_{\frac{-\varepsilon}{2}})F_-}{\varepsilon} + \nabla_N e^h(q)\sqrt{1 + \left|\frac{\Delta_+}{\varepsilon}\right|^2} \right)$$

$$- \frac{\delta}{|\delta|} \cdot \nabla_N e^h(q)\left(\frac{\sqrt{1 + \left|\frac{\Delta_+}{\varepsilon}\right|^2} - \sqrt{1 + \left|\frac{\Delta_-}{-\varepsilon}\right|^2}}{2} \right) + O\left(\left|\frac{\delta}{\varepsilon^2}\right|\right) + O(|\varepsilon|^\alpha)$$

$$= \frac{\delta}{|\delta|} \cdot \left(-\frac{e^h(q_{\frac{\varepsilon}{2}})F_+ - e^h(q_{\frac{-\varepsilon}{2}})F_-}{\varepsilon} + \nabla_N e^h(q)\sqrt{1 + \left|\frac{\Delta_+}{\varepsilon}\right|^2} \right)$$

$$- \frac{\delta}{|\delta|} \cdot \nabla_N e^h(q)\frac{\left(\frac{\Delta_+}{\varepsilon} + \frac{\Delta_-}{-\varepsilon}\right)\left(\frac{\Delta_+}{\varepsilon} - \frac{\Delta_-}{-\varepsilon}\right)}{2\left(\sqrt{1 + \left|\frac{\Delta_+}{\varepsilon}\right|^2} + \sqrt{1 + \left|\frac{\Delta_-}{-\varepsilon}\right|^2}\right)}$$

$$+ O\left(\left|\frac{\delta}{\varepsilon^2}\right|\right) + O(|\varepsilon|^\alpha).$$

$$(21)$$

Let us remark that the term

$$\frac{1}{\varepsilon}\left(e^h(q_{\frac{\varepsilon}{2}})F_+ - e^h(q_{\frac{-\varepsilon}{2}})F_- \right) = \frac{1}{\varepsilon}\left(e^h(q_{\frac{\varepsilon}{2}})F\left(\frac{\Delta_+}{\varepsilon}\right) - e^h(q_{\frac{-\varepsilon}{2}})F\left(\frac{\Delta_-}{-\varepsilon}\right) \right)$$

is a proper difference quotient for the discretisation length ε.

The next result shows that the length of the edges of a three point polygon can be reduced by moving the middle point towards the line connecting the two outer points.

Proposition 1. *Let S_∞^{n-1} be the l^∞-sphere in $\mathbb{R}^{n-1}$,*

$$S_\infty^{n-1} := \left\{ \nu \in \mathbb{R}^{n-1} \,\Big|\, \sup_{j \in \{2,\dots,n\}} |e_j \cdot \nu| \le 1 \text{ and } |e_k \cdot \nu| = 1 \text{ for some } k \in \{2,\dots,n\} \right\}.$$

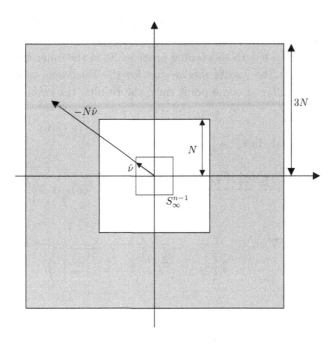

Fig. 2. Geometric configuration as set out in Proposition 1

There exists $N > 1$ and $\varepsilon_0 = \varepsilon_0(N) \in (0, 1/N)$ such that for all $\varepsilon \in (0, \varepsilon_0)$ and all triplets

$$(q_{-\varepsilon}, q, q_\varepsilon) \quad with \quad q_{\pm\varepsilon} = q + (\pm\varepsilon, \Delta_\pm)$$

which satisfy

$$\left| \frac{\Delta_\pm}{\pm\varepsilon} \right| \leq 1 \tag{22}$$

and

$$\frac{\Delta_+ + \Delta_-}{\varepsilon^2} = -\hat{N}\hat{\nu}, \ with \ \hat{N} \in (N, 3N) \ and \ \hat{\nu} \in S_\infty^{n-1}, \ see \ Fig. \ 2, \tag{23}$$

there holds

$$\begin{cases} \Delta\bar{L}(\varepsilon, \delta) > 0 & for \ every \ \delta \ with \ |\delta| \leq \varepsilon^{\alpha+2} \ and \ \frac{\delta}{|\delta|} \cdot \hat{\nu} = 1, \\ \Delta\bar{L}(\varepsilon, \delta) < 0 & for \ every \ \delta \ with \ |\delta| \leq \varepsilon^{\alpha+2} \ and \ \frac{\delta}{|\delta|} \cdot \hat{\nu} = -1. \end{cases}$$

The important implication of this statement is that length-reducing procedures for triplets will not increase the discrete curvature indefinitely. Specifically, if the curvature of a triplet is such that it is in the white inner square in Fig. 2, then the length shortening procedure may increase the curvature (unlike in the Euclidean case). However, if the curvature increases, it will eventually enter the grey region in Fig. 2. Then the proposition shows that a further step "outwards" (increasing the discrete curvature) necessarily increases the length, while the corresponding "inward" step decreases the length. This prevents the discrete curvature to grow without bounds under a length shortening process. So in the inner white region, the Birkhoff procedure for the Jacobi metric and for the Euclidean metric can differ; in fact they have to differ at some point since the results, the respective geodesics, differ.

Proof. From (21) we deduce, as $\left|\frac{\delta}{\varepsilon^2}\right| \leq |\varepsilon|^\alpha$,

$$
\frac{\Delta \bar{L}(\varepsilon, \delta)}{\varepsilon |\delta|} = \frac{\delta}{|\delta|} \cdot \left(-\frac{e^h(q_{\frac{\varepsilon}{2}})F_+ - e^h(q_{\frac{-\varepsilon}{2}})F_-}{\varepsilon} + \nabla_N e^h(q)\sqrt{1 + \left|\frac{\Delta_+}{\varepsilon}\right|^2} \right)
$$

$$
- \frac{\delta}{|\delta|} \cdot \nabla_N e^h(q) \frac{\left(\frac{\Delta_+}{\varepsilon} + \frac{\Delta_-}{-\varepsilon}\right)\left(\frac{\Delta_+}{\varepsilon} - \frac{\Delta_-}{-\varepsilon}\right)}{2\left(\sqrt{1 + \left|\frac{\Delta_+}{\varepsilon}\right|^2} + \sqrt{1 + \left|\frac{\Delta_-}{-\varepsilon}\right|^2}\right)} + O(|\varepsilon|^\alpha). \tag{24}
$$

We rewrite

$$
e^h(q_{\frac{\varepsilon}{2}})F_+ - e^h(q_{\frac{-\varepsilon}{2}})F_- = \frac{e^h(q_{\frac{\varepsilon}{2}}) + e^h(q_{\frac{-\varepsilon}{2}})}{2}(F_+ - F_-)
$$

$$
+ \frac{F_+ + F_-}{2}\left(e^h(q_{\frac{\varepsilon}{2}}) - e^h(q_{\frac{-\varepsilon}{2}})\right). \tag{25}
$$

Firstly, with the identities

$$
\frac{\Delta_\pm}{\pm \varepsilon} = \frac{\Delta_+ - \Delta_-}{2\varepsilon} \pm \frac{\Delta_+ + \Delta_-}{2\varepsilon}, \tag{26}
$$

we infer for the difference of $F_\pm$

$$
(F_+ - F_-) = \frac{\frac{\Delta_+}{\varepsilon}}{\sqrt{1 + \left|\frac{\Delta_+}{\varepsilon}\right|^2}} - \frac{\frac{\Delta_-}{-\varepsilon}}{\sqrt{1 + \left|\frac{\Delta_-}{-\varepsilon}\right|^2}}
$$

$$
= \frac{\frac{\Delta_+ - \Delta_-}{2\varepsilon} + \frac{\Delta_+ + \Delta_-}{2\varepsilon}}{\sqrt{1 + \left|\frac{\Delta_+}{\varepsilon}\right|^2}} - \frac{\frac{\Delta_+ - \Delta_-}{2\varepsilon} - \frac{\Delta_+ + \Delta_-}{2\varepsilon}}{\sqrt{1 + \left|\frac{\Delta_-}{-\varepsilon}\right|^2}}
$$

$$
= \frac{\Delta_+ + \Delta_-}{2\varepsilon}\left(\frac{1}{\sqrt{1 + \left|\frac{\Delta_+}{\varepsilon}\right|^2}} + \frac{1}{\sqrt{1 + \left|\frac{\Delta_-}{-\varepsilon}\right|^2}}\right)
$$

$$
+ \frac{\Delta_+ - \Delta_-}{2\varepsilon}\left(\frac{1}{\sqrt{1 + \left|\frac{\Delta_+}{\varepsilon}\right|^2}} - \frac{1}{\sqrt{1 + \left|\frac{\Delta_-}{-\varepsilon}\right|^2}}\right)
$$

$$
\begin{aligned}
&\frac{\frac{\Delta_+ + \Delta_-}{\varepsilon} \frac{\sqrt{1+\left|\frac{\Delta_+}{\varepsilon}\right|^2} + \sqrt{1+\left|\frac{\Delta_-}{-\varepsilon}\right|^2}}{2} - \frac{\Delta_+ - \Delta_-}{2\varepsilon} \frac{\left|\frac{\Delta_+}{\varepsilon}\right|^2 - \left|\frac{\Delta_-}{-\varepsilon}\right|^2}{\left(\sqrt{1+\left|\frac{\Delta_+}{\varepsilon}\right|^2} + \sqrt{1+\left|\frac{\Delta_-}{-\varepsilon}\right|^2}\right)}}{\sqrt{1+\left|\frac{\Delta_+}{\varepsilon}\right|^2}\sqrt{1+\left|\frac{\Delta_-}{-\varepsilon}\right|^2}} \\
&= A\frac{\Delta_+ + \Delta_-}{\varepsilon},
\end{aligned}
$$

where

$$
A = \frac{\left(\frac{\sqrt{1+\left|\frac{\Delta_+}{\varepsilon}\right|^2} + \sqrt{1+\left|\frac{\Delta_-}{-\varepsilon}\right|^2}}{2}\right)^2 \mathrm{Id} - \left(\frac{\Delta_+ - \Delta_-}{2\varepsilon}\right)\left(\frac{\Delta_+ - \Delta_-}{2\varepsilon}\right)^T}{\sqrt{1+\left|\frac{\Delta_+}{\varepsilon}\right|^2}\sqrt{1+\left|\frac{\Delta_-}{-\varepsilon}\right|^2}\frac{\sqrt{1+\left|\frac{\Delta_+}{\varepsilon}\right|^2} + \sqrt{1+\left|\frac{\Delta_-}{-\varepsilon}\right|^2}}{2}}. \tag{27}
$$

Secondly, (13), (14) and (22) imply

$$
\frac{e^h(q_{\frac{\varepsilon}{2}}) - e^h(q_{\frac{-\varepsilon}{2}})}{e^h(q)} \leq \varepsilon^{1+\alpha} H \cdot 2^{1+\alpha} + \varepsilon\|\nabla h\| \cdot 2.
$$

These two steps imply for (25)

$$
\left| e^h(q_{\frac{\varepsilon}{2}})F_+ - e^h(q_{\frac{-\varepsilon}{2}})F_- - \frac{e^h(q_{\frac{\varepsilon}{2}}) + e^h(q_{\frac{-\varepsilon}{2}})}{2} A\frac{\Delta_+ + \Delta_-}{\varepsilon} \right|
$$
$$
\leq e^h(q)\left(2^{1+\alpha}\varepsilon^{1+\alpha} H + 2\varepsilon\|\nabla h\|\right)\frac{1}{\sqrt{2}},
$$

where we used the bound $|F^\pm| \leq \frac{1}{\sqrt{2}}$ implied by (22).

We return to (24), and with A from (27), together with (23) and $\left|\frac{\Delta_+ - \Delta_-}{\varepsilon}\right| = \left|\frac{\Delta_+}{\varepsilon} + \frac{\Delta_-}{-\varepsilon}\right| \leq 2$, we obtain

$$
\begin{aligned}
\frac{\Delta\bar{L}(\varepsilon,\delta)}{\varepsilon\,|\delta|} &\geq \frac{e^h(q_{\frac{\varepsilon}{2}}) + e^h(q_{\frac{-\varepsilon}{2}})}{2}\left[-\frac{\delta}{|\delta|}A\frac{\Delta_+ + \Delta_-}{\varepsilon^2} - \left(1 + \|\nabla h\| \cdot 2\frac{1}{\sqrt{2}}\right) - \|\nabla h\|\sqrt{2}\right. \\
&\qquad \left. -\|\nabla h\|\frac{1}{2\sqrt{2}}\left|\frac{\Delta_+ + \Delta_-}{\varepsilon}\right|\right] + O(|\varepsilon|^\alpha) \\
&\geq \frac{e^h(q_{\frac{\varepsilon}{2}}) + e^h(q_{\frac{-\varepsilon}{2}})}{2}\left[\frac{\delta}{|\delta|}A\hat{\nu}\cdot\hat{N} - \left(1 + \|\nabla h\| \cdot \frac{2}{\sqrt{2}}\right) - \|\nabla h\|\sqrt{2}\right. \\
&\qquad \left. -\varepsilon\hat{N}\|\nabla h\|\frac{1}{\sqrt{2}}\right] + O(|\varepsilon|^\alpha).
\end{aligned}
$$

We use the inequality

$$
\sqrt{1+|a|^2}\sqrt{1+|b|^2} \geq 1 + a_i b_i
$$

for all $a, b \in \mathbb{R}^{n-1}$ to deduce for a diagonal element of the matrix A

$$e_i^T A e_i \geq \frac{\frac{1}{2}\left(1 + \sqrt{1 + \left|\frac{\Delta_+}{\varepsilon}\right|^2}\sqrt{1 + \left|\frac{\Delta_-}{-\varepsilon}\right|^2} + \frac{e_i^T \Delta_+}{\varepsilon} \cdot \frac{e_i^T \Delta_-}{\varepsilon}\right)}{\sqrt{2}^3} \geq \frac{1}{\sqrt{2}^3}.$$

Thus, we obtain in the case $\frac{\delta}{|\delta|} \cdot \hat{\nu} = 1$

$$\frac{\Delta \bar{L}(\varepsilon, \delta)}{\varepsilon |\delta|} \geq \frac{\frac{e^h(q_{\frac{\varepsilon}{2}}) + e^h(q_{\frac{-\varepsilon}{2}})}{2}}{\sqrt{2}^3}\left\{\hat{N} - \left[\sqrt{2}^3 + \|\nabla h\| \cdot 2\left(2 + 2 + \varepsilon \hat{N}\right)\right]\right\} + O(|\varepsilon|^\alpha).$$

Hence

$$\frac{\Delta \bar{L}(\varepsilon, \delta)}{\varepsilon |\delta|} \geq \frac{\frac{e^h(q_{\frac{\varepsilon}{2}}) + e^h(q_{\frac{-\varepsilon}{2}})}{2}}{2}\left(\hat{N} - 2\left(4 + \|\nabla h\|\left[4 + \hat{N}\varepsilon\right]\right)\right) + O(|\varepsilon|^\alpha).$$

Thus, we may take $N = 3\left(4 + 5\|\nabla h\|\right) \geq 12$ and choose ε_0 smaller than $1/N$. As $\hat{N} \geq N$ and as we may reduce ε_0 possibly further to compensate the error term, we can ensure the positivity of $\frac{\Delta \bar{L}(\varepsilon, \delta)}{\varepsilon |\delta|}$ for all $0 < \varepsilon < \varepsilon_0$.

Analogously, we deduce in the case $\frac{\delta}{|\delta|} \cdot \hat{\nu} = -1$ that there holds

$$\frac{\Delta \bar{L}(\varepsilon, \delta)}{\varepsilon |\delta|} \leq \frac{\frac{e^h(q_{\frac{\varepsilon}{2}}) + e^h(q_{\frac{-\varepsilon}{2}})}{2}}{2}\left(-\hat{N} + 2\left(4 + \|\nabla h\|\left[4 + \hat{N}\varepsilon\right]\right)\right) + O(|\varepsilon|^\alpha) < 0,$$

that is, strict negativity for the opposite sign. $\qquad\square$

3.2. *Birkhoff method for a fixed discretisation*

We recall that, for given q_a and q_b and $\ell = \frac{|q_a - q_b|}{2}$, where $|\cdot|$ is the Euclidean distance on Q (not the Jacobi distance), we have chosen an orthonormal basis for $Q \subset \mathbb{R}^n$ such that $q_b - q_a = 2\ell e_1$. For $q \in Q$, we write $q = (X, Y) \in \mathbb{R} \times \mathbb{R}^{n-1}$.

Definition 2 (Polygon, associated points and differences). Let $M \in \mathbb{N}$ large enough such that $\varepsilon := \frac{\ell}{M} \leq \hat{\varepsilon}$. We define a polygon with $2M + 1$ vertices as

$$q_j = (X_j, Y_j) := \frac{1}{2}(q_b - q_a) + \left(X_j e_1 + \sum_{k=1}^{n-1} Y_k e_{k+1}\right) \tag{28}$$

with $X_j = j\varepsilon$ for $j = -M, \ldots, M$. Note that $q_a = q_{-M} = (X_{-M}, 0) = (-\ell, 0)$ and $q_b = q_M = (X_M, 0) = (\ell, 0)$. The *polygon* γ associated with these points consists of the line segments joining neighbouring points.

We then define the set of interior nodes $\mathcal{J} := \{j \in \mathbb{Z} \mid |j| < M\}$ and set for all $j \in \mathcal{J}$

$$\Delta_j^\pm = Y_{j\pm 1} - Y_j.$$

Remark 1. Note that $\Delta_j^+ = -\Delta_{j+1}^-$ and

$$\Delta_j^- + \Delta_j^+ = Y_{j-1} + Y_{j+1} - 2Y_j.$$

We now show that the fixed boundary points $q_{\pm M}$ and the estimate on the second differences from Proposition 1 ensure that the first differences remain bounded by 1.

Definition 3. Consider a polygon γ as in Definition 2. Let N be given by Proposition 1. We say that $\gamma \in \mathcal{D}_{2N}$ if the second difference quotients of γ satisfy

$$\max_{j \in \mathcal{J}} \max_{i=2,\ldots,n} \left| e_i \cdot \frac{\Delta_j^- + \Delta_j^+}{\varepsilon^2} \right| \leq 2N. \tag{29}$$

Lemma 2. *Let $\ell \leq \ell_0 := \frac{1}{4N\sqrt{n-1}}$, where N is given by Proposition 1. Let γ be a polygon as in Definition 2. If $\gamma \in \mathcal{D}_{2N}$ (see Definition 3), then there holds*

$$\sup_{j \in \mathcal{J}} \left| \frac{\Delta_j^\pm}{\pm\varepsilon} \right| \leq 1 \tag{30}$$

as well as γ is contained in

$$\mathcal{P}_{2N} := \left\{ q = (X,Y) : X \in [-\ell, \ell], |Y| \leq N \left(\ell^2 - |X|^2 \right) \right\}. \tag{31}$$

Proof. We add up the second differences: As $Y_{\pm M} = 0$, any unit vector e_k, $k = 2, \ldots, n$, orthogonal to e_1 satisfies $e_k \cdot Y_{\pm M} = 0$. Hence, for $\Delta_j^+ = Y_{j+1} - Y_j$ it follows that

$$0 = e_k \cdot (Y_M - Y_{-M}) = \sum_{j \in \mathcal{J} \cup \{-M\}} e_k \cdot \Delta_j^+.$$

By the Mean Value Theorem, there exists a $j_- \in \mathcal{J} \cup \{-M\}$ such that $e_k \cdot \Delta_{j_-}^+ \leq 0$. For any $j \in \mathcal{J} \cup \{-M\}$ we thus deduce from the boundedness of the second difference quotients

$$e_k \cdot \Delta_j^+ = e_k \cdot \Delta_{j_-}^+ + \mathrm{sign}\,(j - j_-) \sum_{l=\min\{j,j_-\}}^{\max\{j,j_-\}-1} e_k \cdot \left(\Delta_{l+1}^+ - \Delta_l^+ \right)$$

$$= e_k \cdot \Delta_{j_-}^+ + \mathrm{sign}\,(j - j_-) \sum_{l=\min\{j,j_-\}}^{\max\{j,j_-\}-1} e_k \cdot \left(\Delta_{l+1}^+ + \Delta_{l+1}^- \right)$$

$$\leq |j - j_-| \cdot 2N\varepsilon^2 \leq 2M \cdot 2N\varepsilon^2 = 2\ell \cdot 2N\varepsilon.$$

We proceed analogously for a $j_+ \in \mathcal{J} \cup \{-M\}$ with $e_k \cdot \Delta_{j_+}^+ \geq 0$ and obtain for all $j \in \mathcal{J} \cup \{-M\}$

$$\left| e_k \cdot \frac{\Delta_j^+}{\varepsilon} \right| \leq 4N\ell.$$

As e_k is arbitrary, we find $\left|\frac{\Delta_j^+}{\varepsilon}\right| \le 4N\ell\sqrt{n-1}$. As $\ell \le \ell_0$, we conclude

$$\left|\frac{\Delta_j^+}{\varepsilon}\right| \le 1.$$

As $\Delta_j^+ = -\Delta_{j+1}^-$, the estimate extends to the corresponding 'negative' differences Δ_j^-. Note that an integration of the second differences yields directly the inclusion $\gamma \in \mathcal{P}_{2N}$.   □

We now define the Birkhoff method for fixed $\epsilon > 0$. We start with a polygon γ^0 represented by the vertices q_j^0 that can be written as in (28). We think of the Birkhoff method as an iterative process to update a polygon γ^l to a polygon γ^{l+1} which has a strictly smaller discrete length. It will be shown later that such an update is, for a fixed discretisation, not always possible and the Birkhoff method thus terminates.

Definition 4 (Birkhoff step). We consider a polygon γ^l represented by $2M + 1$ points q_j^l, as in Definition 2. Let $0 < \zeta \le \varepsilon^{2+\alpha}$ and

$$\mathcal{N} := \{\sigma e_i \mid \sigma \in \{\pm 1\} \text{ and } i = 2, \ldots, n\}.$$

Then

(1) we consider sequentially every $j \in \mathcal{J}$. For given j, consider sequentially every $\nu \in \mathcal{N}$ and try to move the interior point q_j^l to $q_j^* = q_j^l + \delta$, with $\delta := \zeta\nu$, to achieve

$$\bar{L}_{[q_{j-1}^l, q_j^*]} + \bar{L}_{[q_j^*, q_{j+1}^l]} < \bar{L}_{[q_{j-1}^l, q_j^l]} + \bar{L}_{[q_j^l, q_{j+1}^l]}.$$

That is, the passage via q_j^* is shorter than via the original q_j^l. In the affirmative case, then we define the update γ^{l+1} as

$$q_j^{l+1} := q_j^* \text{ and } q_k^{l+1} := q_k^l \text{ for } k \ne j.$$

Thus the update has strictly smaller discrete length.

(2) If 1 is not affirmative for any $j \in \mathcal{J}$, then Birkhoff step is called *void*.

The Birkhoff step depends on the sequential order chosen for $\mathcal{J}$ and $\mathcal{N}$. Here and later, we regard this choice as fixed and hence the Birkhoff step is uniquely defined.

It is immediate that if q_j^l is on the grid defined in Section 3.1, then q_j^{l+1} lies on the grid as well; the Birkhoff step thus makes only movements which are compatible with the grid, and thus results in polygons with vertices on the grid.

Definition 5 (Birkhoff method and map). The *Birkhoff method* is the iteration obtained by consecutive Birkhoff steps starting with a polygon γ^0 until the Birkhoff step is void.

The *Birkhoff map* maps a starting polygon γ^0 to the final polygon obtained by the Birkhoff method.

Proposition 2. *Let $\ell_0 > 0$ be given by Lemma 2 with $N > 1$ be given by Proposition 1. Consider an initial polygon γ^0 as in Definition 2 with sufficiently close endpoints, that is $\ell \leq \ell_0$. Furthermore, assume $\gamma_0 \in \mathcal{D}_{2N}$, that is, the second differences are bounded. The updates of the Birkhoff steps obey the same assumptions, that is the remain* graph-like *polygons in the sense of Definition 2 and are contained in $\mathcal{D}_{2N}$. Furthermore, the Birkhoff method terminates after finitely many steps with a final polygon in the strictly smaller set $\mathcal{D}_N$, that is, there holds (recall Definition 3)*

$$\max_{j \in \mathcal{J}} \max_{i=2,\ldots,n} \left| e_i \cdot \frac{\Delta_j^- + \Delta_j^+}{\varepsilon^2} \right| \leq N.$$

We point out and will use later that the second differences bound associated with $\mathcal{D}_N$ is exactly half the bound associated with $\mathcal{D}_{2N}$. This will be crucial to compensate for the doubling of the discrete curvature if one halves the stepsize.

Proof. Firstly, we notice that $\gamma^0 \in \mathcal{P}_{2N}$ by Lemma 2. Consider a Birkhoff step according to Definition 4. (i) If the step is *void*, then γ^0 is the final polygon. The final polygon has to lie in $\mathcal{D}_N$ as otherwise Proposition 1 would ensure the existence of a further affirmative Birkhoff step, that is, an update site q_j^l and an associated direction δ such that $\Delta \bar{L}(\varepsilon, \delta) < 0$.

(ii) Otherwise, consider a single affirmative Birkhoff step acting on a polygon $\gamma^l \in \mathcal{D}_{2N}$, that is, there is a $j \in \mathcal{J}$ and $\nu \in \mathcal{N}$ such that

$$\bar{L}_{\left[q_{j-1}^l, q_j^l + \zeta\nu\right]} + \bar{L}_{\left[q_j^l + \zeta\nu, q_{j+1}^l\right]} < \bar{L}_{\left[q_{j-1}^l, q_j^l\right]} + \bar{L}_{\left[q_j^l, q_{j+1}^l\right]}.$$

For $q := q_j^l$, $q_\pm := q_{j\pm1}^l$ and $\delta = \zeta\nu$, it follows that

$$\Delta \bar{L}(\varepsilon, \delta) = \left(\bar{L}_{[q_-, q+(0,\delta)]} + \bar{L}_{[q+(0,\delta), q_+]} \right) - \left(\bar{L}_{[q_-, q]} + \bar{L}_{[q, q_+]} \right) < 0. \qquad (32)$$

We want to show that the Birkhoff step leaves $\mathcal{D}_{2N}$ invariant.

Case 1: If

$$\max_{j \in \mathcal{J}} \max_{i=2,\ldots,n} \left| e_i \cdot \frac{\Delta_j^{-,l+1} + \Delta_j^{+,l+1}}{\varepsilon^2} \right| \leq 2N,$$

then by definition of $\mathcal{D}_{2N}$ the update γ^{l+1} is contained in $\mathcal{D}_{2N}$, so nothing is to be shown.

Case 2: Now we assume on the contrary that γ^{l+1} is not in the set $\mathcal{D}_{2N}$. That is, there exist a $j \in \mathcal{J}$ and an $i \in \{2, \ldots, n\}$ such that

$$\left| e_i \cdot \frac{\Delta_j^{-,l+1} + \Delta_j^{+,l+1}}{\varepsilon^2} \right| > 2N;$$

thus we can write

$$\frac{\Delta_j^{-,l+1} + \Delta_j^{+,l+1}}{\varepsilon^2} = \frac{\Delta_j^{-,l} + \Delta_j^{+,l}}{\varepsilon^2} - 2\frac{\delta}{\varepsilon^2} =: -\hat{N}\hat{\nu} - 2\frac{\zeta}{\varepsilon^2}\nu.$$

In view of the notation used in Proposition 1, we write

$$-\hat{N}\hat{\nu} = \frac{\Delta^- + \Delta^+}{\varepsilon^2},$$

where $\Delta_\pm := \Delta_j^{\pm,l}$ with $\left|\frac{\Delta_\pm}{\pm\varepsilon}\right| \leq 1$ by Lemma 2. Furthermore, there holds $\hat{N} \in (N, 2N]$, $\hat{\nu} \in S_\infty^{n-1}$ and $\delta = \zeta\nu$, $\nu \in \mathcal{N}$. Note that $\hat{\nu} \cdot \nu$ is restricted to the values $-1, 0, 1$ by the definitions of S_∞^{n-1} and $\mathcal{N}$. By assumption, however, γ^{l+1} has left the set $\mathcal{D}_{2N}$, hence $\hat{\nu} \cdot \nu = 1$. Proposition 1 then implies $\Delta\bar{L}(\varepsilon, \delta) > 0$. This contradicts (32), so Case 2 is in fact impossible.

Using the inclusion $\mathcal{D}_{2N} \subset S$, we conclude that for a finite discretisation length ε, the number of distinguished polygons represented on the discrete grid is finite. Each affirmative Birkhoff step is strictly reducing the length, thus does not allow us to visit the same polygon twice. Hence the method terminates after finitely many affirmative Birkhoff steps. As already argued in (i), the final polygon has to lie in $\mathcal{D}_N$ as otherwise one could show the existence of a further affirmative Birkhoff step.   □

4. Birkhoff refinement

This section consists of three parts. In the first part, we define a sequence of polygons γ_k, which are the the final polygons of the Birkhoff method for the discretisation length ε_k. As $\varepsilon_k \to 0$, we will show that the curves γ_k converge to a curve γ. In the second part, we recall a weak formulation of the geodesic equation. In the third part, we show that the limit γ satisfies this weak geodesic equation, hence γ is smooth and a stationary curve for the Jacobi length functional.

4.1. Refinement and convergence

Definition 6 (Refinement). Let M_0 be large enough such that $\varepsilon_0 := \frac{\ell}{M_0} \leq \hat{\varepsilon}$ and $\zeta_0 := \varepsilon_0^3$. Let us define the starting polygon γ_0 as the ε_0–discretisation of the straight line segment with endpoints q_a, q_b.

For $k = 1, 2, \ldots$, we want to halve the discretisation length, that is, $\varepsilon_k := \frac{\ell}{M_k}$ for $M_k := 2M_{k-1}$. Let $\mathcal{J}_k := \{j \in \mathbb{Z} \mid |j| < M_k\}$. We embed the polygon γ_{k-1} into the finer grid by introducing new vertices at the midpoints of the connecting line segments. In the notation of Definition 2, this means the embedded polygon $\bar{\gamma}_k$ has the vertices

$$\bar{q}_j^k = q_{\frac{j}{2}}^{k-1}, \qquad \text{for even } j \in \mathcal{J}_k \cup \{\pm M_k\},$$

$$\bar{q}_j^k = \tfrac{1}{2}\left(q_{\frac{j-1}{2}}^{k-1} + q_{\frac{j+1}{2}}^{k-1}\right) \text{ for odd } j \in \mathcal{J}_k.$$

We define $\zeta_k := \frac{1}{6}\zeta_{k-1}$. For this choice, the vertices of $\bar{\gamma}_k$ lie on the finer grid $(\varepsilon_k \mathbb{Z}, \zeta_k \mathbb{Z}^{n-1})$. Then γ_k is the Birkhoff map of $\bar{\gamma}_k$ for the discretisation length ε_k.

To simplify the notation, we write on each level k

$$\Delta_j^k := \Delta_j^+, \quad \text{which implies} \quad -\Delta_{j-1}^k = \Delta_j^-.$$

We now want to prove that the Birkhoff refinement laid out above will lead to a converging sequence of polygons if we start the iteration with properly chosen initial points. Specifically, let us first assume that the total energy E, and thus Q, is fixed, and let us choose a nonempty compact set $P \subset Q$. Then let N be as in the proof of Proposition 1, $N = 3\left(4 + 5\|\nabla e^h\|\right)$ with the norm taken over the set P.

Definition 7. Any pair of endpoints (q_a, q_b) is *admissible* if they meet the following two conditions, which depend on N:

(1) $\ell = \frac{|q_a - q_b|}{2} \leq \ell_0$, with $\ell_0 := \frac{1}{4N\sqrt{n-1}}$ as in Lemma 2;
(2) the set $\mathcal{D}_{2N}$ defined in Lemma 2 is contained in P.

Theorem 1. *For an admissible pair* (q_a, q_b), *we consider the sequence of graphs* $(x, f_k(x))$ *with* $x \in (-\ell, \ell)$ *representing the polygons* γ_k *obtained by the Birkhoff refinement. As* $\gamma_k \in \mathcal{D}_N$, *the centred differences satisfy for all* $k \geq 0$ *the combined estimates*

$$\max_{j \in \mathcal{J}_k \cup \{-M_k\}} \left| \frac{\Delta_j^k}{\varepsilon_k} \right| \leq 1, \tag{33}$$

$$\max_{j \in \mathcal{J}_k} \max_{i=2,\ldots,n} \left| e_i \cdot \frac{\Delta_j^k - \Delta_{j-1}^k}{\varepsilon_k^2} \right| \leq N. \tag{34}$$

Proof. Observe that the initial polygon γ_0 is contained in $\mathcal{D}_N$, since all its finite differences vanish.

By induction let $k = 1, 2, \ldots$. According to Definition 6, we embed γ_{k-1} into the finer grid of size ε_k, which is half the size of ε_{k-1}. The recursive definition of ζ_k implies $\zeta_k \leq \varepsilon_k^{2+\alpha}$, with $\alpha = \frac{\log 6}{\log 2} - 2 \in (0, 1)$.

With respect to ε_k, the embedded polygons γ_{k-1} are in $\mathcal{D}_{2N}$; this follows since all newly introduced vertices have a vanishing second difference quotient $\frac{\Delta_j^k - \Delta_{j-1}^k}{\varepsilon_k^2}$ at odd nodes $j \in \mathcal{J}_k$, whereas all quotients with even $j \in \mathcal{J}_k$ are doubled in size,

$$\frac{\Delta_j^k - \Delta_{j-1}^k}{\varepsilon_k^2} = 2\frac{\Delta_{\frac{j}{2}}^{k-1} - \Delta_{\frac{j}{2}-1}^{k-1}}{\varepsilon_{k-1}^2}.$$

From Proposition 2, it follows that the final polygon γ_k obtained from the Birkhoff map is in fact in the smaller set $\mathcal{D}_N$. This is crucial for our argument, since it shows that one application of the Birkhoff map after a refinement retains the same discrete curvature bound as before.

This shows that every polygon γ_k with step-size ε_k belongs to $\mathcal{D}_N$. By Lemma 2 it obeys also a bound on the first difference quotients, which is uniform in k. $\quad\square$

Corollary 1. *For an admissible pair* (q_a, q_b), *we consider a sequence of graphs* $(x, f_k(x))$ *with* $x \in (-\ell, \ell)$ *representing the polygons* γ_k *obtained by the Birkhoff refinement. Then the functions* f_k *converge in* $C^\beta \left((-\ell, \ell); \mathbb{R}^{n-1} \right)$ *to a limit* $f \in C^{0,1} \left((-\ell, \ell); \mathbb{R}^{n-1} \right)$. *We write* $\gamma = (x, f(x))$ *for the limit graph.*

Proof. As $\gamma_k \in C^{0,1}([0,1]; S)$, we deduce the claimed convergence in $C^\beta([0,1]; S)$ for any $\beta \in (0,1)$ by Arzelà-Ascoli. $\square$

Let us remark that the graphs of all polygons γ_k and hence the limit γ belong to the subset $\mathcal{P}_{2N}$ of Q defined in Lemma 2.

4.2. *Variational formulation for a geodesic*

Let us consider a geodesic which can be represented as a graph. In this subsection, we derive a weak formulation for its governing equation. Thus, we consider a curve represented as a graph $\gamma(x) = (x, f(x))$, $x \in (-\ell, \ell)$. The length of this curve in the Jacobi metric $g_{ij}^{\text{Jac}}(q) = e^{2h}(q)\delta_{ij}$ is given by

$$L[\gamma] = \int_{-\ell}^{\ell} \sqrt{g_{ij}^{\text{Jac}}(\gamma(x))\dot{\gamma}^i \dot{\gamma}^j} \, \mathrm{d}x = \int_{-\ell}^{\ell} e^{h(\gamma(x))} \sqrt{1 + \left| \tfrac{\mathrm{d}}{\mathrm{d}x} f \right|^2} \, \mathrm{d}x.$$

For a *orthogonal* perturbation $\gamma_\varepsilon = (x, f + \varepsilon y)$, where y has compact support in $(-\ell, \ell)$, we deduce via integration by parts

$$\left. \frac{d}{d\varepsilon} \right|_{\varepsilon=0} L[\gamma_\varepsilon]$$

$$= \int_{-\ell}^{\ell} \left[y(x) \cdot \nabla_N e^{h(\gamma(x))} \sqrt{1 + \left| \tfrac{\mathrm{d}}{\mathrm{d}x} f(x) \right|^2} + e^{h(\gamma(x))} \frac{\tfrac{\mathrm{d}}{\mathrm{d}x} f(x) \cdot \tfrac{\mathrm{d}}{\mathrm{d}x} y(x)}{\sqrt{1 + \left| \tfrac{\mathrm{d}}{\mathrm{d}x} f(x) \right|^2}} \right] \mathrm{d}x.$$

Thus, a graph $\gamma(x) = (x, f(x))$ with $f \in H^1(-\ell, \ell)$ is stationary for the length functional L_γ if the variational derivative $\left. \frac{d}{d\varepsilon} \right|_{\varepsilon=0} L_{\gamma_\varepsilon}$ vanishes for all functions $y \in H_0^1(-\ell, \ell)$. We will use this weak formulation, rather than the more common one obtained by a further integration by parts.

4.3. *Characterisation of the limit of the Birkhoff refinement*

Using a smoother interpolation of the grid points representing the polygons of the Birkhoff refinement we obtain a better convergence and a smoother characterisation of its limit than the previous result of Corollary 1.

Theorem 2. *For an admissible pair* (q_a, q_b), *consider the sequence of polygons* $\gamma_k = (x, f_k(x))$ *obtained by the Birkhoff refinement. The functions* f_k *converge in* $H^1(-\ell, \ell)$ *to a limit* f. *Furthermore,* $f \in C^{1,1}(-\ell, \ell)$ *and the limit graph*

$\gamma(x) = (x, f(x))$ *satisfies*

$$0 = \int_{-\ell}^{\ell} \left[y(x) \cdot \nabla_N e^{h(\gamma(x))} \sqrt{1 + \left| \frac{d}{dx} f(x) \right|^2} + e^{h(\gamma(x))} \frac{\frac{d}{dx} f(x) \cdot \frac{d}{dx} y(x)}{\sqrt{1 + \left| \frac{d}{dx} f(x) \right|^2}} \right] dx. \quad (35)$$

for every function $y \in H_0^1(-\ell, \ell)$.

Proof. Analogously to the definition of $\Delta \bar{L}(\varepsilon, \delta)$ in (21), we define

$$\Delta \bar{L}_j(\varepsilon_k, \zeta_k \sigma e_i) := \left(\bar{L}[q_{j-1}, q_j + \zeta_k \sigma e_i] + \bar{L}[q_j + \zeta_k \sigma e_i, q_{j+1}] \right) - \left(\bar{L}[q_{j-1}, q_j] + \bar{L}[q_j, q_{j+1}] \right).$$

It is convenient to write for $j \in \mathcal{J}_k$ in analogy to (19)

$$q_{j+\frac{1}{2}}^k := \frac{1}{2} \left(q_j^k + q_{j+1}^k \right),$$

$$X_{j+\frac{1}{2}}^k := \frac{1}{2} \left(X_j^k + X_{j+1}^k \right).$$

In further analogy to (20), let

$$F_j^k := \frac{\frac{\Delta_j^k}{\varepsilon_k}}{\sqrt{1 + \left| \frac{\Delta_j^k}{\varepsilon_k} \right|^2}}. \quad (36)$$

We want to examine $\Delta \bar{L}(\varepsilon, \delta)$ of (21) evaluated at a node q_j for $\varepsilon = \varepsilon_k$ and $\delta = \zeta_k \sigma e_i$, where $\sigma \in \{\pm 1\}$ and $i \in \{2, \ldots, n\}$. Theorem 1 provides the necessary estimates for the difference quotients, so that

$$\nabla_N e^h(q_j^k) \frac{\left(\frac{\Delta_j^k}{\varepsilon_k} + \frac{-\Delta_{j-1}^k}{-\varepsilon_k} \right) \left(\frac{\Delta_j^k}{\varepsilon_k} - \frac{-\Delta_{j-1}^k}{-\varepsilon_k} \right)}{2 \left(\sqrt{1 + \left| \frac{\Delta_j^k}{\varepsilon_k} \right|^2} + \sqrt{1 + \left| \frac{-\Delta_{j-1}^k}{-\varepsilon_k} \right|^2} \right)} = O(\varepsilon_k),$$

since $\dfrac{\frac{\Delta_j^k}{\varepsilon_k} + \frac{\Delta_{j-1}^k}{\varepsilon_k}}{\sqrt{1 + \left| \frac{\Delta_j^k}{\varepsilon_k} \right|^2} + \sqrt{1 + \left| \frac{\Delta_{j-1}^k}{\varepsilon_k} \right|^2}} \leq 1$ and $\dfrac{\Delta_j^k - \Delta_{j-1}^k}{\varepsilon_k} = O(\varepsilon_k)$ by (33)–(34). Further, by Definition 6,

$$O \left(\left| \frac{\zeta_k}{\varepsilon_k^2} \right| \right) = O \left(|\varepsilon_k|^\alpha \right).$$

Thus, we can write

$$\frac{\Delta \bar{L}_j(\varepsilon_k, \zeta_k \sigma e_i)}{\sigma \varepsilon_k |\zeta_k|} + O(|\varepsilon_k|^\alpha)$$

$$= e_i \cdot \left(-\frac{e^h(q_{j+\frac{1}{2}}^k) F_j^k - e^h(q_{j-\frac{1}{2}}^k) F_{j-1}^k}{\varepsilon_k} + \nabla_N e^h(q_j) \sqrt{1 + \left| \frac{\Delta_j^k}{\varepsilon_k} \right|^2} \right).$$

We identify the right hand side as a product of a function w evaluated on the piecewise constant interpolation γ_k^{pc} times a function z evaluated on the piecewise linear interpolation γ_k^{pl}, that is, for all $x \in \left(X_{j-\frac{1}{2}}^k, X_{j+\frac{1}{2}}^k \right)$ there holds

$$
\frac{\Delta \bar{L}_j(\varepsilon_k, \zeta_k \sigma e_i)}{\sigma \varepsilon_k |\zeta_k|} + O(|\varepsilon_k|^\alpha)
$$
$$
= e_i \cdot \left[D_{\varepsilon_k} \left(e^h \left(x, \gamma_k^{\mathrm{pc}}(x) \right) F \left(\tfrac{\mathrm{d}}{\mathrm{d}x} \gamma_k^{\mathrm{pl}}(x) \right) \right) + \nabla_N e^h \left(x, \gamma_k^{\mathrm{pc}}(x) \right) P_F \left(\tfrac{\mathrm{d}}{\mathrm{d}x} \gamma_k^{\mathrm{pl}}(x) \right) \right],
$$
$$(37)$$

where D_{ε_k} is the centred difference quotient

$$
D_\varepsilon y(x) := \frac{y(x + \frac{\varepsilon}{2}) - y(x - \frac{\varepsilon}{2})}{\varepsilon}
$$

and

$$
F(\xi) := \frac{\xi}{\sqrt{1 + |\xi|^2}}, \quad P_F(\xi) := \sqrt{1 + |\xi|^2}.
$$

We remark that the right-hand of (37) side is globally defined for $x \in (-\ell, \ell)$, and piecewise constant. Given an arbitrary fixed function $y \in H_0^1(-\ell, \ell)$, we can multiply (37) with y and integrate by parts to obtain

$$
\sum_{j=-M+1}^{M-1} \int_{X_{j-\frac{1}{2}}}^{X_{j+\frac{1}{2}}} y(x) \left[\frac{\Delta \bar{L}_j(\varepsilon_k, \zeta_k \sigma e_i)}{\sigma \varepsilon_k |\zeta_k|} + O(|\varepsilon_k|^\alpha) \right] \mathrm{d}x
$$
$$
= e_i \cdot \int_{-\ell+\frac{\varepsilon_k}{2}}^{\ell-\frac{\varepsilon_k}{2}} y(x) \left[D_{\varepsilon_k} \left(e^h \left(x, \gamma_k^{\mathrm{pc}}(x) \right) F \left(\tfrac{\mathrm{d}}{\mathrm{d}x} \gamma_k^{\mathrm{pl}}(x) \right) \right) \right.
$$
$$
\left. + \nabla_N e^h \left(x, \gamma_k^{\mathrm{pc}}(x) \right) P_F \left(\tfrac{\mathrm{d}}{\mathrm{d}x} \gamma_k^{\mathrm{pl}}(x) \right) \right] \mathrm{d}x \qquad (38)
$$
$$
= e_i \cdot \int_{-\ell+\frac{\varepsilon_k}{2}}^{\ell-\frac{\varepsilon_k}{2}} \left[-\left(D_{\varepsilon_k} y(x) \right) e^h \left(x, \gamma_k^{\mathrm{pc}}(x) \right) F \left(\tfrac{\mathrm{d}}{\mathrm{d}x} \gamma_k^{\mathrm{pl}}(x) \right) \right.
$$
$$
\left. + y(x) \nabla_N e^h \left(x, \gamma_k^{\mathrm{pc}}(x) \right) P_F \left(\tfrac{\mathrm{d}}{\mathrm{d}x} \gamma_k^{\mathrm{pl}}(x) \right) \right] \mathrm{d}x.
$$

We observe that the right-hand side does not depend on σ. Now recall that stoppage of the algorithm on level k implies that $\Delta \bar{L}_j(\varepsilon_k, \zeta_k \sigma e_i) \geq 0$ for both choices of the sign of $\sigma = \pm 1$.

Let us assume for the moment that piecewise constant interpolation γ_k^{pc} and the piecewise linear interpolation γ_k^{pl} converge strongly to the same limit γ as $k \to \infty$ in the sense that

$$
\| \gamma_k^{\mathrm{pc}} - \gamma \|_{L^2} \to 0, \qquad (39)
$$

and

$$
\left\| \tfrac{\mathrm{d}}{\mathrm{d}x} \gamma_k^{\mathrm{pl}} - \tfrac{\mathrm{d}}{\mathrm{d}x} \gamma \right\|_{L^2} \to 0. \qquad (40)
$$

Then the argument can be finished as follows. Observe that $\Delta \bar{L}_j(\varepsilon_k, \zeta_k \sigma e_i)$ is non-negative for both choices of $\sigma = \pm 1$, whereas the last line of (38) does not depend on the chosen σ anymore. On each interval $\left(X_{j-\frac{1}{2}}, X_{j+\frac{1}{2}}\right)$, we first choose σ to have the same sign as $\int_{X_{j-\frac{1}{2}}}^{X_{j+\frac{1}{2}}} y(x)\,\mathrm{d}x$. Then the sum on the left-hand side is non-negative. We pass to the limit $k \to \infty$ in (38) and find

$$
\begin{aligned}
0 \leq \lim_{k\to\infty} \sum_{j=-M+1}^{M-1} \int_{X_{j-\frac{1}{2}}}^{X_{j+\frac{1}{2}}} y(x) \frac{\Delta \bar{L}_j(\varepsilon_k, \zeta_k \sigma e_i)}{\sigma \varepsilon_k |\zeta_k|}\,\mathrm{d}x \\
= e_i \cdot \int_{-\ell}^{\ell} \left[-\left(\tfrac{\mathrm{d}}{\mathrm{d}x} y(x)\right) e^h\left(x, \gamma(x)\right) F\left(\tfrac{\mathrm{d}}{\mathrm{d}x}\gamma(x)\right)\right. \\
\left. + y(x) \nabla_N e^h\left(x, \gamma(x)\right) P_F\left(\tfrac{\mathrm{d}}{\mathrm{d}x}\gamma(x)\right)\right]\,\mathrm{d}x;
\end{aligned}
\tag{41}
$$

here the convergence of the difference quotient $D_{\varepsilon_k} y(x)$ follows from [11, Lemma 7.24].

Similarly, we then choose σ to have always the opposite sign and obtain the reversed inequality. Together this yields

$$
\begin{aligned}
0 = e_i \cdot \int_{-\ell}^{\ell} \left[-\left(\tfrac{\mathrm{d}}{\mathrm{d}x} y(x)\right) e^h\left(x, \gamma(x)\right) F\left(\tfrac{\mathrm{d}}{\mathrm{d}x}\gamma(x)\right)\right. \\
\left. + y(x) \nabla_N e^h\left(x, \gamma(x)\right) P_F\left(\tfrac{\mathrm{d}}{\mathrm{d}x}\gamma(x)\right)\right]\,\mathrm{d}x.
\end{aligned}
$$

As we can test all normal directions e_i, with $i = 2, \ldots, n$, in this fashion, we obtain the vectorial identity

$$
\begin{aligned}
0 = \int_{-\ell}^{\ell} \left[-\left(\tfrac{\mathrm{d}}{\mathrm{d}x} y(x)\right) e^h\left(x, \gamma(x)\right) F\left(\tfrac{\mathrm{d}}{\mathrm{d}x}\gamma(x)\right)\right. \\
\left. + y(x) \nabla_N e^h\left(x, \gamma(x)\right) P_F\left(\tfrac{\mathrm{d}}{\mathrm{d}x}\gamma(x)\right)\right]\,\mathrm{d}x
\end{aligned}
\tag{42}
$$

in $\mathbb{R}^{n-1}$. By substituting the definitions of F and P_F, we recover the claim. $\qquad\square$

4.3.1. *Proof of* (39) *and* (40)

The previously assumed convergence of the interpolants is established in the following arguments.

Lemma 3 (Estimates for interpolants). *Given a function $\gamma^{\mathrm{q}} \in H^2$, let γ^{pc} be the piecewise constant interpolation on an equidistant grid of size ε, and similarly γ^{pl} the piecewise linear interpolation. Then the following error estimates hold*

$$
\|\gamma^{\mathrm{pc}} - \gamma^{\mathrm{q}}\|_{L^2} \leq C\varepsilon^2 \left\|\tfrac{\mathrm{d}^2}{\mathrm{d}x^2}\gamma^{\mathrm{q}}\right\|_{L^2},
$$

and

$$
\left\|\tfrac{\mathrm{d}}{\mathrm{d}x}\gamma^{\mathrm{pl}} - \tfrac{\mathrm{d}}{\mathrm{d}x}\gamma^{\mathrm{q}}\right\|_{L^2} \leq C\varepsilon \left\|\tfrac{\mathrm{d}^2}{\mathrm{d}x^2}\gamma^{\mathrm{q}}\right\|_{L^2}.
$$

Proof. This is a standard argument in Finite Elements, see for example [4, Theorem 0.8.7 and Section 4].   □

For a fixed refinement level k, let q_j be the point set associated to γ_k, the output of the Birkhoff map at level k. Now, we first construct a quadratic interpolation γ_k^{pq} of $\{q_j^k\}$ with the special property that its piecewise constant and piecewise linear interpolation coincide with the interpolations γ_k^{pc} and γ_k^{pl} introduced before. Specifically, in our situation the quadratic interpolation can be chosen in such a way that $\left\|\frac{d^2}{dx^2}\gamma_k^{pq}\right\|_{L^2}$ is bounded independently of k.

Explicitly, to construct γ_k^{pq}, any two neighbouring nodes $q_j^k = (X_j^k, Y_j^k)$ and $q_{j+1}^k = (X_{j+1}^k, Y_{j+1}^k) = (X_j^k + \varepsilon_k, Y_j^k + \Delta_j^k)$ are connected via two quadratic splines $s_\pm$ on $\left(X_j^k, X_{j+\frac{1}{2}}^k\right)$ and $\left(X_{j+\frac{1}{2}}^k, X_{j+1}^k\right)$, with matching conditions

$$s_-(X_j^k) = Y_j^k, \qquad s_-'(X_j^k) = b_- := \frac{\Delta_j^k + \Delta_{j-1}^k}{2\varepsilon_k},$$

$$s_+(X_{j+1}^k) = Y_{j+1}^k, \qquad s_+'(X_{j+1}^k) = b_+ := \frac{\Delta_{j+1}^k + \Delta_j^k}{2\varepsilon_k},$$

$$s_-(X_{j+\frac{1}{2}}^k) = s_+(X_{j+\frac{1}{2}}^k), \quad s_-'(X_{j+\frac{1}{2}}^k) = s_+'(X_{j+\frac{1}{2}}^k).$$

On the boundary, we vary the definition slightly for $j = -M_k$ and $j = M_k - 1$ respectively,

$$s_-(X_{-M_k}^k) = Y_{-M_k}^k, \qquad s_-'(X_{-M_k}^k) = b_- := \frac{\Delta_j^k}{\varepsilon_k}$$

and

$$s_+(X_{M_k}^k) = Y_{M_k}^k, \qquad s_+'(X_{M_k}^k) = b_+ := \frac{\Delta_{M_k-1}^k}{\varepsilon_k}.$$

We recall that the polygons $\gamma_k = (x, f_k(x))$ obtained by the Birkhoff refinement satisfy the finite difference estimates (33)–(34). For the chosen assignments $b_\pm$ of the first derivatives at the nodes, the piecewise quadratic interpolation γ_k^{pq} is continuously differentiable throughout. Further, it can be computed that the bound on the second difference quotients implies a bound

$$\left\|\frac{d^2}{dx^2}\gamma_k^{pq}\right\|_{L^\infty} \le C, \tag{43}$$

which holds uniformly in k. Hence the sequence $\{\gamma_k^{pq}\}$ is uniformly bounded in $C^{1,1}$, and we infer convergence in $C^{1,\beta}$ to $\gamma \in C^{1,1}$ by Arzelà-Ascoli. This implies that

$$\|\gamma_k^{pq} - \gamma\|_{L^2} \to 0, \tag{44}$$

and

$$\left\|\frac{d}{dx}\gamma_k^{pq} - \frac{d}{dx}\gamma\right\|_{L^2} \to 0 \tag{45}$$

as $k \to \infty$. Choosing γ_k^{pq} as the function γ^q in Lemma 3, we infer for all k that

$$\|\gamma_k^{pc} - \gamma\|_{L^2} \leq \|\gamma_k^{pc} - \gamma_k^{pq}\|_{L^2} + \|\gamma_k^{pq} - \gamma\|_{L^2}$$
$$\leq C\varepsilon_k^2 \left\|\frac{d^2}{dx^2}\gamma_k^{pq}\right\|_{L^2} + \|\gamma_k^{pq} - \gamma\|_{L^2}$$

and

$$\left\|\frac{d}{dx}\gamma_k^{pl} - \frac{d}{dx}\gamma\right\|_{L^2} \leq \left\|\frac{d}{dx}\gamma_k^{pc} - \frac{d}{dx}\gamma_k^{pq}\right\|_{L^2} + \left\|\frac{d}{dx}\gamma_k^{pq} - \frac{d}{dx}\gamma\right\|_{L^2}$$
$$\leq C\varepsilon_k \left\|\frac{d^2}{dx^2}\gamma_k^{pq}\right\|_{L^2} + \left\|\frac{d}{dx}\gamma_k^{pq} - \frac{d}{dx}\gamma\right\|_{L^2}.$$

All expressions tend to 0 by (43)–(45). This proves the previously assumed claims (39) and (40).

5. Numerical investigations

Here we present experimental convergence rates and estimates for the computational effort for a simple benchmark problem. First, we obtain an explicit solution for a special metric g^{Jac}, which will serve as comparison for the numerical approximations on different discretisation levels ε_k.

5.1. *Special analytic solution*

From the weak formulation of the geodesic equation for graphs (35), we obtain by integration by parts the following strong version

$$\nabla_N e^h \sqrt{1 + |f'|^2} - \left(\frac{d}{dx}e^h + f' \cdot \nabla_N e^h \sqrt{1 + |f'|^2}\right) \frac{f'}{\sqrt{1 + |f'|^2}}$$
$$= e^h \frac{(1 + |f'|^2)I - f'f'^T}{\sqrt{1 + |f'|^2}^3} f'',$$

which has to be satisfied by a geodesic connecting the two points $(\pm\frac{\ell}{2}, 0)$ being a graph $\gamma(x) = (x, f(x))$, $f(\pm\frac{\ell}{2}) = 0$.

If we assume further that h does not depend on x, the last equation simplifies to

$$0 = \frac{\left((1 + |f'|^2)I - f'f'^T\right)}{\sqrt{1 + |f'|^2}^3}\left(\nabla_N h\left(1 + |f'|^2\right) - f''\right).$$

That is, we need to solve the simpler equation

$$f'' = \nabla_N h\left(1 + |f'|^2\right).$$

Let us restrict ourselves to linear h, not depending on the 'horizontal' co-ordinate x, that is, $h = h(y) = -\alpha n^T y$ with $\alpha > 0$ and $n \in \mathbb{R}^{n-1}$ a unit vector, $|n| = 1$.

Then $\nabla_N h = -\alpha n$, hence, $f'' = -\alpha \left(1 + |f'|^2\right) n$. With the ansatz $f' = \phi' n$, the equation can be rewritten as

$$\phi'' n = -\alpha n \left(1 + |\phi'|^2\right),$$

hence we are left with the scalar equation

$$\phi'' = -\alpha \left(1 + |\phi'|^2\right)$$

for ϕ' alone. Its solution is $\phi'(x) = \tan(C - \alpha x)$. Thus, integrating once again we deduce $\phi(x) = \frac{1}{\alpha} \ln \cos(-C + \alpha x) + D$. Matching the boundary conditions, we find that the geodesic connecting the two points $(\pm\frac{\ell}{2}, 0)$ is

$$\gamma(x) = (x, f(x)), \quad x \in (-\tfrac{\ell}{2}, \tfrac{\ell}{2}), \quad f(x) = n\phi(x)$$

with

$$\phi(x) = \frac{1}{\alpha} \ln \frac{\cos \alpha x}{\cos \alpha \frac{\ell}{2}}.$$

We remark that this representation requires $\alpha\frac{\ell}{2} < \frac{\pi}{2}$, or equivalently $\alpha < \frac{\pi}{\ell}$. Note that for $\alpha \to 0$, the metric approaches the constant Euclidean metric $h \equiv 0$, and we obtain the straight line segment $(x, f \equiv 0)$ in the limit.

For the other extreme, we remark that if $\alpha \to \frac{\pi}{\ell}$, the geodesic converges to the set of two parallel lines $\{(\pm\frac{\ell}{2}, y) \mid y \geq 0\}$ connected at infinity. In fact, this set of constant and finite length and will be the minimising configuration for all values of $\alpha \geq \frac{\pi}{\ell}$.

5.2. *Computational effort*

We use the analytic solution found above as a benchmark test for convergence of the method. Let

$$g_{ij}^{\text{Jac}}(x, y) := e^{2h(y)} \delta_{ij} \quad \text{with} \quad h(y) = -\alpha y.$$

We now present computations in $\mathbb{R}^2$ with $\ell = 2$, so connecting $(\pm 1, 0)$. Let us consider three different metrics by choosing three values $\alpha_1 = 0.65$, $\alpha_2 = 0.9$, and $\alpha_3 = 1.1$, all less than $\frac{\pi}{2}$. Larger values of α emphasise the difference of g compared to the flat Euclidean metric, hence one expects a larger curvature in the geodesic, which bends further away from the horizontal straight connection between the boundary points $(\pm 1, 0)$. We also explore how the performance of the algorithm depends on this geometrical feature.

We begin by presenting numerical statistics of the geodesic computation. The plots in Figure 3 show the computational error of the polygonal approximation compared to the explicit geodesic (measured in the L^2 norm) in the left panel and the computational effort, measured in the number of affirmative Birkhoff steps, in the right panel.

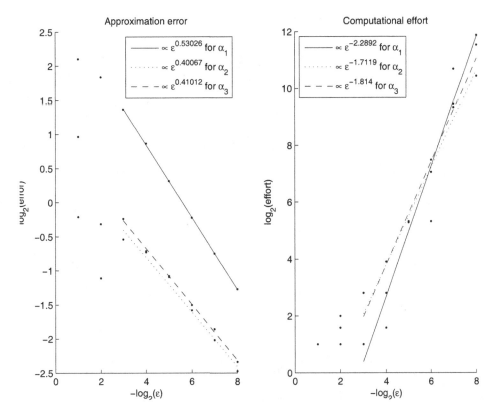

Fig. 3. Exponential metric, $n = 2$: $k = 1, \ldots, 8$. Left panel: The rate of the error proportional to ε^m. Right panel: The effort grows approximately with the rate ε^{-2}.

The error decreases as $\varepsilon_k = 2^{-k}$ decreases for $k = 1, \ldots, 8$. In fact, the calculation shows the error is proportional to ε^m, where the exponent m varies from 0.40 to 0.53, depending on the choice of α in the metric.

The effort increases as ε decreases, and the simulation shows that the effort turns out to be (inversely) proportional to $1/\varepsilon^m$, where m ranges from 1.71 to 2.2, as α is taken from $\alpha_1 = 0.65$, $\alpha_2 = 0.9$, $\alpha_3 = 1.1$.

Hence, for geodesics of larger curvature variation, the Birkhoff procedure starts closer but converges with smaller rate in ε, whereas the effort increases with largely uniform rate in ε of about $1/\varepsilon^2$.

Figure 4 shows how the polygons on the discrete level ε_k approach the exact solution

$$\gamma(x) = \left(x, \frac{1}{\alpha} \ln \frac{\cos \alpha x}{\cos \alpha} \right), \quad x \in (-1, 1).$$

We show the approximation for the three different values for α simultaneously in one graph. In each case, the solid line is the exact geodesic, the dotted line is the polygonal approximation for $e_3 = 1/8$, with the thicker dots indicating the

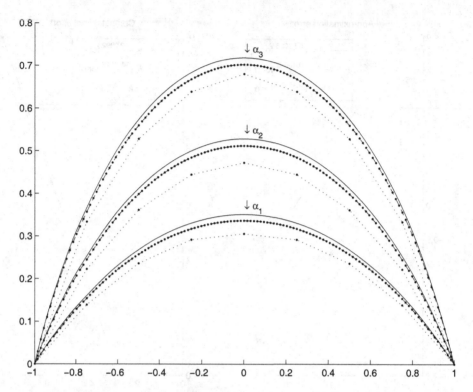

Fig. 4.   Exponential metric, $n = 2$:   approximate solutions compared to the analytic solution (solid line).

stencil points. The intermediate polygon with densely distributed stencils is the approximation for $e_7 = 1/128$.

Acknowledgments

J. Z. gratefully acknowledges the financial support of the EPSRC through an Advanced Research Fellowship (GR/S99037/1), grant EP/K027743/1 and the Leverhulme Trust (RPG-2013-261). Both authors are grateful to Michael Ortiz for pointing out the relevance of the Jacobi principle. They benefited from helpful discussions during the first annual meeting of the EPSRC network "Mathematical Challenges of Molecular Dynamics: A Chemo-Mathematical Forum" (EP/F03685X/1) and thank the anonymous reviewers for helpful comments.

References

[1] Ajit Banerjee and Noah P. Adams. Dynamics of classical systems based on the principle of stationary action. *The Journal of Chemical Physics*, 92(12):7330–7339, 1990.
[2] V. Benci, D. Fortunato, and F. Giannoni. On the existence of geodesics in static

Lorentz manifolds with singular boundary. *Ann. Scuola Norm. Sup. Pisa Cl. Sci. (4)*, 19(2):255–289, 1992.

[3] George D. Birkhoff. *Dynamical systems*. With an addendum by Jurgen Moser. American Mathematical Society Colloquium Publications, Vol. IX. American Mathematical Society, Providence, R.I., 1966.

[4] Susanne C. Brenner and L. Ridgway Scott. *The mathematical theory of finite element methods*, volume 15 of *Texts in Applied Mathematics*. Springer, New York, third edition, 2008.

[5] Maria Cameron, Robert V. Kohn, and Eric Vanden-Eijnden. The string method as a dynamical system. *J. Nonlinear Sci.*, 21(2):193–230, 2011.

[6] Weinan E, Weiqing Ren, and Eric Vanden-Eijnden. String method for the study of rare events. *Phys. Rev. B*, 66:052301, Aug 2002.

[7] Weinan E, Weiqing Ren, and Eric Vanden-Eijnden. Energy landscape and thermally activated switching of submicron-sized ferromagnetic elements. *Journal of Applied Physics*, 93(4):2275–2282, 2003.

[8] Weinan E, Weiqing Ren, and Eric Vanden-Eijnden. Minimum action method for the study of rare events. *Comm. Pure Appl. Math.*, 57(5):637–656, 2004.

[9] Weinan E, Weiqing Ren, and Eric Vanden-Eijnden. Finite temperature string method for the study of rare events. *J. Phys. Chem. B*, 109(14):6688–6693, 2005. PMID: 16851751.

[10] Daan Frenkel and Berend Smit. *Understanding Molecular Simulation*, volume 1 of *Computational Science Series*. Academic Press, Inc., Orlando, FL, USA, second edition, 2002.

[11] David Gilbarg and Neil S. Trudinger. *Elliptic partial differential equations of second order*. Classics in Mathematics. Springer-Verlag, Berlin, 2001. Reprint of the 1998 edition.

[12] Richard E. Gillilan and Kent R. Wilson. Shadowing, rare events, and rubber bands. a variational Verlet algorithm for molecular dynamics. *J. Chem. Phys.*, 97(3):1757–1772, 1992.

[13] William B. Gordon. The existence of geodesics joining two given points. *J. Differential Geometry*, 9:443–450, 1974.

[14] Jürgen Jost. *Riemannian geometry and geometric analysis*. Universitext. Springer-Verlag, Berlin, third edition, 2002.

[15] Anatole Katok and Boris Hasselblatt. *Introduction to the modern theory of dynamical systems*, volume 54 of *Encyclopedia of Mathematics and its Applications*. Cambridge University Press, Cambridge, 1995. With a supplementary chapter by Katok and Leonardo Mendoza.

[16] Herbert B. Keller. *Numerical methods for two-point boundary value problems*. Dover Publications Inc., New York, 1992. Corrected reprint of the 1968 edition.

[17] Roberto Olender and Ron Elber. Calculation of classical trajectories with a very large time step: Formalism and numerical examples. *J. Chem. Phys.*, 105(20):9299–9315, 1996.

[18] Paul H. Rabinowitz. Periodic solutions of Hamiltonian systems. *Comm. Pure Appl. Math.*, 31(2):157–184, 1978.

[19] Paul H. Rabinowitz. Periodic solutions of Hamiltonian systems: a survey. *SIAM J. Math. Anal.*, 13(3):343–352, 1982.

[20] Hartmut Schwetlick and Johannes Zimmer. Calculation of long time classical trajectories: Algorithmic treatment and applications for molecular systems. *J. Chem. Phys.*, 130(12):124106, 2009.

[21] Hartmut Schwetlick and Johannes Zimmer. The computation of long time Hamilto-

nian trajectories for molecular systems via global geodesics. In *ENUMATH Proceedings*, 2012.

[22] H. Seifert. Periodische Bewegungen mechanischer Systeme. *Math. Z.*, 51:197–216, 1948.

[23] Josef Stoer and Roland Bulirsch. *Numerische Mathematik. 2.* Springer-Lehrbuch. [Springer Textbook]. Springer-Verlag, Berlin, third edition, 1990. Eine Einführung—unter Berücksichtigung von Vorlesungen von F. L. Bauer. [An introduction, with reference to lectures by F. L. Bauer].

Chapter 13

Variational Symmetries and Pluri-Lagrangian Systems

Yuri B. Suris

Institut für Mathematik, MA 7-2, Technische Universität Berlin,
Str. des 17. Juni 136, 10623 Berlin, Germany
E-mail: suris@math.tu-berlin.de

We analyze the relation of the notion of a pluri-Lagrangian system, which recently emerged in the theory of integrable systems, to the classical notion of variational symmetry, due to E. Noether.

To Professor Armin Leutbecher on the occasion of his 80-th birthday, with the very best wishes and in the hope to please his flawless mathematical taste

Contents

1. Introduction

In the last decade, a new understanding of integrability of discrete systems as their multi-dimensional consistency has been a major breakthrough [8, 9, 22]. This led to classification of discrete 2-dimensional integrable systems (ABS list) [1], which turned out to be rather influential. According to the concept of multi-dimensional consistency, integrable two-dimensional systems can be imposed in a consistent way on all two-dimensional sublattices of a lattice $\mathbb{Z}^p$ of arbitrary dimension. This means that the resulting multi-dimensional system possesses solutions whose restrictions to any two-dimensional sublattice are generic solutions of the corresponding two-dimensional system. To put this idea differently, one can impose the two-dimensional equations on any quad-surface in $\mathbb{Z}^p$ (i.e., a surface composed of elementary squares), and transfer solutions from one such surface to another one, if they are related by a sequence of local moves, each one involving one three-dimensional cube, like the moves shown of Fig. 1.

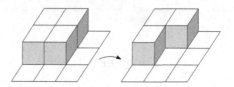

Fig. 1.   Local move of a quad-surface involving one three-dimensional cube

A further fundamental conceptual development was initiated in [19] and deals with variational (Lagrangian) formulation of multi-dimensionally consistent systems. Solutions of any ABS equation on any quad-surface $\Sigma \subset \mathbb{Z}^p$ are critical points of a certain action functional $S_\Sigma = \int_\Sigma \mathcal{L}$ obtained by integration of a suitable discrete Lagrangian 2-form $\mathcal{L}$. It was observed in [19] that the critical value of the action remains invariant under local changes of the underlying quad-surface, like those on Fig. 1, or, in other words, that the 2-form $\mathcal{L}$ is closed on solutions of quad-equations, and it was suggested to consider this as a defining feature of integrability. Results of [19], found on the case-by-case basis for some equations of the ABS list, have been extended to the whole list and were given a more conceptual proof in [10]. Subsequently, this research was pushed in various directions [2, 11, 12, 14, 15, 20, 21, 29]. These developments led us to introduce the following notion [11, 13].

Definition 1. (d-dimensional pluri-Lagrangian system) Let $\mathcal{L}[u]$ be a d-form on the p-dimensional space $\mathbb{R}^p$ (called multi-time, $p > d$), smoothly depending on a field $u : \mathbb{R}^p \to U$ with values in some vector space U and on partial derivatives of u up to a certain finite (but unspecified) order.

- To an arbitrary oriented d-dimensional manifold $\Sigma \subset \mathbb{R}^p$ with boundary $\partial\Sigma$, there corresponds the *action functional*, which assigns to $u|_\Sigma$ the number

$$S_\Sigma[u] = \int_\Sigma \mathcal{L}[u].$$

- We say that the field $u_\Sigma : \Sigma \to U$ is a *critical point* of S_Σ, if

$$\left.\frac{d}{d\epsilon}\right|_{\epsilon=0} S_\Sigma[u_\Sigma + \epsilon v_\Sigma] = 0$$

for any variation $v_\Sigma : \Sigma \to U$ such that $v_\Sigma|_{\partial\Sigma} = 0$.
- We say that the field $u : \mathbb{R}^p \to U$ solves the *pluri-Lagrangian problem* for the Lagrangian d-form $\mathcal{L}$ if, *for any oriented d-dimensional manifold with boundary* $\Sigma \subset \mathbb{R}^p$, the restriction $u|_\Sigma$ is a critical point of the corresponding action S_Σ.

A general theory of one-dimensional pluri-Lagrangian systems, continuous and discrete, has been developed in [27]. It was shown that they are related to systems of commuting Hamiltonian flows, resp. to systems of commuting symplectic maps. The general structure of multi-time Euler-Lagrange equations for discrete two-dimensional pluri-Lagrangian problems was studied in [13]. It was shown that the corresponding 2-form $\mathcal{L}$ is closed on general solutions of the multi-time Euler-Lagrange equations.

As argued in the latter paper, the original idea of [19] has significant precursors. These include:

- *Theory of pluriharmonic functions* and, more generally, of pluriharmonic maps [16, 24, 26]. By definition, a pluriharmonic function of several complex variables $f : \mathbb{C}^p \to \mathbb{R}$ minimizes the Dirichlet functional $E_\Gamma = \int_\Gamma |(f \circ \Gamma)_z|^2 dz \wedge d\bar{z}$ along any holomorphic curve in its domain $\Gamma : \mathbb{C} \to \mathbb{C}^p$. Differential equations governing pluriharmonic functions (and maps) are heavily overdetermined. Therefore it is not surprising that they belong to the theory of integrable systems.
- *Baxter's Z-invariance* of solvable models of statistical mechanics [3, 4]. This concept is based on invariance of the partition functions of solvable models under elementary local transformations of the underlying planar graphs. It is well known (see, e.g., [7]) that one can identify planar graphs underlying these models with quad-surfaces in $\mathbb{Z}^p$. On the other hand, the classical mechanical analogue of the partition function is the action functional. This suggests the relation of Z-invariance to the concept of closedness of the Lagrangian 2-form, at least at the heuristic level. This relation has been made mathematically precise for a number of models, through the quasiclassical limit [5, 6].
- The classical notion of *variational symmetry*, going back to the seminal work of E. Noether [23], turns out to be directly related to the idea of the closedness of the Lagrangian form in the multi-time.

The first of these precursors motivates the term "pluri-Lagrangian systems". The intention of the present note is to explain the relation of this notion to the third of the above mentioned precursors, namely to the notion of variational symmetries. For this aim, we recall the necessary definitions in section 2, illustrating them with one of the most familiar examples, the sine-Gordon equation and its variational symmetry given by the modified KdV equation. Then, in section 3 we establish the relation of these classical notions with the idea of closedness of the pluri-Lagrangian form. Finally, in section 4 we present the Euler-Lagrange equations for two-dimensional pluri-Lagrangian problems of second order, and establish the pluri-Lagrangian structure of the sine-Gordon equation.

2. Variational symmetries

We start with reminding the notion of a variational symmetry of Lagrangian differential equations, introduced (yet without this modern terminology) in the seminal paper by E. Noether [23]. See also a detailed historical discussion in [18]. Our presentation is based on [25].

We consider the differential algebra of functions u^α $(\alpha = 1, \ldots, q)$ of d independent variables x^i $(i = 1, \ldots, d)$. It has generators u_I^α, $I = (i_1, \ldots, i_d)$ being a multiindex, with the meaning

$$u_I^\alpha = \left(\frac{\partial}{\partial x^1}\right)^{i_1} \cdots \left(\frac{\partial}{\partial x^d}\right)^{i_d} u^\alpha.$$

The derivation D_j, understood as a total derivative w.r.t. x^j, acts on generators according to $D_j u_I^\alpha = u_{I+e_j}^\alpha$. Thus, for any differential function f we have

$$D_j f = \frac{\partial f}{\partial x^j} + \sum u_{I+e_j}^\alpha \frac{\partial f}{\partial u_I^\alpha}.$$

We now define more general derivations (generalized vector fields). In what follows, we only consider "vertical" (or evolutionary) generalized vector fields, i.e., those acting on dependent variables only. This is done, on one hand, for simplicity of notation, and, on the other hand, because in the discrete case only these are relevant, due to the absence of changes of independent variable.

Definition 2. (Evolutionary vector field) An evolutionary vector field generated by the set of q differential functions $\varphi[u] = (\varphi^1[u], \ldots, \varphi^q[u])$ is given by

$$D_\varphi = \sum_{\alpha, I} \varphi_I^\alpha \frac{\partial}{\partial u_I^\alpha}, \quad \varphi_I^\alpha = D_I \varphi^\alpha = D_1^{i_1} \ldots D_d^{i_d} \varphi^\alpha. \tag{1}$$

Usually [25, eq. (5.6)], this is called an (infinite) prolongation of the evolutionary vector field

$$\sum_\alpha \varphi^\alpha \frac{\partial}{\partial u^\alpha}.$$

The following is an adaptation of Definition 5.51 from [25].

Definition 3. (Variational symmetry) A generalized evolutionary vector field (1) is called a *variational symmetry* of an action functional

$$S[u] = \int L[u] dx^1 \ldots dx^d,$$

if the action of D_φ on the Lagrange function L is a complete divergence, that is, if there exist functions $M_1[u], \ldots, M_d[u]$ such that

$$D_\varphi L = \text{Div } M = \sum_{i=1}^d D_i M_i. \tag{2}$$

The intention of this definition is clear: the integral of a complete divergence (with fixed boundary values) vanishes, so D_φ preserves the value of the action functional. The following statement (see [25, Theorem 5.53]) justifies the previous definition.

Theorem 1. *If an evolutionary vector field D_φ is a variational symmetry of the action functional S, then it is a generalized symmetry of the Euler-Lagrange equations $\delta L/\delta u = 0$, that is, $D_\varphi(\delta L/\delta u) = 0$ on each smooth solution of $\delta L/\delta u = 0$.*

Example. Sine-Gordon equation, $d = 2$, $q = 1$:

$$u_{xy} = \sin u, \tag{3}$$

is the Euler-Lagrange equation for

$$L[u] = \frac{1}{2} u_x u_y - \cos u. \tag{4}$$

We show that the (prolongation of the) evolutionary vector field $\varphi \, \partial/\partial u$ with

$$\varphi[u] = u_{xxx} + \frac{1}{2} u_x^3 \tag{5}$$

is a variational symmetry for the sine-Gordon equation. The corresponding computation is mentioned in [25, p. 336], but is not presented there in detail, being replaced by a less direct method. We show that

$$D_\varphi L = D_x N + D_y M \tag{6}$$

with the following differential functions:

$$M[u] = \frac{1}{2} \varphi u_x - \frac{1}{8} u_x^4 + \frac{1}{2} u_{xx}^2, \tag{7}$$

$$N[u] = \frac{1}{2} \varphi u_y - \frac{1}{2} u_x^2 \cos u - u_{xx}(u_{xy} - \sin u). \tag{8}$$

Indeed, we compute:

$$D_\varphi L = \frac{1}{2}(\varphi_y u_x + \varphi_x u_y) + \varphi \sin u, \tag{9}$$

and

$$
\begin{aligned}
D_y M + D_x N &= \frac{1}{2}\varphi_y u_x + \frac{1}{2}\varphi u_{xy} - \frac{1}{2} u_x^3 u_{xy} + u_{xx} u_{xxy} \\
&\quad + \frac{1}{2}\varphi_x u_y + \frac{1}{2}\varphi u_{xy} - u_x u_{xx} \cos u + \frac{1}{2} u_x^3 \sin u \\
&\quad - u_{xxx}(u_{xy} - \sin u) - u_{xx}(u_{xxy} - u_x \cos u) \\
&= \frac{1}{2}(\varphi_y u_x + \varphi_x u_y) + \varphi \sin u + \left(\varphi - \frac{1}{2} u_x^3 - u_{xxx}\right)(u_{xy} - \sin u). \tag{10}
\end{aligned}
$$

Comparing (9) and (10), we see that identity (6) is satisfied under the choice of the differential function φ as in (5). By Noether's theorem, existence of a variational symmetry leads to a corresponding conservation law for the sine-Gordon equation given in [25, p. 336]. For convenience of the reader, we reproduce this computation here:

$$
\begin{aligned}
\varphi\frac{\delta L}{\delta u} &= \varphi\left(\frac{\partial L}{\partial u} - D_x\frac{\partial L}{\partial u_x} - D_y\frac{\partial L}{\partial u_y}\right) \\
&= \varphi\frac{\partial L}{\partial u} + \varphi_x\frac{\partial L}{\partial u_x} + \varphi_y\frac{\partial L}{\partial u_y} - D_x\left(\varphi\frac{\partial L}{\partial u_x}\right) - D_y\left(\varphi\frac{\partial L}{\partial u_y}\right) \\
&= D_\varphi L - D_x\left(\varphi\frac{\partial L}{\partial u_x}\right) - D_y\left(\varphi\frac{\partial L}{\partial u_y}\right) \\
&= D_x\left(N - \varphi\frac{\partial L}{\partial u_x}\right) + D_y\left(M - \varphi\frac{\partial L}{\partial u_y}\right) \\
&= D_x\left(N - \frac{1}{2}\varphi u_y\right) + D_y\left(M - \frac{1}{2}\varphi u_x\right) \\
&= -D_x\left(\frac{1}{2}u_x^2\cos u + u_{xx}(u_{xy} - \sin u)\right) + D_y\left(-\frac{1}{8}u_x^4 + \frac{1}{2}u_{xx}^2\right).
\end{aligned}
$$

3. Variational symmetries and closedness of multi-time Lagrangian forms

Now, we would like to promote an alternative point of view. In the standard approach, reproduced in the previous section, equation (2) is a certain (differential-) algebraic property of the vector field D_φ. However, this way of thinking about this equation ignores one of the main interpretations of the notion of "symmetry", namely the interpretation as a commuting flow. In this interpretation, one introduces a *new independent variable* z corresponding to the "flow" of the generalized vector field D_φ,

$$u_z^\alpha = \varphi^\alpha[u], \tag{11}$$

and considers *simultaneous solutions* of the Euler-Lagrange equations $\delta L/\delta u = 0$ and of the flow (11) as functions of $p = d + 1$ independent variables $x^1, \ldots, x^d, z$. Then equation (2) reads

$$D_z L - \sum_{i=1}^{d} D_i M_i = 0. \tag{12}$$

The key observation is that equation (12) is nothing but the *closedness condition* $d\mathcal{L} = 0$ of the following d-form in the $(d + 1)$-dimensional space:

$$\mathcal{L} = L[u]\, dx^1 \wedge \ldots \wedge dx^d - \sum_{i=1}^{d}(-1)^i M_i[u]\, dz \wedge dx^1 \wedge \ldots \wedge \widehat{dx^i} \wedge \ldots \wedge dx^d. \tag{13}$$

Thus, we are led to define the *extended action functional*

$$S_\Sigma = \int_\Sigma \mathcal{L}, \tag{14}$$

where Σ is some d-dimensional surface (with boundary) in the $(d + 1)$-dimensional space of independent variables $x^1, \ldots, x^d, z$. In particular, the action S_Σ over the hypersurface $\Sigma \subset \{z = \text{const}\}$ is the original action S. Equation (12) means that the extended action does not depend on local changes of the d-dimensional integration surface Σ preserving boundary. Of course, this statement only holds *on simultaneous solutions* of the Euler-Lagrange equations $\delta L/\delta u = 0$ and of the flow (11).

Example. To clearly see this in our above example of the sine-Gordon equation, we re-write the previous computations in our new notation. We consider u as a function of $d + 1 = 3$ variables x, y, z, and replace φ by u_z. We have:

$$\mathcal{L} = L[u] \, dx \wedge dy - M[u] \, dz \wedge dx + N[u] \, dz \wedge dy, \tag{15}$$

where

$$L[u] = \frac{1}{2} u_x u_y - \cos u, \tag{16}$$

$$M[u] = \frac{1}{2} u_x u_z - \frac{1}{8} u_x^4 + \frac{1}{2} u_{xx}^2, \tag{17}$$

$$N[u] = \frac{1}{2} u_y u_z - \frac{1}{2} u_x^2 \cos u - u_{xx}(u_{xy} - \sin u). \tag{18}$$

Then the previous computation tells us that

$$L_z - (M_y + N_x) = -\left(u_z - \frac{1}{2} u_x^3 - u_{xxx}\right)(u_{xy} - \sin u). \tag{19}$$

Thus, the form $\mathcal{L}$ is closed as soon as $u_z = \frac{1}{2} u_x^3 + u_{xxx}$. This shows us once again that the modified KdV equation

$$u_z = u_{xxx} + \frac{1}{2} u_x^3 \tag{20}$$

is a variational symmetry of the sine-Gordon equation (3).

The remarkable factorized form of the right-hand side of (19) shows that it also vanishes as soon as $u_{xy} = \sin u$. This suggests that the above relation could be reversed, namely, that the sine-Gordon equation should be a variational symmetry of the modified KdV equation, as well. Two facts apparently stand in the way of this interpretation: first, modified KdV equation is not Lagrangian, and, second, sine-Gordon equation is not evolutionary. Nevertheless, this interpretation is still possible. To show this, we first observe that the function $M[u]$ from (17) can be considered as a Lagrangian for the action S_Σ over the hypersurface $\Sigma \subset \{y = \text{const}\}$. The corresponding Euler-Lagrange equation $\delta M/\delta u = 0$ is

$$u_{zx} - \frac{3}{2} u_x^2 u_{xx} - u_{xxxx} = 0, \tag{21}$$

the differentiated form of modified KdV. It is this equation for which we want to declare the derivation D_y as a variational symmetry. To overcome the difficulty that D_y is not an evolutionary vector field (i.e., that u_y is not defined by our differential equations), we observe that we only need to define the action of D_y on the Lagrangian M. However, the latter function does not contain u alone, but only its derivatives (of degree 1 w.r.t z and of higher degrees w.r.t. x). For such functions the formula

$$D_y f = u_{yz} \frac{\partial f}{\partial u_z} + \sum_{I:\, i_1 \geq 1} u_{Iy} \frac{\partial f}{\partial u_I}$$

works perfectly as an evolutionary vector field. Indeed, one can use the equation

$$u_{yz} = u_{xxxy} + \frac{3}{2} u_x^2 u_{xy} = (\sin u)_{xx} + \frac{3}{2} u_x^2 \sin u = u_{xx} \cos u + \frac{1}{2} u_x^2 \sin u,$$

which is obtained from (20) by differentiation upon use of the sine-Gordon equation, as well as relations $u_{Iy} = D_{I-e_1} \sin u$ for multiindices I with $i_1 \geq 1$.

4. Pluri-Lagrangian structure of the sine-Gordon equation

Motivated by considerations of the previous section, we consider the pluri-Lagrangian problem with the 2-form (15). To derive the corresponding multi-time Euler-Lagrange equations turns out to be a remarkably difficult task. We will not give the complete derivation here, but restrict ourselves to formulating (without a proof which will be given elsewhere [28]) the following statement which covers our main example in this note, namely the sine-Gordon equation.

Theorem 2. *Consider a pluri-Lagrangian problem of dimension $d = 2$ and of order 2, with the 2-form*

$$\mathcal{L} = \sum_{1 \leq i < j \leq p} L_{ij}[u] dx^i \wedge dx^j,$$

where $L_{ij} = -L_{ji}$ are differential functions depending on the second jet of the field $u = u(x^1, \ldots, x^p)$, so that $L_{ij}[u] = L_{ij}(u, u_k, u_{km})$, where $u_k = \partial u / \partial x^k$ and $u_{km} = \partial^2 u / \partial x^k \partial x^m$. The system of multi-time Euler-Lagrange equations consists of:

$$\frac{\delta L_{ij}}{\delta u} = 0, \tag{22}$$

$$\frac{\delta L_{ij}}{\delta u_k} = 0 \quad \text{for all } k \notin \{i, j\}, \tag{23}$$

$$\frac{\delta L_{ij}}{\delta u_j} = \frac{\delta L_{ik}}{\delta u_k} \quad \text{for all } j, k \neq i, \tag{24}$$

$$\frac{\delta L_{ij}}{\delta u_{km}} = 0 \quad \text{for all } k, m \notin \{i, j\}, \tag{25}$$

$$\frac{\delta L_{ij}}{\delta u_{jm}} = \frac{\delta L_{ik}}{\delta u_{km}} \quad \text{for all } j, k \neq i, \tag{26}$$

and

$$\frac{\delta L_{ij}}{\delta u_{ij}} + \frac{\delta L_{jk}}{\delta u_{jk}} + \frac{\delta L_{ki}}{\delta u_{ki}} = 0. \tag{27}$$

Here the following notations are used:

$$\frac{\delta L_{ij}}{\delta u} := \frac{\partial L_{ij}}{\partial u} - D_i \frac{\partial L_{ij}}{\partial u_i} - D_j \frac{\partial L_{ij}}{\partial u_j} + D_i^2 \frac{\partial L_{ij}}{\partial u_{ii}} + D_i D_j \frac{\partial L_{ij}}{\partial u_{ij}} + D_j^2 \frac{\partial L_{ij}}{\partial u_{jj}}, \quad (28)$$

$$\frac{\delta L_{ij}}{\delta u_k} := \frac{\partial L_{ij}}{\partial u_k} - D_i \frac{\partial L_{ij}}{\partial u_{ik}} - D_j \frac{\partial L_{ij}}{\partial u_{jk}}, \quad (29)$$

$$\frac{\delta L_{ij}}{\delta u_{km}} := \frac{\partial L_{ij}}{\partial u_{km}}. \quad (30)$$

These are the standard variational derivatives with respect to the coordinate plane (x^i, x^j) *to which the component* L_{ij} *of the 2-form* $\mathcal{L}$ *corresponds.*

Equations (22), (23), (25) can be derived by considering action over coordinate planes as surfaces Σ, while equations (24), (26), (27) are derived by considering general surfaces Σ and are less obvious.

Example. Sine-Gordon equation. We have:

$$L = L_{12} = \frac{1}{2} u_x u_y - \cos u, \quad (31)$$

$$M = L_{13} = \frac{1}{2} u_z u_x - \frac{1}{8} u_x^4 + \frac{1}{2} u_{xx}^2, \quad (32)$$

$$N = -L_{23} = \frac{1}{2} u_z u_y - \frac{1}{2} u_x^2 \cos u - u_{xx}(u_{xy} - \sin u). \quad (33)$$

For these Lagrangians, the above system of multi-time Euler-Lagrange equations reduces to

$$u_{xy} = \sin u, \quad (34)$$

$$u_{xz} = \frac{3}{2} u_x^2 u_{xx} + u_{xxxx}, \quad (35)$$

$$u_{yz} = u_{xx} \cos u + \frac{1}{2} u_x^2 \sin u, \quad (36)$$

$$u_{xxy} = u_x \cos u, \quad (37)$$

$$u_z = \frac{1}{2} u_x^3 + u_{xxx}. \quad (38)$$

Indeed, for $p = 3$ the above system consists of 19 equations. With the present choice of L_{ij}, the majority of these equations are satisfied identically. The non-trivial equations are derived from the following ones:

- equation (34) is obtained from two equations, namely, from $\delta L/\delta u = 0$ and from $\delta N/\delta u_{xx} = 0$;
- equation (35) is obtained from $\delta M/\delta u = 0$;
- equation (36) is obtained from $\delta N/\delta u = 0$;
- equation (37) is obtained from $\delta N/\delta u_x = 0$;
- equation (38) is obtained from

$$\frac{\delta N}{\delta u_y} + \frac{\delta M}{\delta u_x} = \left(\frac{\partial N}{\partial u_y} - D_y \frac{\partial N}{\partial u_{yy}} - D_z \frac{\partial N}{\partial u_{yz}} \right) + \left(\frac{\partial M}{\partial u_x} - D_z \frac{\partial M}{\partial u_{xz}} - D_x \frac{\partial M}{\partial u_{xx}} \right) = 0.$$

It remains to notice that equations (35)–(37) are corollaries of (34) and (38), derived by differentiation.

Theorem 3. *Multi-time Euler-Lagrange equations for the pluri-Lagrangian problem with the 2-form (15) with the components (16)–(18) consist of the sine-Gordon equation (34) and of the modified KdV equation (38). On simultaneous solutions of these equations, the 2-form $\mathcal{L}$ is closed.*

It is remarkable that multi-time Euler-Lagrange equations are capable of producing evolutionary equations.

Considering further variational symmetries of the sine-Gordon equation which are commuting flows of the modified KdV hierarchy (see, e.g., [25, p. 336] for for a variational symmetry of order 5) would lead to a pluri-Lagrangian 2-form in a space of an arbitrary number of independent variables, encoding the whole hierarchy.

5. Conclusions

In subsequent publications, we will address the following problems:

- To derive multi-time Euler-Lagrange equations for pluri-Lagrangian problems of arbitrary order, i.e., for forms $\mathcal{L}$ depending on jets of arbitrary order.
- To extend the classical De Donder-Weyl theory of calculus of variations to the pluri-Lagrangian context.
- To elaborate on the pluri-Lagrangian structure of classical integrable hierarchies, like the KdV or, more generally, Gelfand-Dickey hierarchies. In the monograph [17] (which is, in my opinion, one of the best sources on the Lagrangian field theory) whose program, according to the foreword, is "that the book is about hierarchies of integrable equations rather than about individual equations", it is the Lagrangian part (chapters 19, 20) that only deals with individual equations. The reason for this is apparently the absence of the concept of pluri-Lagrangian systems.
- To establish a general relation of pluri-Lagrangian structure to more traditional notions of integrability.
- To study the general relation of pluri-Lagrangian structure to Z-invariance of statistical-mechanical problems, via quasi-classical limit, as exemplified in [5, 6].

This research is supported by the DFG Collaborative Research Center TRR 109 "Discretization in Geometry and Dynamics".

References

[1] V.E. Adler, A.I. Bobenko, Yu.B. Suris. *Classification of integrable equations on quad-graphs. The consistency approach*, Commun. Math. Phys., 233 (2003), 513–543.

[2] J. Atkinson, S.B. Lobb, F.W. Nijhoff. *An integrable multicomponent quad-equation and its lagrangian formulation*, Theor. Math. Phys., **173** (2012), 1644–1653.

[3] R.J. Baxter. *Solvable eight-vertex model on an arbitrary planar lattice*, Philos. Trans. R. Soc. London, Ser. A **289** (1978) 315–346.

[4] R.J. Baxter. *Free-fermion, checkerboard and Z-invariant lattice models in statistical mechanics*. Proc. R. Soc. Lond. A **404** (1986) 1–33.

[5] V.V. Bazhanov, V.V. Mangazeev, S.M. Sergeev. *Faddeev-Volkov solution of the Yang-Baxter equation and discrete conformal geometry*, Nucl. Phys. B **784** (2007), 234–258.

[6] V.V. Bazhanov, V.V. Mangazeev, S.M. Sergeev. *A master solution of the quantum Yang-Baxter equation and classical discrete integrable equations*, Adv. Theor. Math. Phys. **16** (2012) 65–95.

[7] A.I. Bobenko, Ch. Mercat, Yu.B. Suris. *Linear and nonlinear theories of discrete analytic functions. Integrable structure and isomonodromic Green's function*. J. Reine Angew. Math. **583** (2005), 117–161.

[8] A.I. Bobenko, Yu.B. Suris. *Integrable systems on quad-graphs*, Intern. Math. Research Notices, **2002**, Nr. 11 (2002), 573–611.

[9] A.I. Bobenko, Yu.B. Suris. *Discrete Differential Geometry: Integrable Structures*, Graduate Studies in Mathematics, Vol.98, AMS, 2008.

[10] A.I. Bobenko, Yu.B. Suris. *On the Lagrangian structure of integrable quad-equations*, Lett. Math. Phys. **92** (2010), 17–31.

[11] A.I. Bobenko, Yu.B. Suris. *Discrete pluriharmonic functions as solutions of linear pluri-Lagrangian systems*, Commun. Math. Phys. **336** (2015), No. 1, 199–215.

[12] R. Boll, M. Petrera, Yu.B. Suris. *Multi-time Lagrangian 1-forms for families of Bäcklund transformations. Toda-type systems*, J. Phys. A: Math. Theor. **46** (2013), No. 27, 275204, 26 pp.

[13] R. Boll, M. Petrera, Yu.B. Suris. *What is integrability of discrete variational systems?*, Proc. Royal Soc. A **470** (2014), No. 2162, 20130550, 15 pp.

[14] R. Boll, M. Petrera, Yu.B. Suris. *Multi-time Lagrangian 1-forms for families of Bäcklund transformations. Relativistic Toda-type systems*, J. Phys. A: Math. Theor. **48** (2015), No. 8, 085203, 28 pp.

[15] R. Boll, Yu.B. Suris. *On the Lagrangian structure of 3D consistent systems of asymmetric quad-equations*, J. Phys. A: Math. Theor. **45** (2012), No. 11, 115201, 18 pp.

[16] F. Burstall, D. Ferus, F. Pedit, and U. Pinkall. *Harmonic tori in symmetric spaces and commuting Hamiltonian systems on loop algebras*. Ann. Math. **138** (1993) 173–212.

[17] L. Dickey. *Soliton equations and Hamiltonian systems*. 2nd edition. World Scientific, 2003.

[18] Y. Kosmann-Schwarzbach. *The Noether theorems. Invariance and conservation laws in the 20th century*. Springer, 2011.

[19] S. Lobb, F.W. Nijhoff. *Lagrangian multiforms and multidimensional consistency*, J. Phys. A: Math. Theor. **42** (2009) 454013.

[20] S.B. Lobb, F.W. Nijhoff. *Lagrangian multiform structure for the lattice Gel'fand-Dikij hierarchy*, J. Phys. A: Math. Theor. **43** (2010) 072003.

[21] S.B. Lobb, F.W. Nijhoff, G.R.W. Quispel. *Lagrangian multiform structure for the lattice KP system*, J. Phys. A: Math. Theor. **42** (2009) 472002.

[22] F.W. Nijhoff. *Lax pair for the Adler (lattice Krichever-Novikov) system*, Phys. Lett. A **297** (2002), 49–58.

[23] E. Noether. *Invariante Variationsprobleme*, Nachrichten von der Gesellschaft der Wissenschaften zu Göttingen, Math.-Phys. Kl. (1918), 235–257.

[24] Y. Ohnita, G. Valli. *Pluriharmonic maps into compact Lie groups and factorization into unitons*, Proc. London Math. Soc. **61** (1990) 546–570.

[25] P. Olver. *Applications of Lie groups to differential equations*. Graduate Texts in Mathematics, Vol. 107. 2nd edition, Springer, 1993.

[26] W. Rudin. *Function theory in polydiscs*. Benjamin (1969).

[27] Yu.B. Suris. *Variational formulation of commuting Hamiltonian flows: multi-time Lagrangian 1-forms*, J. Geom. Mech. **5** (2013), pp. 365–379.

[28] Yu.B. Suris, M. Vermeeren. *On the Lagrangian structure of integrable hierarchies*, arXiv:1510.03724 [math-ph] (2015), http://arxiv.org/abs/1510.03724.

[29] S. Yoo-Kong, S. Lobb, F. Nijhoff. *Discrete-time Calogero-Moser system and Lagrangian 1-form structure*. J. Phys. A: Math. Theor. **44** (2011) 365203.

Printed in the United States
By Bookmasters